No. 1962
$22.95

electronics MATH

R. JESSE PHAGAN

Dedicated To My Mother

FIRST EDITION
FIRST PRINTING

Printed in the United States of America

Library of Congress Cataloging in Publication Data

Phagan, R. Jesse.
Electronics math.

Includes index.
1. Electronics—Mathematics. I. Title.
TK7835.P48 1986 621.381'0151 85-27804
ISBN 0-8306-0962-8
ISBN 0-8306-1962-3 (pbk.)

Edited by Donald A. Morris
in memoriam

Contents

Introduction

Electronics is a fast changing field in the world today, and will continue to change at a very rapid rate for many years to come. It is exciting to be a part of this change, and many more people are becoming excited by the prospects of the electronics field.

To most people, electronics still remains a bit of magic. It is fine to keep some people in the dark, but a person who truly wants to learn about the field needs more than just magic as an answer to how things work.

Electronics can be a difficult subject to study, but it doesn't really have to be that way. Many of the electronics theory books go to great lengths to try and explain the theory behind a certain subject. They will ignore the necessary math until the end of the chapter, where there are lots of questions that cannot be solved without calculations. As a result, the student feels he has not learned the subject because he cannot answer the questions. Other students will spend so much time concentrating on the math that the theory is lost.

Electronics Math is intended to change the way people have been learning electronics. I believe a great deal of effort needs to be placed on learning the math involved in electronics. However, understanding the theory of how things work is just as important. This book makes every effort to show step-by-step procedures for understanding the mathematics necessary to solve problems. At the same time, theories are also explained.

Everything in this book is from the technician's point of view and, as a result, many things have been simplified. Electronics will always be magic to some, but students using this book will find they understand the math much better, and the theory of operation will also be easier to understand.

The subject matter of this book concentrates on the basic circuits, such as dc circuits and circuits containing inductors and capacitors. There is also a separate chapter dealing with the sine wave.

The chapters presented here are considered the building blocks of electronic circuits. Many students have trouble understanding electronics at all levels of study because they do not understand the basics.

Chapter 1

10^3 = kilo 10^{-3} = milli

10^6 = mega

10^{-6} = micro 10^9 = giga

Scientific and Engineering Notation

There are times when it is necessary to deal with very large or very small numbers, but working with numbers like these is somewhat cumbersome using ordinary arithmetic. The term scientific notation suggests science and technical fields have a definite need for such a mathematical system. In electronics, it is especially appropriate to use a system that makes using numbers easier. In electronics and other engineering fields, a modification of scientific notation, called engineering notation is used.

SIGNIFICANT FIGURES

Mathematics, with its many rules and precise calculations gives the impression that it is an accurate as well as precise tool. This is true of the math itself, however, the numbers that are used for those calculations is questionable to their degree of accuracy. Therefore, it makes sense that if the numbers are questionable, then the accuracy of the calculations is questionable. For example, when a meter is read, the meter has a tolerance and the person reading it has a tolerance. The value of resistors used in calculations will often have a 10 percent tolerance.

With all the different tolerances being introduced to our calculations, there needs to be a way of determining which digits of a number are reliable and accurate and which figures are questionable, or for that matter, not needed due to their unreliability. The way a number can be written and show the importance of each figure is called significant figures.

Often when writing numbers in electronics, the most commonly used number of figures is three. This is a common occurrence, not a rule.

Rules for Determining Significant Figures

To determine the significant figures of a number, follow these guidelines:

Rule 1 Non-zero numbers (the digits 1 through 9) are always significant figures.
Rule 2 A zero used as a place holder between two non-zero digits is considered to be a significant figure.
Rule 3 A zero to the right of the last digit to the right of a decimal is considered a significant figure.
Rule 4 A zero used only as a place holder, either on the right or the left side of the decimal point, if it is not between two non-zero digits, is *not* considered a significant figure.
Rule 5 When a number is greater than one, a decimal point can be placed following the number to show that all of the digits are significant, including zeros.

Examples

1234 has four significant figures
1230 has three significant figures
1200 has two significant figures
1034 has four significant figures
1030 has three significant figures
1030. has four significant figures
.00123 has three significant figures
0.0123 has three significant figures
0.0104 has three significant figures
.2050 has four significant figures
25.0 has two significant figures
25.5 has three significant figures
53.05 has four significant figures
50.50 has three significant figures
005 has one significant figure
00500 has one significant figure

ROUNDING NUMBERS

The process of rounding numbers is used to make working with numbers easier. It can be seen that the digit in a number with the least significance often has little power in changing the value of a number. For example, there is very little difference between $100 and $102.

Whenever it is necessary to round numbers, first determine how many significant figures will appear in the final number. The next digit to the right will determine the outcome of rounding.

Rules for Rounding Numbers

Rule 1 Determine how many significant figures are to be used in the final number. The next number to the right determines the outcome of rounding.
Rule 2 If the number to the right is 5 or more, drop this digit and all the digits to the right. *Add 1* to the last figure kept. Replace the digits dropped with zeros to maintain the place value of the figures remaining.
Rule 3 If the number to the right is less than 5, drop this digit and all the digits to the right. Replace the digits dropped with zeros to maintain the place value of the figures remaining.

Examples (rounded to three significant figures)

806956 = 807000	.1009 = .101
875.5 = 876	.0000003553 = .000000355
9083.2 = 9080	80.78 = 80.8
.00387 = .00387	104 = 100
.008737 = .00874	98 = 100

Practice Problems

Round to three significant figures.

1. 5609
2. 5605
3. 5603
4. 29888
5. 359,999
6. 568.8
7. 573.09
8. 333.1
9. 12.09
10. 23.15
11. 54.51909
12. 56.99103
13. 8.2
14. 8.4300
15. 9.00199
16. 00.00333333333
17. 0.9000999
18. .0009315
19. 0.009999
20. 0.19999999
21. .6666666666
22. 0.005555555
23. 0.7447474747
24. 0.0909090909
25. 0.0125125125

POSITIVE AND NEGATIVE NUMBERS

Scientific notation expresses a number as a power of 10 in either positive or negative form. All numbers have a value of either positive or negative. The positive and negative signs are written in front of the number and becomes part of the number. For example; positive three is written $+3$ or simply 3, negative three is written -3. If no sign is written, we will always assume the number to be positive.

Keep in mind fractions and decimals will still have the same relationships, even if the number is negative.

Addition and Subtraction with Signed Numbers

Adding and subtracting positive and negative numbers is similar to adding and subtracting whole numbers. Study the rules and examples that follow.

Rules for Adding and Subtracting Signed Numbers

Rule 1 If the signs of the two numbers being added are alike; add the numbers and keep the sign. Examples

$$8 + 2 = 10 \qquad 4 + 3 = 7 \qquad .2 + .5 = .7$$

Note In the examples shown above, both signs of the numbers are positive and the numbers are added. Result is positive. In the examples shown below, both signs of the numbers are negative and the numbers are added with the result being negative.

$-8 + -2 = -10 \qquad -4 + -3 = -7 \qquad -.2 + -.5 = -.7$

Rule 2 If the signs of the two numbers being added are unlike, subtract the numbers and take the sign of the larger number.

Examples

$-8 + 2 = -6 \qquad 8 + -2 = 6 \qquad -.2 + .5 = .3$

Note In the examples shown above, the signs of the two numbers are unlike. The numbers are subtracted, least from the greatest, and the result takes the sign of the greater number.

Rule 3 When subtracting two numbers, change the sign of the number being subtracted to its opposite, change the subtraction sign to addition, and then follow the rules for addition. Examples

$7 - 3 = 7 + -3 = 4 \qquad -7 - 3 = -7 + -3 = -10$
$7 - -3 = 7 + +3 = 10 \qquad -7 - -3 = -7 + +3 = -4$

Note When subtracting more than one number always start at the left and work to the right.

Practice Problems

Complete the operation.

1. $-6 + 4$
2. $7 + 3$
3. $8 + -2$
4. $9 - 5$
5. $-9 - -5$
6. $10 - -6$
7. $-3 + 9$
8. $-2 + -5$
9. $-1 -6$
10. $-1 - 1$
11. $-3 - -3$
12. $1 + 2 - 4$
13. $-3 - 5 + -6 - -3$
14. $10 - 6 + 3 - -10 - 3 + 6$
15. $0.3 - 0.5 - 0.8 + 1.5$
16. $2.5 - 3.2 + 6.8 + -.8$
17. $0 + 8 - 0.01 - 0 - 0.03$
18. $-4.02 + 8.12 + 3.14 - 18.0$
19. $4.5 - 10.6 - 3.02 + .01$
20. $-\frac{3}{4} - \frac{5}{8}$

Multiplication and Division with Signed Numbers

When multiplying and dividing signed numbers disregard the signs and multiply or divide. Then follow the rules shown below.

Rule 1 Disregard the signs and multiply (or divide) the numbers. If the signs are like, the product (or quotient) is positive.

Examples

$4 \times 2 = 8 \qquad -4 \times -2 = 8$
$8 \div 2 = 4 \qquad -8 \div -4 = 2$

Rule 2 If the signs are unlike, the product (or quotient) is negative.
Examples

$4 \times -2 = -8 \qquad -4 \times 2 = -8$
$8 \div -2 = -4 \qquad -8 \div 2 = -4$

The remainder of this chapter deals with the rules and operations of scientific notation.

Practice Problems

Complete the operation.

Note Parentheses written together ( ) ( ) signify multiplication. Division is represented by the use of the division symbol ÷ or using the fraction bar.

1. $(2)(3)$
2. -4×-5
3. 5×-3
4. -6×3
5. $(-2)(3)$
6. $(-1)(-2)$
7. $0 \times 3 \times -5$
8. $.4 \times .5$
9. 0.3×-0.6
10. -2^2
11. -2^3
12. $20 \div 4$
13. $25 \div -5$
14. $-30 \div 6$
15. $-45 \div -9$
16. $\frac{-8}{2}$
17. $\frac{2}{-12}$
18. $\frac{2 \times -3}{-1}$
19. $-\frac{3 \times -4}{-2 \times -2}$
20. $\frac{(-3)(-3)(-3)}{(-1)(-1)(-1)}$

SCIENTIFIC NOTATION

Scientific notation is a mathematical tool used to make numbers much easier to work with. Even with the use of a calculator, it is still helpful to use scientific notation when dealing with the types of numbers found in electronics.

Scientific notation is the writing of a number as a number between 1 and 10, times a power of 10. First it is necessary to see how numbers that are multiples of ten are expressed as powers of ten.

Table 1-1. Powers of 10.

Number	Power of 10	Read as
1,000,000	10^6	ten to the sixth power
100,000	10^5	ten to the fifth power
10,000	10^4	ten to the fourth power
1,000	10^3	ten to the third power
100	10^2	ten to the second power
10	10^1	ten to the first power
1	10^0	ten to the zero power
.1	10^{-1}	ten to the negative first power
.01	10^{-2}	ten to the negative second power
.001	10^{-3}	ten to the negative third power
.0001	10^{-4}	ten to the negative fourth power
.00001	10^{-5}	ten to the negative fifth power
.000001	10^{-6}	ten to the negative sixth power

The following key points should be noted.

- Numbers that are greater than 1 have a positive power of 10.
- Numbers that are less than 1 have a negative power of 10.
- 10 to the zero power (10^0) equals 1. Any number to the zero power equals 1.

Rules for Writing Numbers in Scientific Notation

Rule 1 When a number is greater than 1, express it in scientific notation by moving the decimal point to the *left* enough places so the number is written between 1 and 10. Count the number of decimal places the point was moved and that is the positive power of 10.

Examples

$85,500 = 8.65 \times 10^4$ $563,000,000 = 5.63 \times 10^8$

$230 = 2.30 \times 10^2$ $7,602 = 7.602 \times 10^3$

$7.3 = 7.3 \times 10^0$ $20 = 2.0 \times 10^1$

Rule 2 When a number is less than 1, express is in scientific notation by moving the decimal point to the *right* enough places so the number is written between 1 and 10. Count the number of decimal places the point was moved to determine the negative power of 10.

Examples

$.1 = 1.0 \times 10^{-1}$ $0.023 = 2.3 \times 10^{-2}$

$0.000356 = 3.56 \times 10^{-4}$ $0.03004 = 3.004 \times 10^{-2}$

$.000098 = 9.8 \times 10^{-5}$ $.00000030 = 3.0 \times 10^{-7}$

Sometimes numbers are already written as a number times some power of 10. When this is the case, it is usually seen that the number is not expressed between 1 and 10. It is then necessary to rewrite the number to follow the form of scientific notation.

Example

$365 \times 10^4 = 3.65 \times 10^6$

The easiest way to solve this type of problem is to first write the number in its proper form, with a power of 10 corresponding to the initial amount of the decimal move. The next step would be to add the exponents according to the rules of signed numbers.

Step 1 $365 = 3.65 \times 10^2$ write the number in proper form
Step 2 $10^2 + 10^4 = 10^6$ add the powers of 10
Step 3 3.65×10^6 combine steps 1 and 2 for the final answer

More Examples

$560{,}000 \times 10^{-5} = 5.6 \times 10^{0}$
Step 1 $560{,}000 = 5.6 \times 10^{5}$
Step 2 $10^{5} + 10^{-5} = 10^{0}$

$890{,}000{,}000 \times 10^{-3} = 8.9 \times 10^{5}$
Step 1 $890{,}000{,}000 = 8.9 \times 10^{8}$
Step 2 $10^{8} + 10^{-3} = 10^{5}$

$0.00035 \times 10^{8} = 3.5 \times 10^{4}$
Step 1 $0.00035 = 3.5 \times 10^{-4}$
Step 2 $10^{-4} + 10^{8} = 10^{4}$

$0.00000057 \times 10^{-4} = 5.7 \times 10^{-11}$
Step 1 $0.00000057 = 5.7 \times 10^{-7}$
Step 2 $10^{-7} + 10^{-4} = 10^{-11}$

Sometimes it is necessary to remove the power of 10 and return to the original number, without scientific notation. When this is necessary, there are two basic rules to follow.

Rule 1 If the power of 10 is positive, move the decimal to the right, the number of places stated by the power of 10.

Example

$8.6 \times 10^{6} = 8{,}600{,}000.$
$56.7 \times 10^{3} = 56{,}000$

Rule 2 If the power of 10 is negative, move the decimal to the left, the number of places stated by the power of 10.

$0.0034 \times 10^{-5} = 0.000000034$
$56.3 \times 10^{-1} = 5.63$

Practice Problems

Write each in the form of scientific notation.

1. 876,000
2. 1,030,000,000
3. 32,000
4. 25
5. 5.8
6. .03
7. 0.00056
8. 00.00405
9. .0000001000
10. .200
11. $13{,}000 \times 10^{5}$
12. 520×10^{4}
13. 0.0045×10^{7}
14. 0.000039×10^{3}
15. $0.000000056 \times 10^{5}$
16. $52{,}000 \times 10^{-3}$
17. $3{,}200 \times 10^{-6}$
18. $.00046 \times 10^{-5}$
19. 0.00705×10^{-2}
20. 0.0000004×10^{-7}

Practice Problems

Write each as a standard numeral.

1. 4.8×10^5
2. 7.85×10^2
3. 8.9×10^0
4. 34.6×10^1
5. 457×10^3
6. 1.0×10^{-1}
7. 2.01×10^{-3}
8. 00.003×10^{-5}
9. 100×10^{-2}
10. $100,000 \times 10^{-6}$
11. 0.00035×10^5
12. 0.00000400×10^6
13. 709×10^{-3}
14. 5600×10^3
15. $.00000065 \times 10^8$
16. 9.8×10
17. 1.09×10^{-4}
18. 0.00000078×10^0
19. 10×10^0
20. 1×10^1

ENGINEERING NOTATION

Engineering notation is a variation of scientific notation. In electronics and many other scientific and technical fields there are certain powers of 10 that are used more often than others. Names have been placed on certain powers of 10 to make the use of them even easier.

The names of the powers are given in multiples of three. For example: 10^3, 10^6, 10^9, 10^{-3}, 10^{-6}, 10^{-9}, etc. Refer to Table 1-2. Notice how the powers of 10 that are positive also have corresponding powers that are negative.

The engineering notation names have prefixes that are written in front of, and attached to the unit name. For example, when dealing with volts as the unit of measure, engineering notation prefixes could be millivolts, microvolts, kilovolts, or megavolts. Refer to Table 1-2 for a list of names and powers of 10 commonly used in electronics.

Writing Numbers Using Engineering Notation

Convert 48,000 watts to engineering notation. The first step would be to write the number in modified scientific notation, that is to say, use a power of 10 that is a multiple of 3. Then Step 2 would be to replace the power of 10 with the multiplier name. Refer to Table 1-2.

Step 1 48,000 watts = 48×10^3 watts
Step 2 10^3 = kilo . . . therefore . . . 48 kilowatts

Table 1-2. Engineering Notation.

Multiply By	Power of 10	Multiplier Name	Symbol
1,000,000,000,000	10^{12}	tera	T
1,000,000,000	10^9	giga	G
1,000,000	10^6	mega	M
1,000	10^3	kilo	k
1	10^0	basic unit	(no multiplier)
0.001	10^{-3}	milli	m
0.000 001	10^{-6}	micro	μ
0.000 000 001	10^{-9}	nano	n
0.000 000 000 001	10^{-12}	pico	p

Convert 305,000,000 hertz to engineering notation.

Step 1 305×10^6 hertz or 0.305×10^9 hertz (writing the number with a power of 10).
Step 2 305 megahertz or 0.305 gigahertz.

Note When it is not specified which multiplier name to use, either can be selected, depending on the application in the problem.

Convert 0.0025 amps to engineering notation.

Step 1 2.5×10^{-3} amps
Step 2 2.5 milliamps

Sometimes it is necessary to take a number that is already written in scientific notation and change the multiplier name to make it more convenient to use. The easiest way to do this is by removing the multiplier given and replacing it with the proper power of 10. At this point, most students find it best to rewrite the number, without the use of scientific notation. In other words, this returns the number back to the basic unit. Once it is returned to the basic unit, it is simply a matter of moving the decimal the correct number of places to have a power of 10 equal to the multiplier desired.

Examples:

Convert 2500 milliamps to amps.
Step 1 2500 milliamps = 2500×10^{-3} amps
Step 2 2500×10^{-3} = 2.5 amps

Convert 475 kilovolts to megavolts.
Step 1 475 kilovolts = 475×10^3 volts
Step 2 475×10^3 volts = 475,000 volts
Step 3 0.475×10^6 volts = 0.475 megavolts

Convert .255 μA (microamps) to mA (milliamps).
Step 1 .255 μA = $.255 \times 10^{-6}$ amps
Step 2 $.255 \times 10^{-6}$ = .000000255 amps
Step 3 .000000255 = $.000255 \times 10^{-3}$ amps
Step 4 .000255 milliamps

Convert .001 pf (picofarads) to μF (microfarads).
Step 1 .001 pf = $.001 \times 10^{-12}$ farads
Step 2 $.001 \times 10^{-12}$ = .000 000 000 000 001 farads
Step 3 .000 000 001 $\times 10^{-6}$ farads
Step 4 .000 000 001 microfarads

Practice Problems

Change each to the units shown.

1. 2,400,000 Ω	____________	kΩ	____________	MΩ	(ohms)
2. 353,000 Hz	____________	kHz	____________	MHz	(hertz)
3. 2500 W	____________	mW	____________	kW	(watts)
4. 25 V	____________	mV	____________	kV	(volts)

5. .01 H	____________	μH	____________	mH	(henries)
6. 1.5 kΩ	____________	Ω	____________	MΩ	(ohms)
7. 25 MHz	____________	kHz	____________	GHz	(hertz)
8. .03 GHz	____________	MHz	____________	Hz	(hertz)
9. 56 MV	____________	kV	____________	V	(volts)
10. 75 kW	____________	W	____________	MW	(watts)
11. 25 mA	____________	μA	____________	A	(amps)
12. 1500 μA	____________	A	____________	mA	(amps)
13. 1000 μF	____________	pf	____________	f	(farads)
14. .001 pF	____________	μF	____________	nf	(farads)
15. .025 mV	____________	μV	____________	V	(volts)
16. 7500 mW	____________	W	____________	μW	(watts)
17. 500 V	____________	kV	____________	mV	(volts)
18. 2,400 mV	____________	V	____________	kV	(volts)
19. 1 A	____________	kA	____________	mA	(amps)
20. 10 V	____________	mV	____________	kV	(volts)

The following key point should be noted.

- When converting from one multiplier to another, the actual value of the number does not change, even though the decimal place moves.

Multiplication and Division with Engineering Notation

In electronics, there are a great deal of calculations, most of which involve the use of powers of 10. Calculations can be performed through the use of scientific notation or through the use of engineering notation, with multiplier names. Using multiplier names is much more common and therefore, the method that will be concentrated on here.

There are a few basic rules to learn, as there is in any form of mathematics.

Rules for Multiplication and Division

Rule 1 To multiply numbers that contain powers of 10, multiply the numbers and add the powers of 10 exponents.

Example

Multiply 20 kilo × 5 milli

Step 1 multiply the numbers $20 \times 5 = 100$
Step 2 add the powers of 10 exponents
kilo = 10^3
milli = 10^{-3}
$10^3 + 10^{-3} = 10^0 = 1$
Step 3 combine step 1 (the number) with step 2 (the power of 10)
final answer = 100

- When a positive multiplier is multiplied by a negative multiplier of the same size (kilo and milli) the result is canceled unit multipliers. The final answer will be in basic units.

Multiply 100 × 25 milli

Step 1 multiply the numbers 100 × 25 = 2500
Step 2 add the powers of 10 exponents
100 is in basic units = 10^0
milli = 10^{-3}
$10^0 + 10^{-3} = 10^{-3}$
Step 3 combine the number with the power of 10
2500 milli
Step 4 adjust the decimal in the number to better fit the multiplier
2500 milli = 2.5 (basic unit)

Rule 2 To divide numbers that contain powers of 10; divide the numbers and subtract the powers of 10 exponents.

Example

Divide: $\dfrac{100}{25 \text{ milli}}$

Step 1 divide the numbers 100 ÷ 25 = 4
Step 2 subtract the powers of 10 exponents
100 is in basic units = 10^0
milli = 10^{-3}
$10^0 - 10^{-3} = 10^3$
Always subtract the denominator (bottom) from the numerator (top).
Step 3 combine step 1 (the number) with step 2 (the power of 10)
final answer = 4 kilo

- When a positive multiplier is divided into basic units, the result is a negative multiplier of equal size (kilo divided into basic units results in milli). Also, basic divided by negative gives a positive.

Rule 3 Whenever the power of 10 is moved from the numerator to the denominator or the denominator to the numerator, move the power of 10 by changing the sign of the exponent.

Example (to show moving of power only)

$$\frac{1}{\text{micro}} = \frac{1}{10^{-6}} = \frac{10^6}{1} = \text{mega}$$

$$\text{kilo} = 10^3 = \frac{1}{10^{-3}} = \frac{1}{\text{milli}}$$

Note Although, at this time it may be difficult to see a use for rule 3, shown above, it often comes in quite useful. The same operation can be performed by using the rules for division by allowing the 1 shown in the examples to be 10^0.

When working with multiplier names, it is necessary to keep in mind that there are units to deal with. Rather than dealing with the units in this chapter, it is felt the student will have better success with the multiplier names if that is all we have to deal with. In electronics, multiplication, or division of two units often leads to a new unit altogether. An excellent example of this is Ohm's law. E (volts) = I (amps) × R (ohms).

Practice Problems

Complete the operation. Write the answer in engineering notation with the most convenient multiplier. Show the correct power of 10 in answer.

1. 100 kilo × 1 milli
2. 20 mega × 3 milli
3. 10 × 1.5 kilo
4. 4 kilo × 2 micro
5. 5 × 6 nano
6. $\frac{10}{2 \text{ kilo}}$
7. $\frac{50}{25 \text{ milli}}$
8. $\frac{20 \text{ micro}}{5}$
9. $\frac{25 \text{ giga}}{5 \text{ mega}}$
10. 2 × 3.14 × 1 kilo × .01 milli
11. 2 × 3.14 × 10 mega × 10 milli
12. $\frac{1}{2 \times 3.14 \times 1 \text{ mega} \times 1 \text{ micro}}$
13. $\frac{1}{2 \times 3.14 \times 1 \text{ kilo} \times 10 \text{ pico}}$
14. $\frac{2 \text{ giga} \times 3 \text{ kilo}}{6 \text{ mega} \times 1 \text{ milli}}$
15. $\frac{5 \text{ milli} \times 4 \text{ micro}}{2 \text{ nano} \times 2 \text{ kilo}}$

Addition and Subtraction with Engineering Notation

Addition and subtraction with numbers having powers of 10 is an arithmetic function found quite often in the study of electronics.

There are two common ways of adding and subtracting. The first method is by keeping the power of 10 for both numbers the same. The second method is by converting both of the numbers back to basic units and using regular arithmetic. Both methods should be learned since they are both used at different times.

Rules for Adding and Subtracting with Engineering Notation

Rule 1 When the powers of 10 are the same; add or subtract the numbers, keeping the same power of 10.

Examples:

2 kilo + 3 kilo = 5 kilo
4 milli + 6 milli = 10 milli
300 + 5 kilo = .3 kilo + 5 kilo = 5.3 kilo (it is necessary to make the units the same)

Rule 2 If the powers of 10 are not the same; it is often easier to convert the numbers to basic units and use regular arithmetic.

Examples:

100 milli + 1 = .1 + 1 = 1.1 basic units
1.5 kilo + .06 mega = 1500 + 60,000 = 61,500 = 61.5 kilo

2.7 mega + 10 = 2,700,000 + 10 = 2,700,010 = 2.7 mega

Note In this last example, the very small basic unit of 10 was dropped. Using the rules for rounding to three significant figures, this is acceptable.

Practice Problems

Complete the operation. Write the answer in engineering notation with the most convenient multiplier. Show the correct power of 10. Round to three significant figures if needed.

1. 3 kilo + 41 kilo
2. 305 + 609
3. 75 kilo + 1200
4. .85 mega + 150 kilo
5. .75 mega − 25 kilo
6. 2.2 kilo − 850
7. 10 + 2500 milli
8. 25 kilo + 10 milli
9. 85 milli + 1000 micro
10. .90 milli + 150 micro
11. 8 milli − 2 milli
12. 10 milli − 10 micro
13. .1 milli − 90 micro
14. 250 micro + 1500 nano
15. 25 nano + 100 pico
16. .01 micro + .001 micro
17. 15 pico + 25 pico
18. .01 micro + 1500 pico
19. 1.0 + 0.1 milli
20. 201 + 20.1

CHAPTER SUMMARY

This first chapter dealt mostly with writing numbers in either scientific notation or engineering notation. Keep in mind that although engineering notation is used most of the time in electronics, and other technical areas, the multiplier names are used to replace powers of 10.

The following key points should be noted.

- Numbers that are greater than 1 have a positive power of 10.
- Numbers that are less than 1 have a negative power of 10.
- 10 to the zero power (10^0) equals 1. Any number to the zero power equals 1.
- When converting from one multiplier to another, the actual value of the number does not change, even though the decimal place moves.
- When a positive multiplier is multiplied by a negative multiplier of the same size (kilo and milli) the result is canceled unit multipliers. The final answer will be in basic units.
- When a positive multiplier is divided into basic units, the result is a negative multiplier of equal size (kilo divided into basic units results in milli). Also, basic divided by negative gives a positive.

Chapter 2

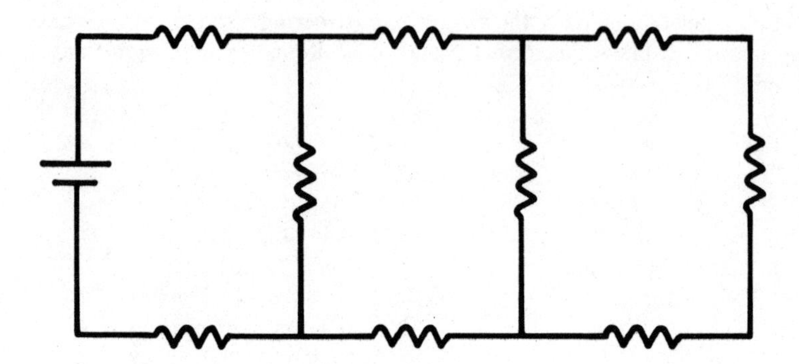

DC Circuit Math

Direct current circuit math is considered the most basic of all mathematics in electronics. However, many students progress to the more difficult subject areas and find themselves with an inadequate background in basic circuit math. It has been found that due to the learning backgrounds of most students, memorizing new material is the most widely used method for learning. The process of memorizing is helpful in the beginning, but in the field of electronics, it is very important for the technician to have a thorough understanding of the subject material.

Resistive circuits are used in every aspect of electronic circuits. It is probably the most important of all subject matter to understand fully.

OHM'S LAW . . . $E = I \times R$

Ohm's law states that the voltage equals the current times the resistance. Many of the formulas in electronics have two, or more, units multiplied, or divided, to produce a completely different unit.

- Stated by Ohm's law: voltage = amps × ohms

All the rules of multipliers apply whenever units are used. Review the rule covered in Chapter 1, if necessary.

Exactly what the formula means and how it is used is discussed later in this chapter. This section of the chapter deals with becoming proficient with performing the arithmetic of the formula and rearranging it to suit the needs of the particular application.

Because the formula has three letters, called unknowns, the formula can be used to solve for any of the three unknowns. In order to use the equation, however, two of the three values must be given. Then, with two values given, the third value is found using the equation.

Start with the basic formula: E = I × R. Use algebra to change the formula as needed.

To find I:

Step 1 $E = I \times R$ formula **Formula 2-1**

Step 2 $\frac{E}{R} = \frac{I \times R}{R}$ divide both sides by R.

Step 3 $\frac{E}{R} = I$ cancel the R on the right side.

$I = \frac{E}{R}$ **Formula 2-1A**

To find R:

Step 1 $E = I \times R$ formula **Formula 2-1**

Step 2 $\frac{E}{I} = \frac{I \times R}{I}$ divide both sides by I.

Step 3 $\frac{E}{I} = R$ cancel the I on the right side.

$R = \frac{E}{I}$ **Formula 2-1B**

Summaries of Ohm's Law Formulas

$E = I \times R$ This formula is used when current (I) and resistance (R) are known, solve for voltage (E). **Formula 2-1**

$I = \frac{E}{R}$ This formula is used when voltage (E) and resistance (R) are known, solve for current (I). **Formula 2-1A**

$R = \frac{E}{I}$ This formula is used when voltage (E) and current (I) are known, solve for resistance (R). **Formula 2-1B**

Note In each of these formulas, the letter E is used to represent voltage. The letter V may be used rather than the letter E.

The practice problems have been designed for the student to grasp full use of the Ohm's law formulas. It is recommended that the student spend time learning the formulas before trying to work the practice problems. The Ohm's law formulas will be used repeatedly in the electrical/electronics field.

Later in the chapter, a fuller understanding of how the formulas work to solve electrical circuits will be developed.

Shortcut symbols the student should be aware of are shown below.

voltage: V for volts, E for EMF (electro-motive force)
current: I for current, A for amps, mA for milliamps, μA for microamps
resistance: ohms or Ω, Greek letter omega, kΩ for kilohms, MΩ for megohms,
R for resistance.

Examples With Ohm's Law Formulas

$E = I \times R$ **Formula 2-1**

Given: I = 5 amps, R = 20 ohms
Find: E

Step 1 $E = I \times R$ formula
Step 2 E = 5 amps × 20 ohms . . . substitute the known values in place of the letters, then multiply
E = 100 volts

$I = \frac{E}{R}$ **Formula 2-1A**

Given: E = 250 volts, R = 2.5 kilohms

Find: I

Step 1 $I = \frac{E}{R}$ formula

Step 2 $I = \frac{250 \text{ volts}}{2.5 \text{ kilohms}}$ substitute the known values in place of the letters, then divide

I = 100 milliamps

$R = \frac{E}{I}$ **Formula 2-1B**

Given: E = 10 volts, I = 5 milliamps
Find: R

Step 1 $R = \frac{E}{I}$ formula

Step 2 $R = \frac{10 \text{ volts}}{5 \text{ milliamps}}$ substitute the known values in place of the letters, then divide

R = 2 kilohms

Practice Problems

Use Ohm's law to solve for the unknown quantity. Show the answers with the proper units and use engineering notation, whenever possible.

1. I = 2 amps, R = 10 ohms, find E:
2. R = 100 ohms, I = .5 amps, find E:
3. R = 1 kilohms, I = 20 milliamps, find E:
4. I = 100 mA, R - 3 kΩ, find E:
5. R = 1.2 MΩ, I = 10 μA, find E:
6. E = 10 volts, R = 100 ohms, find I:
7. R = 50 Ω, E = 50 V, find I:
8. R = 1 kohm, E = 10 V, find I:
9. E = 12 volts, R = 1.2 kΩ, find I:
10. E = .5 volts, R = 50 ohms, find I:
11. E = 10 volts, I = 1 amp, find R:
12. I = 5 amps, E = 50 volts, find R:
13. I = 10 mA, E = 20 volts, find R:
14. E = 30 volts, I = 15 μA, find R:
15. V = 50 volts, I = 10 mA, find R:
16. V = 40 volts, R = 400 ohms, find I:
17. I = 15 amps, R = 2 ohms, find E:
18. R = 1500 ohms, E = 1.5 kV, find I:
19. I = 2 mA, E = 2.4 kV, find R:
20. E = 24 volts, I = 24 amps, find R:

POWER FORMULAS $P = I \times E$

The formula shown is considered to be the basic power formula. It is used to calculate the power dissipated in a dc circuit when the voltage and current are known. Of course, if the voltage and current are known, then resistance can also be calculated.

- In any dc circuit there are four quantities that can be calculated; voltage, current, resistance, and power.

In this section, formulas based on both Ohm's law and the power formula that will allow calculations of all four quantities, with any two of the four given, will be developed.

$P = I \times E$ **Formula 2-2**

Power (watts) = Current (amps) × Voltage (volts)

The formula as it is shown here will allow the calculation of power if current and voltage are given. The formula can be rearranged to solve for either of the other two quantities.

Step 1 $P = I \times E$ original formula **Formula 2-2**

Step 2 $\frac{P}{E} = \frac{I \times E}{E}$ divide both sides by E

$\frac{P}{E} = I$ **Formula 2-2A**

Step 1 $P = I \times E$ original formula **Formula 2-2**

Step 2 $\frac{P}{I} = \frac{I \times E}{I}$ . . . divide both sides by I

$$\frac{P}{I} = E$$ **Formula 2-2B**

Combining Ohm's Law Formulas with the Power Formulas

If Ohm's law formulas are combined with the power formulas, it is possible to solve for any of the four quantities (P, I, R, and E), regardless of which two values are given.

Find P, substitute $I \times R$ for E

Step 1 $P = I \times E$ and $E = I \times R$ original formulas
Step 2 $P = I \times (I \times R)$ . . . substitute the $I \times R$ of Ohm's law into the power formula for E.
Step 3 $P = I^2R$ . . . combine like terms. **Formula 2-3**

Formula 2-3 shown here can be used to calculate power when current and resistance are known. Rearranging this formula makes it possible to solve for either current or resistance when the other two are known.

Find I

Step 1 $P = I^2R$ formula

Step 2 $\frac{P}{R} = \frac{I^2R}{R}$ divide both sides by R.

Step 3 $\frac{P}{R} = I^2$ . . . cancel the R on the right side.

Step 4 $I = \sqrt{\frac{P}{R}}$ taking the square root of both sides **Formula 2-3A**

Find R

Step 1 $P = I^2R$ . . . formula.

Step 2 $\frac{P}{I^2} = \frac{I^2R}{I^2}$ divide both sides by I^2.

Step 3 $\frac{P}{I^2} = R$ cancel the I^2 on the right side **Formula 2-3B**

Find P, substitute $\frac{E}{R}$ for I

Step 1 $P = I \times E$ and $I = \frac{E}{R}$ formulas

Step 2 $P = \frac{E}{R} \times E$ . . . substitute the E/R for the I

Step 3 $P = \frac{E^2}{R}$ . . . combine like terms. **Formula 2-4**

The formula as it is shown here can be used to calculate the power when the voltage and resistance are given. Rearranging the formula makes it possible to solve for either voltage or resistance when power is known.

Find E,

Step 1 $P = \frac{E^2}{R}$ formula

Step 2 $P \times R = \frac{E^2 \times R}{R}$ . . . multiply both sides by R

Step 3 $P \times R = E^2$ cancel the R on the right side.

Step 4 $E = \sqrt{P \times R}$ . . . take the square root of both sides **Formula 2-4A**

Table 2-1. Ohm's Law and Power Formulas.

Given	Find	Formula	Alternate Formulas
I, R	E	$E = I \times R$	
I, R	P	$P = I^2 R$	$E = I \times R$ then $P = I \times E$
E, R	I	$I = \frac{E}{R}$	
E,R	P	$P = \frac{E^2}{R}$	$I = \frac{E}{R}$ then $P = I \times E$
I, E	R	$R = \frac{E}{I}$	
I, E	P	$P = I \times E$	
P, E	I	$I = \frac{P}{E}$	
P, E	R	$R = \frac{E^2}{P}$	$I = \frac{P}{E}$ then $R = \frac{E}{I}$
P, I	E	$E = \frac{P}{I}$	
P, I	R	$R = \frac{P}{I^2}$	$E = \frac{P}{I}$ then $R = \frac{E}{I}$
P, R	I	$I = \sqrt{\frac{P}{R}}$	
P, R	E	$E = \sqrt{P \times R}$	$I = \sqrt{\frac{P}{R}}$ then $E = I \times R$

Find R,

Step 1 $P = \frac{E^2}{R}$ formula

Step 2 $P \times R = \frac{E^2 \times R}{R}$ . . . multiply both sides by R.

Step 3 $P \times R = E^2$ cancel R on right side.

Step 4 $\frac{P \times R}{P} = \frac{E^2}{P}$ divide both sides by P.

Step 5 $R = \frac{E^2}{P}$ cancel the P on the left side. **Formula 2-4B**

Between Ohm's law formula and power formula, there are 12 different combinations. Notice, in each combination there are only three variables, therefore, each variation of the original formula needs only two known values.

Practice Problems

Use Ohm's law and the power formulas to complete the table.

	Voltage	Current	Resistance	Power
1.	E = ______	I = 15 amps	R = 2 ohms	P = ______
2.	E = ______	I = 2 amps	R = ______	P = 5 watts
3.	E = 100 volts	I = .01 amps	R = ______	P = ______
4.	E = 250 volts	I = ______	R = 5 ohms	P = ______
5.	E = 15 volts	I = ______	R = ______	P = 100 mW
6.	E = ______	I = ______	R = 1 kΩ	P = 10 mW
7.	E = ______	I = 25 mA	R = 10 kΩ	P = ______
8.	E = ______	I = 2 mA	R = ______	P = 25 mW
9.	E = 50 volts	I = 5 mA	R = ______	P = ______
10.	E = 5 volts	I = ______	R = 2500 ohms	P = ______
11.	E = 0 volts	I = ______	R = 100 ohms	P = ______
12.	E = 10 volts	I = ______	R = ______	P = 10 watts
13.	E = ______	I = 10 μa	R = 10 kΩ	P = ______
14.	E = ______	I = 100 mA	R = ______	P = 1 watt
15.	E = 25 volts	I = 1 amp	R = ______	P = ______
16.	E = 20 volts	I = ______	R = 200 ohms	P = ______
17.	E = 1 volt	I = ______	R = ______	P = 1 watt
18.	E = ______	I = ______	R = 1000 ohms	P = .1 watt
19.	E = ______	I = 1000 mA	R = 10,000 Ω	P = ______
20.	E = 10 volts	I = 100 mA	R = ______	P = ______

DIRECT/INDIRECT RELATIONSHIPS IN FORMULAS

Now that there has been practice with the use of the formulas, it is time to look at how the units of voltage, current, resistance, and power react with each other. In other words, when looking at any of the Ohm's law or power formulas, what happens to the third quantity if one is held constant and the other changed?

There are two types of relationships that are of particular concern, the direct relationship and the indirect relationship.

Direct Relationships

A direct relationship can be defined as follows: when a quantity is increased, the other also increases; when one is decreased, the other decreases.

Examples:

$E = I \times R$ Relationships are always compared on opposite sides of the equals sign. If R is held constant, when there is an increase in current, there will be an increase in voltage. If I is held constant, when there is an increase in resistance there will be an increase in voltage.

$P = I \times E$ If E is held constant, there will be an increase in P when I is increased. If I is held constant, there will be an increase in P when E is increased.

By these two examples we can establish some key points.

- I is directly related to P.
- E is directly related to I.
- E is directly related to R.
- E is directly related to P.

These key points help to show the effect on different quantities in the event it is necessary to change or compare the values of different quantities in a circuit.

Indirect Relationships

An indirect relationship can be defined as follows: when a quantity is increased, the other decreases; when one is decreased, the other increases. It should be noted, with the direct relationships, the two quantities that are directly related were both in the numerator of the equation. It stands to reason that, with indirect relationships, one quantity will be in the numerator and the other in the denominator.

Examples:

$I = \frac{E}{R}$ Relationships are always compared on opposite sides of the equals sign. If E is held constant, an increase in R will cause the current to decrease. A decrease in R will cause the current to increase.

Notice with the example shown here the E and the I are both in the numerator. Their relationship does not change.

- R is indirectly related to I.

Due to the fact that resistance is directly related to voltage and indirectly related to current, this makes the relationship of resistance and power difficult to see from the formulas.

- R is indirectly related to P.

SERIES CIRCUITS

A series circuit is defined as having only one path for current to flow. In order to have any current flow, in any circuit, there must be a complete path for the current to return to the same power supply. That is to say, for each electron leaving a power source, one must return. A series circuit has four key points.

- In a series circuit, there is only one current path. Current is the same throughout

a series circuit. This is the definition of a series circuit.
- Voltage drops at each resistance directly related to the size of the resistance.
- The sum of the voltage drops in a series circuit must be equal to the supply voltage.
- The sum of the powers dissipated in a series circuit must be equal to the total power.

Refer to Fig. 2-1. This is a series circuit because current has only one path to follow. There can be any number of resistors, provided there is only one path for current to flow, it will still be a series circuit. Notice in the drawing that there is an arrow indicating a direction to current flow. The direction is from minus to plus, starting from the minus, or negative side of the battery (smaller of the two battery terminals), flowing through the load, returning to the positive (plus) side of the battery. As long as current can return to the power source it is considered a complete circuit.

One word about direction of current flow. This book, and most modern textbooks, refer to current flow as being from negative to positive. This is called electron current flow. Technicians need to be aware that current can be described as flowing in the opposite direction, from plus to minus. This is called conventional current flow.

- Electron current flow is from negative to positive. Conventional is from positive to negative.

Calculations

Calculate current, often referred to as total current. See Fig. 2-1.

Step 1 $I = \frac{E}{R}$ write the equation

Step 2 $I = \frac{10 \text{ volts}}{100 \text{ ohms}}$ substitute values

$I = .1$ amps or 100 mA

Calculate power. True power is only dissipated in a resistance. In a later chapter, apparent power will be discussed. Power is the work done, or in the case of electricity, the heat produced.

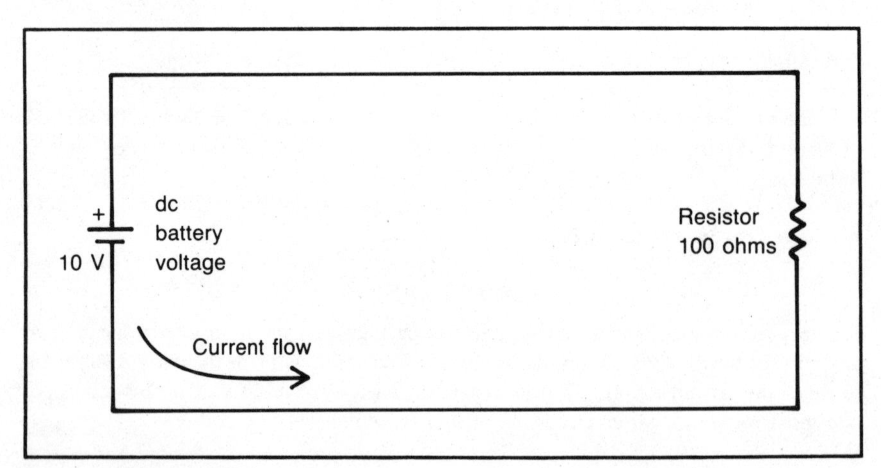

Fig. 2-1. Series circuit.

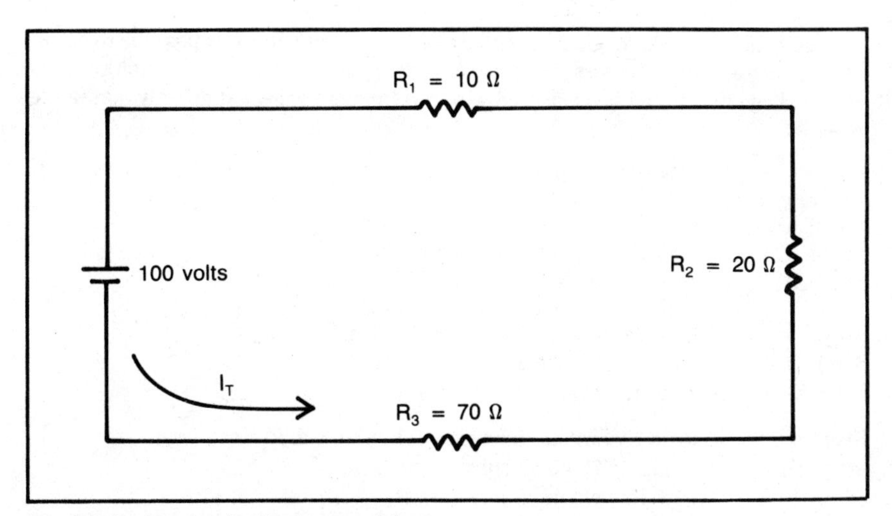

Fig. 2-2. Series circuit with three resistors.

Step 1 P = I × E write the equation
Step 2 P = .1 amps × 100 ohms substitute values
P = 10 watts

Note In the power calculations shown above, the value of current used is the result of a previous calculation. If a mistake had been made in the current calculation, power would also be wrong. Be careful or use given values in all calculations.

Figure 2-2 shows three resistors in series, connected to a single power supply. The direction of current flow is from negative to positive and current must flow through each resistor to return to the power source. The current is labeled I_T to represent total current. Keep in mind, current is the same throughout a series circuit.

In Fig. 2-2 or, for that matter, any circuit that contains more than one resistor in series requires more calculations than just the use of Ohm's law. It is first necessary to determine the total resistance of the circuit. Resistors in series add, giving the formula:

resistors in series $R_T = R_1 + R_2 + R_3 + \ldots$ **Formula 2-5**

Using the resistance total formula for series circuits, calculate the total resistance of Fig. 2-2.

Calculate total resistance.

Step 1 $R_T = R_1 + R_2 + R_3$ formula
Step 2 R_T = 10 ohms + 20 ohms + 70 ohms substitute circuit values
Step 3 R_T = 100 ohms perform the addition

Now that the circuit resistance is known, Ohm's law can be used to find the circuit current.

Step 1 $I = \frac{E}{R}$ formula

Step 2 $I = \frac{100 \text{ volts}}{100 \text{ ohms}}$ substitute circuit values

Step 3 I_T = 1 amp

Using the current through the resistor and the resistor value it is possible to calculate the voltage dropped across each resistor, often called the IR drop. The voltage drop is the voltage that would be measured with a voltmeter connected directly across the resistor.

$E = I \times R$ formula to be used

Step 1 $E_{R1} = I_T \times R_1$ formula for voltage across R_1
Step 2 $E_{R1} = 1$ amp $\times$ 10 ohms substitute values
Step 3 $E_{R1} = 10$ volts voltage drop across R_1

Step 1 $E_{R2} = I_T \times R_2$ formula for voltage across R_2
Step 2 $E_{R2} = 1$ amp $\times$ 20 ohms substitute values
Step 3 $E_{R2} = 20$ volts voltage drop across R_2

Step 1 $E_{R3} = I_T \times R_3$ formula for voltage drop across R_3
Step 2 $E_{R3} = 1$ amp $\times$ 70 ohms substitute values
Step 3 $E_{R3} = 70$ volts voltage drop across R_3

Summary of voltage drops:

$E_{R1} = 10$ volts
$E_{R2} = 20$ volts
$E_{R3} = 70$ volts

If all the voltage drops were calculated correctly, they should add to the applied voltage.

Step 1 $E_{R1} + E_{R2} + E_{R3} =$ applied voltage, E_T
Step 2 $10 + 20 + 70 = 100$ volts checks with applied voltage

- The sum of the voltage drops in a series circuit equals the applied voltage.

Calculate the power dissipated by each resistor.

$P = I \times E$ formula to be used

Step 1 $P_{R1} = I_T \times E_{R1}$ power in R_1
Step 2 $P_{R1} = 1$ amp $\times$ 10 volts substitute values
Step 3 $P_{R1} = 10$ watts power in R_1

Step 1 $P_{R2} = I_T \times E_{R2}$ formula for power in R_2
Step 2 $P_{R2} = 1$ amp $\times$ 20 volts substituting values
Step 3 $P_{R2} = 20$ watts power in R_2

Step 1 $P_{R3} = I_T \times E_{R3}$ formula for power in R_3
Step 2 $P_{R3} = 1$ amp $\times$ 70 volts substituting values
Step 3 $P_{R3} = 70$ watts power in R_3

Summary of power dissipation.

$P_{R1} = 10$ watts
$P_{R2} = 20$ watts
$P_{R3} = 70$ watts

If all the powers have been calculated correctly, the power dissipated by the resistors should be equal to the power supplied by the power supply.

Step 1 $P_{R1} + P_{R2} + P_{R3} = P_T$
Step 2 $10 + 20 + 70 = 100$ watts

The sum of the powers can be checked by using the power formula with total current and applied voltage.

$P = I \times E$ formula to be used

Step 1 $P_T = I_T \times E_T$ power total equals the total current times the total applied voltage
Step 2 $P_T = 1$ amp $\times$ 100 volts substituting values
Step 3 $P_T = 100$ watts checks with the sum of the powers

- The total power dissipated in a series circuit is equal to the sum of the individual powers.

There are other observations that should be made about the series circuit and the calculations just made. Note the following key points.

- In a series circuit, the largest resistance will drop the most voltage.
- In a series circuit, the largest resistance will dissipate the most power.

There are two other conditions that can happen to a circuit, a short circuit and an open circuit. Both of these conditions are usually considered defects or faults.

- A short circuit has a resistance of zero. This condition has unlimited current, zero voltage drop and zero power.
- An open circuit has a resistance of infinity. This condition causes zero current, applied voltage dropped at open, zero power.

General Guidelines for Solving Series Circuits

The following general procedure is intended as a general guideline when solving the series circuit. It is not intended to consider every possible condition of known and unknown values in a circuit.

Step 1 Find R_T. Find the total resistance first is almost always the best way to proceed. If one value of resistance is not known, then the current must be known, or be able to be calculated from some other information.
Step 2 Calculate the total current, I_T. Keep in mind, current is the same throughout a series circuit.
Step 3 Find the individual voltage drops across each resistor. Use I_T and the resistance values.
Step 4 Check the calculated voltage drops by adding them together to see if they equal the applied voltage.
Step 5 Calculate the power dissipated in each resistor. Use the current, I_T, and the individual voltage drops.
Step 6 The individual powers should be checked by adding and see if they equal the total power. Total power is calculated using total current and applied voltage.

Note Once the total circuit resistance is found, all the other calculations will fall into place. Sometimes, one of the resistor values is not known and the circuit will have to be worked differently.

Practice Problems

Use the schematic diagrams shown to find the unknown quantities.

1. Find: I_T, R_T, E_{R1}, E_{R2}, P_{R1}, P_{R2}

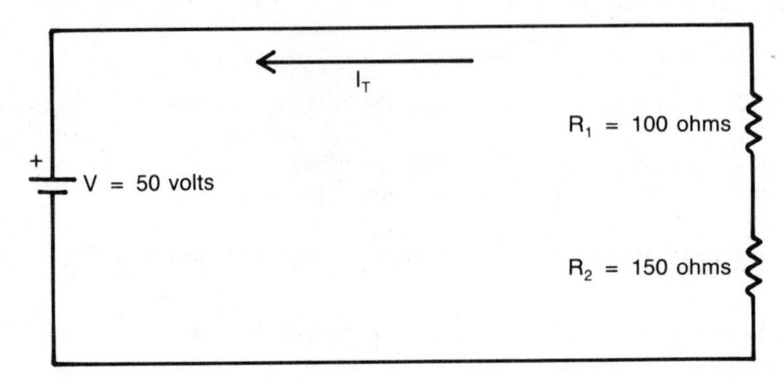

2. Find: I_T, R_T, E_{R1}, E_{R2}, E_{R3}, P_{R1}, P_{R2}, P_{R3}

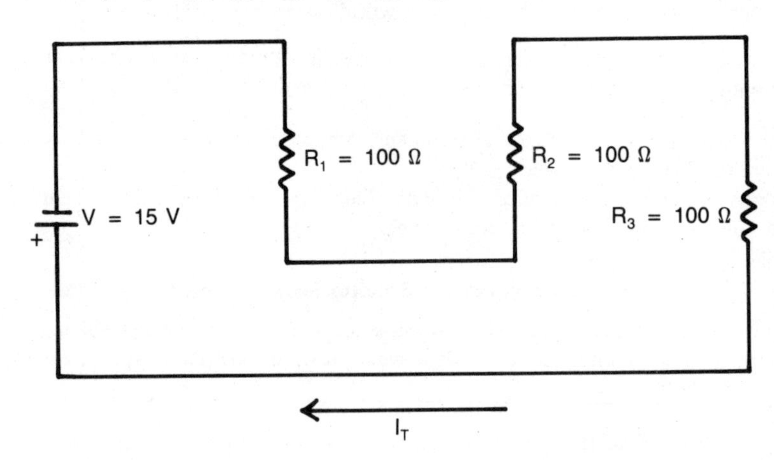

3. Find: R_T, I_T

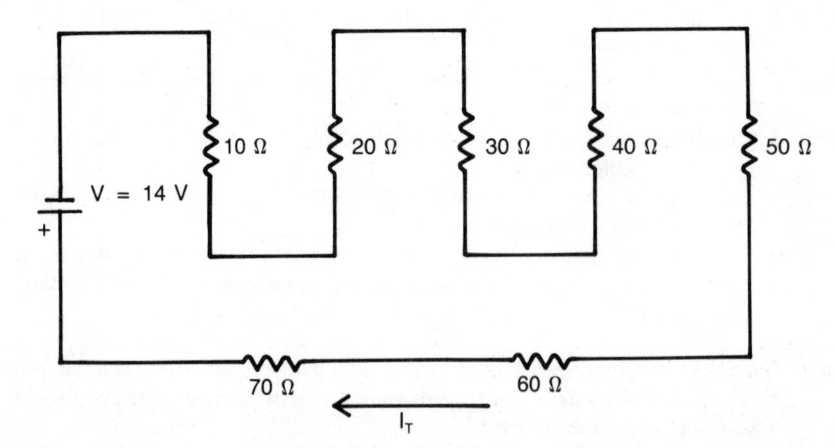

4. Given: E_{R2} = 90 volts; Find: I_T, V

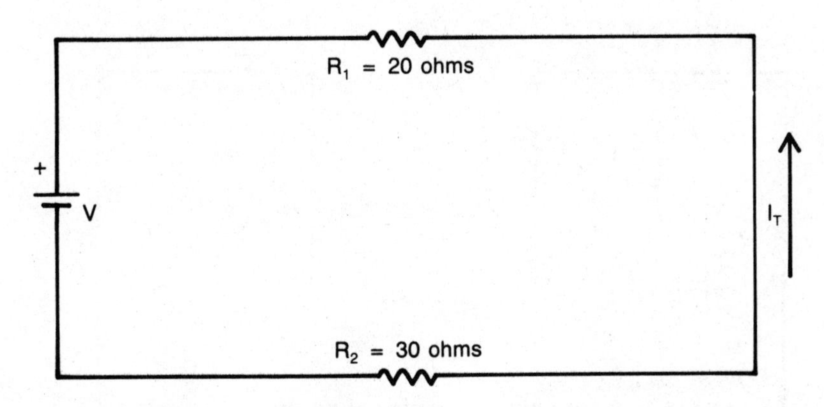

5. Given: E_{R3} = 60 volts; Find: R_T, I_T, V

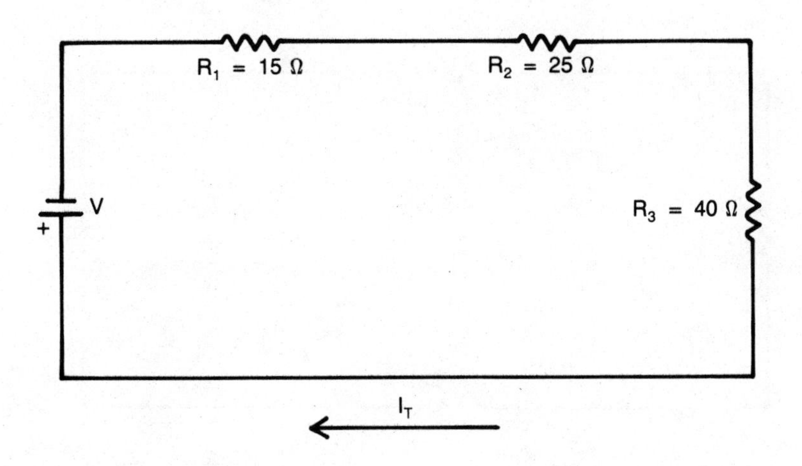

6. Given: I_T = 2 mA; Find: R_T, R_1

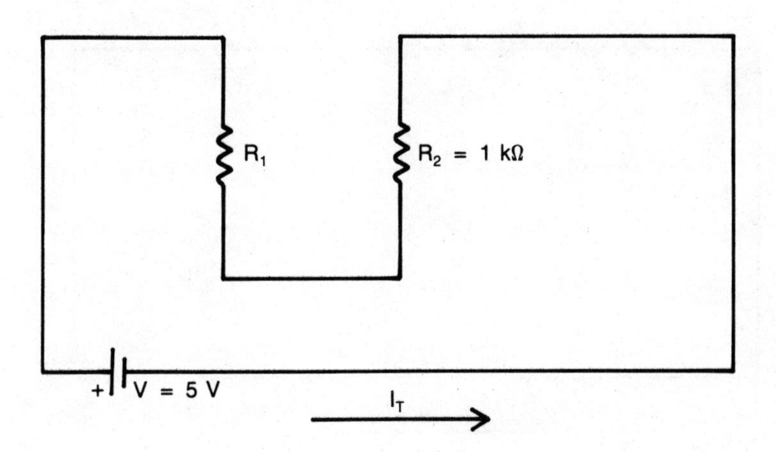

7. Given: I_T = 100 mA; Find: R_T, R_3, V
E_{R3} = 10 volts

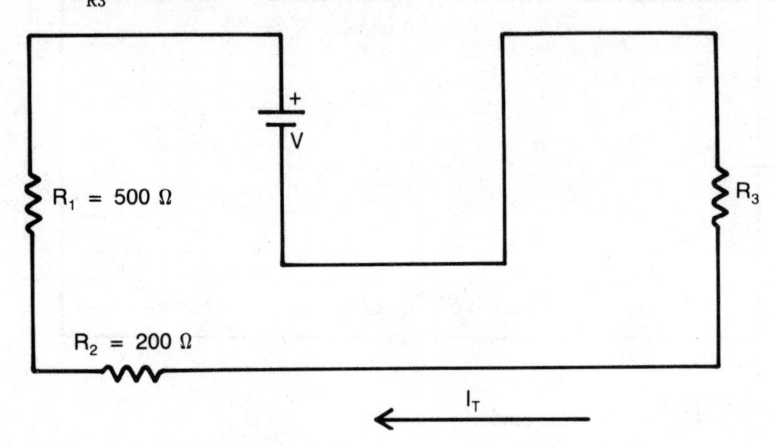

8. Given: P_{R3} = 200 mW; Find: I_T, R_2, E_{R1}

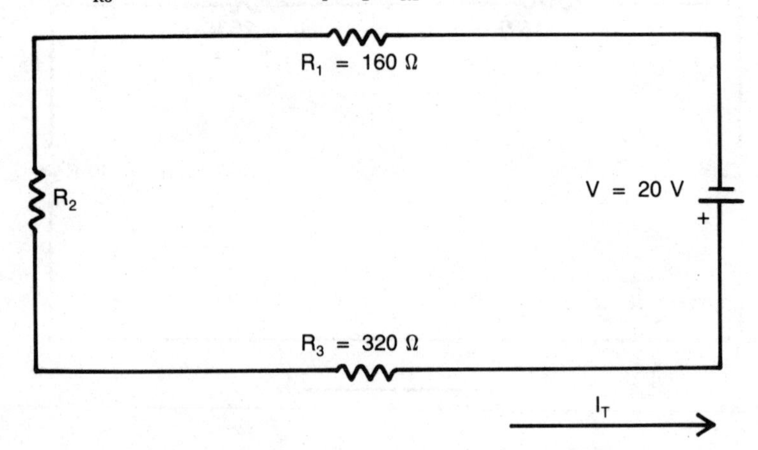

9. Given: P_T = 12.8 watts; Find: V, I_T

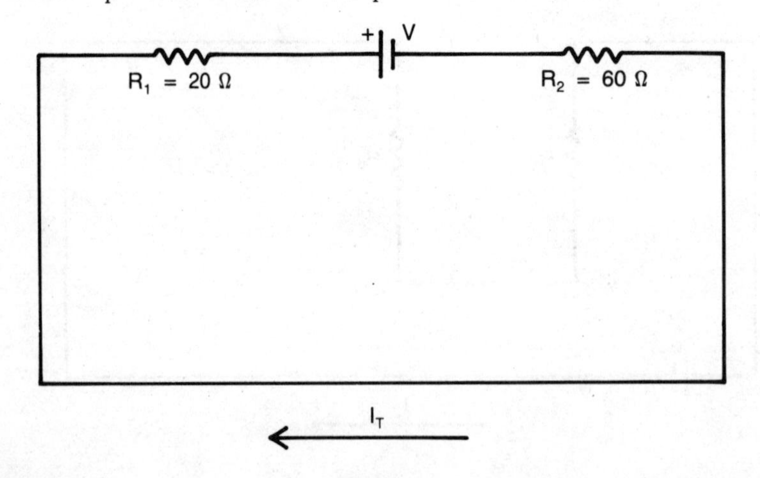

10. Given: P_T = 240 watts, P_{R1} = 40 watts; Find: V, I_T, R_2

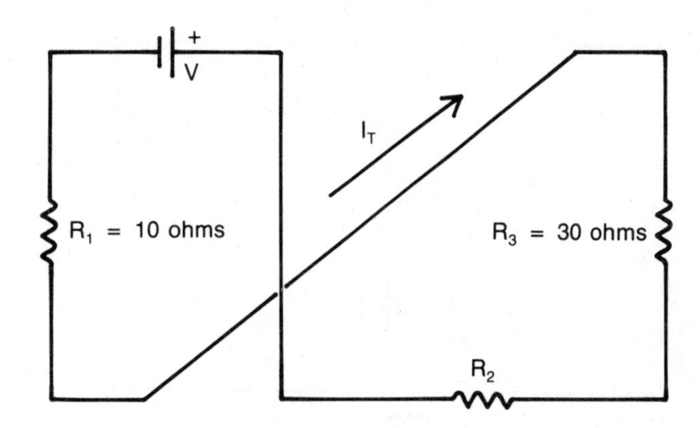

PARALLEL CIRCUITS

A parallel circuit is defined as having more than one path for current to flow. In a simple parallel circuit there is a piece of wire connecting all the resistors on one side and another piece of wire to connect the resistors on their other side. Since a piece of wire has no voltage dropped across it, voltage to all the resistors is the same. Remember, in a series circuit, each resistor had a different voltage, directly proportional to the size of the resistor. In a parallel circuit, it is the current that will be divided to each resistor branch, indirectly proportional to the size of the resistance.

- In a parallel circuit, the current divides to the individual branches indirectly proportional to the size of the resistors.
- Total current in a parallel circuit is equal to the sum of the individual branch currents.
- Voltage is the same throughout a parallel circuit.
- The sum of the powers dissipated in a parallel circuit is equal to the total power.

Figure 2-3 shows a parallel circuit with two resistive branches. Notice how the arrows in the drawing that show the current leaving the power supply is the sum of the individual branch currents. The branch currents are calculated using the voltage across the resistor and the resistance value, keeping in mind the voltage is the same across both branches.

Calculating Current in a Parallel Circuit

Each individual branch can be treated as an individual series circuit. Calculate the branch current as would be done in any series circuit.

R_1 branch current.

Step 1 $I_1 = \frac{E}{R_1}$ formula

Step 2 $I_1 = \frac{10\ V}{25\ \Omega}$ substitute circuit values

Step 3 I_1 = .4 amps performing division gives the current in R_1 branch

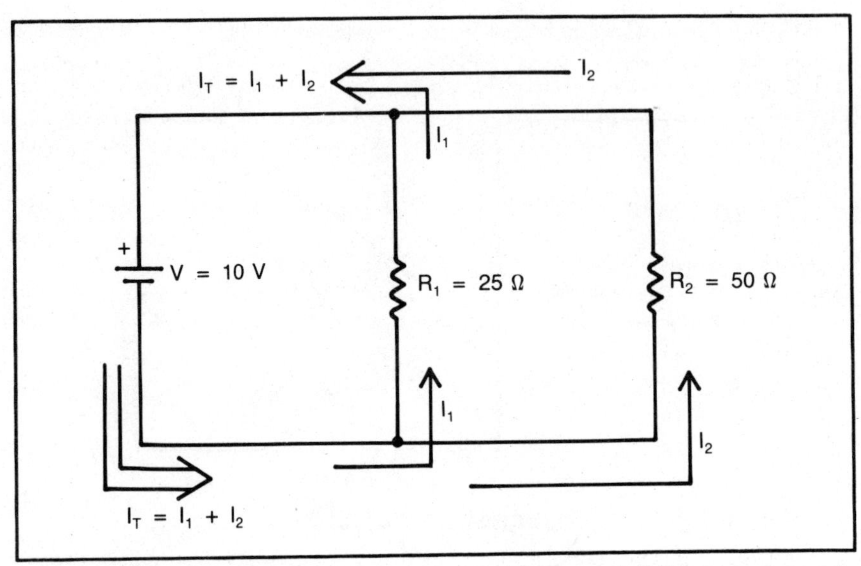

Fig. 2-3. Parallel circuit with two branches showing how current divides to each branch.

R_2 branch current.

Step 1 $I_2 = \frac{E}{R_2}$ formula

Step 2 $I_2 = \frac{10\ V}{50\ \Omega}$ substitute circuit values

Step 3 $I_2 = .2$ amps performing division gives the current in R_2 branch

$I_T = I_1 + I_2$ total current

Step 1 $I_T = .4$ amps + .2 amps
Step 2 $I_T = .6$ amps

Calculating Total Resistance in a Parallel Circuit

Calculating the current in a simple parallel circuit is fairly easy, using Ohm's law. To calculate the total resistance of the simple circuit shown in Fig. 2-3, it can be done using total current and applied voltage.

R_T is the total resistance of the circuit.

Step 1 $R_T = \frac{E}{I_T}$ formula

Step 2 $R_T = \frac{10\ V}{.6\ A}$ substitute applied voltage and total current

Step 3 $R_T = 16.7$ ohms divide

Figure 2-4 shows an equivalent circuit from the calculated values to replace the original circuit shown in Fig. 2-3.

Total resistance can also be called the equivalent resistance of the parallel circuit. When it is connected to the power supply, the total current will be exactly equal to the original circuit. Equivalent circuits are very helpful when the circuits are much more complicated.

Power calculations are performed exactly the same as they are done in series circuits.

The method used to calculate total resistance, shown in Step 4, is fine if total current can be calculated first. In a more complex circuit, it would be impossible to first calculate total current. Therefore, it is necessary to have a way of calculating the total resistance of the parallel circuit independently of current. That leads to a new formula.

resistors in parallel $\frac{1}{R_T} = \frac{1}{R_1} + \frac{1}{R_2} + \frac{1}{R_3} + \ldots$ **Formula 2-6**

The formula shown above is called the reciprocal formula. This formula can be used regardless of how many resistors are connected in parallel.

R_T using the reciprocal formula for Fig. 2-3.

Step 1 $\frac{1}{R_T} = \frac{1}{R_1} + \frac{1}{R_2}$ formula

Step 2 $\frac{1}{R_T} = \frac{1}{25} + \frac{1}{50}$ substitute circuit values

Step 3 $\frac{1}{R_T} = .04 + .02$ changing the fractions to decimals using the calculator is much easier than trying to deal with fractions

Step 4 $\frac{1}{R_T} = .06$ perform the addition. R_T is still in reciprocal form

Step 5 $R_T = 16.7$ ohms take the reciprocal

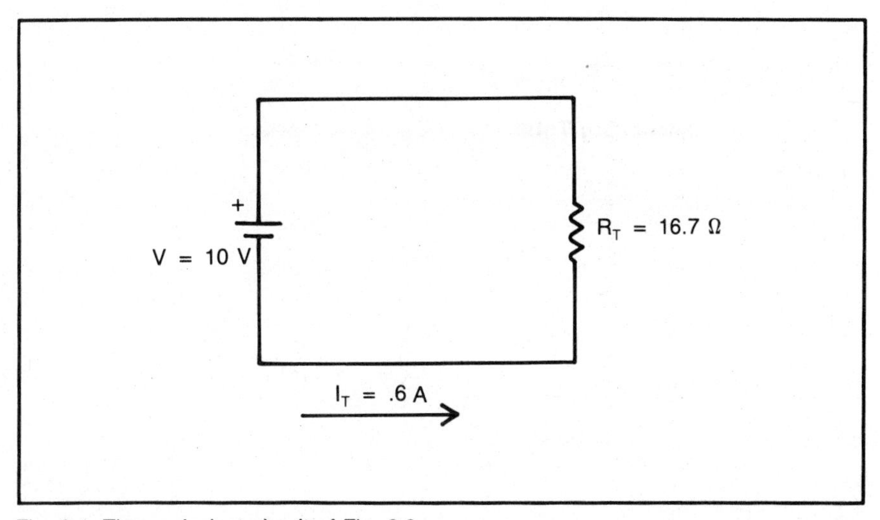

Fig. 2-4. The equivalent circuit of Fig. 2-3.

The method shown in Step 5 arrives at the same answer as using the total current. The method shown here must be used with complicated circuits. If there are only two resistors in the circuit, a shortcut formula can be used.

shortcut formula for two resistors in parallel $R_T = \dfrac{R_1 \times R_2}{R_1 + R_2}$ **Formula 2-7**

The shortcut formula will not be demonstrated here because it is very straightforward. Keep in mind, the shortcut formula is limited to only two resistors. The reciprocal formula is used for any number of resistors in parallel. With the electronic calculator to change fractions into decimals, it is probably the easiest method.

Calculations for parallel circuits must be done separately from the calculations for a series circuit.

The total resistance of a parallel circuit is almost always the first thing that will be found, especially in a more complex circuit. Often, the R_T will be used to find other calculations. If a mistake is made in R_T all other calculations will also be wrong. One quick check is, the resistance of each branch is higher than the total resistance.

- The total resistance of a parallel circuit is smaller than the smallest resistance of any branch.

Figure 2-5 shows three parallel branches. This circuit will be used to demonstrate the effect of having equal resistance in each of the parallel branches. Notice in branch B there are two resistors. This branch is actually a series circuit, by itself. Two resistors in series simply add, so the middle branch has a resistance of 60 ohms, as does the other two branches.

Calculate R_T for Fig. 2-5.

Step 1 $\dfrac{1}{R_T} = \dfrac{1}{R_A} + \dfrac{1}{R_B} + \dfrac{1}{R_C}$ formula

Step 2 $\dfrac{1}{R_T} = \dfrac{1}{60} + \dfrac{1}{60} + \dfrac{1}{60}$ substitute values, branch B is a series circuit equal to 30 + 30 = 60 ohms

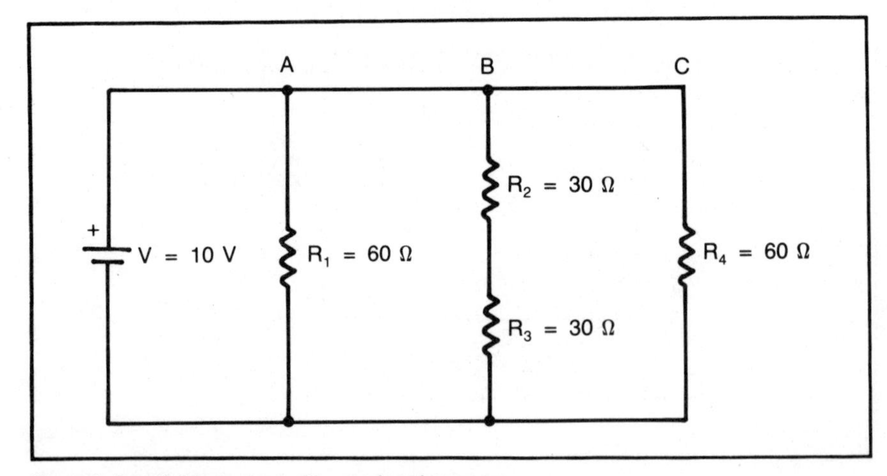

Fig. 2-5. Parallel branches with equal resistances.

Step 3 $\frac{1}{R_T} = .01667 + .01667 + .01667$ change fractions to decimals using the calculator

Step 4 $\frac{1}{R_T} = .05$ add

Step 5 $R_T = 20$ ohms

Notice from the calculations, the value of R_T is one-third the value of one resistor. This problem could have been solved by dividing the resistance of one branch by the number of branches having equal resistance.

- To find the total resistance of equal resistances in parallel, divide the resistance of one branch by the number of branches with equal resistance.

Calculate the value of total current.

Step 1 $I_T = \frac{V}{R_T}$ formula

Step 2 $I_T = \frac{10\ V}{20\ \Omega}$ substitute values

Step 3 $I_T = .5$ amps

Calculate the value of branch currents.

Step 1 $I = \frac{E}{R}$ formula

Step 2 $I = \frac{10\ V}{60\ \Omega}$ substitute values for any branch

Step 3 $I = .1667$ amps any branch current

- In a parallel circuit, if branches have equal resistance, the current through each branch will be equal.
- The total current divided by the number of branches with equal resistances will give the individual branch currents.

Solving a Parallel Circuit with One Unknown Resistance

Refer to Fig. 2-6. In this circuit, there are three branches, with the resistance of only two known. The total current and the applied voltage is given, which means the first step would be to calculate the total resistance. From there, the unknown branch resistance can be calculated. See Fig. 2-6.

Solution

Calculate the total resistance of the parallel circuit.

Step 1 $R_T = \frac{V}{I_T}$ formula

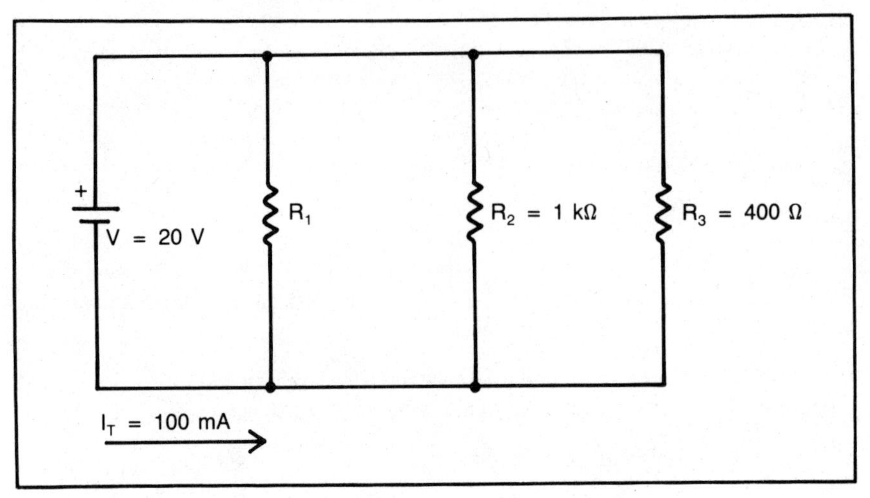

Fig. 2-6. Solving a parallel circuit when one resistance is unknown.

Step 2 $R_T = \dfrac{20\ V}{100\ mA}$ substitute values

Step 3 R_T = 200 ohms

Using the R_T just calculated, find R_1.

Step 1 $\dfrac{1}{R_T} = \dfrac{1}{R_1} + \dfrac{1}{R_2} + \dfrac{1}{R_3}$ formula

Step 2 $\dfrac{1}{200} = \dfrac{1}{R_1} + \dfrac{1}{1000} + \dfrac{1}{400}$ substitute values

Step 3 $.005 = \dfrac{1}{R_1} + .001 + .0025$ convert to decimals

Step 4 $\dfrac{1}{R_1} = .005 - (.001 + .0025)$ rearranging the equation

Step 5 $\dfrac{1}{R_1} = .005 - .0035$ inside parenthesis first

Step 6 $\dfrac{1}{R_1} = .0015$ complete the subtraction

Step 7 R_1 = 667 take the reciprocal

General Guidelines for Solving Parallel Circuits

The following is intended to be a general guideline when solving parallel circuits. Each circuit will have to be worked according to what is given and what is to be found. Having a good working knowledge of the formulas is the best way to solve circuits.

Step 1 Find R_T. It is almost always best to solve a circuit by first finding the total resistance of the circuit. Keep in mind, the total resistance of a parallel circuit is always smaller than the smallest branch resistance.

Step 2 Find I_T. Use the applied voltage and the total resistance. Total current can also be found by adding each of the branch currents.

Step 3 Find branch currents. Keep in mind, voltage is the same across all branches of a parallel circuit, while the current will divide to the individual branches, indirectly proportional to the branch resistance. Largest resistance, smallest current.

Step 4 Power is the same as with series circuits. The current through a resistor, times the voltage across a resistor. Total power is the sum of the individual powers.

Practice Problems

Use the schematic diagrams shown to find the unknowns.

1. Find: R_T, I_T, I_A, I_B, I_C (Branch currents)

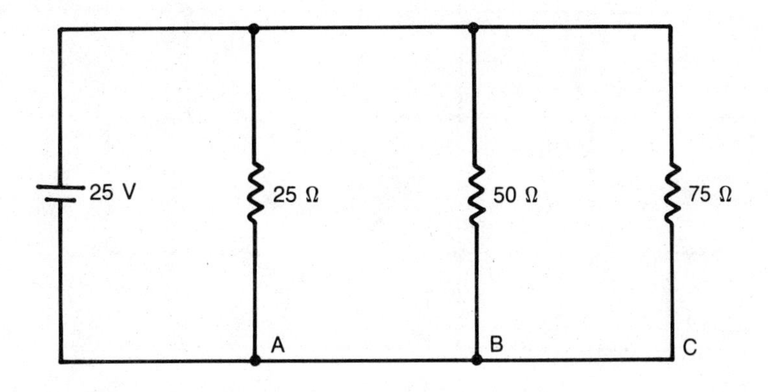

2. Find: R_T, I_T

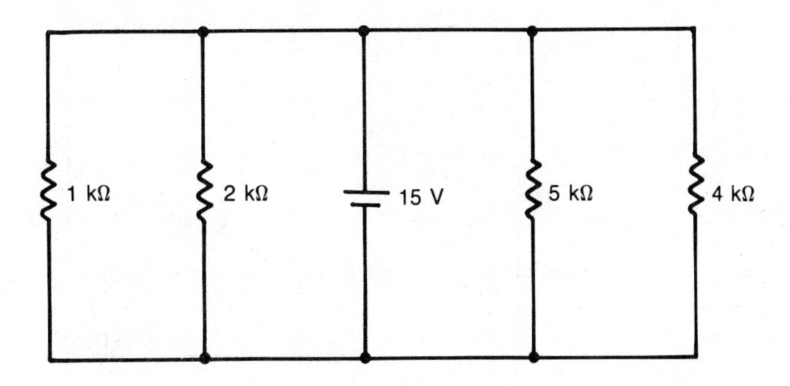

3. Find: R_T

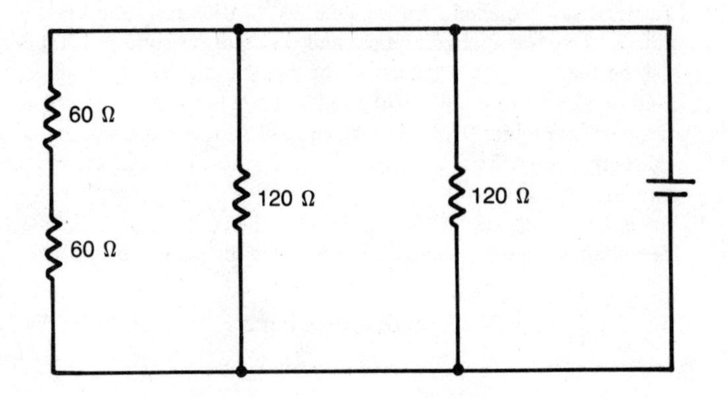

4. Find: R_T

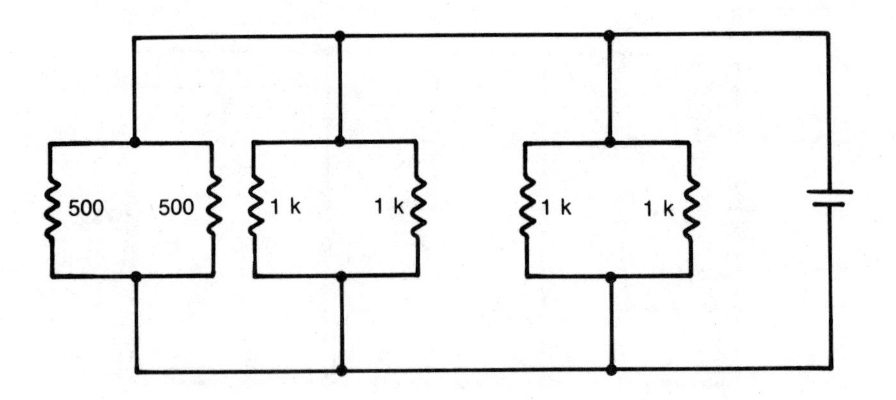

5. Given: $R_T = 150\ \Omega$, Find: R_3

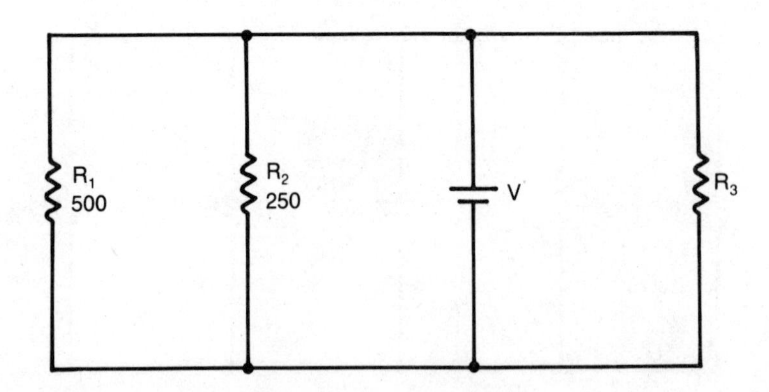

6. Given: $R_T = 5$ ohms, Find: R_4

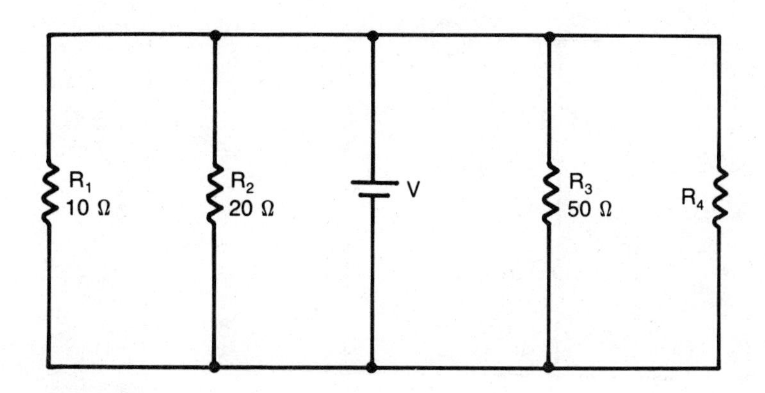

7. Find: R_3, I_T, R_T

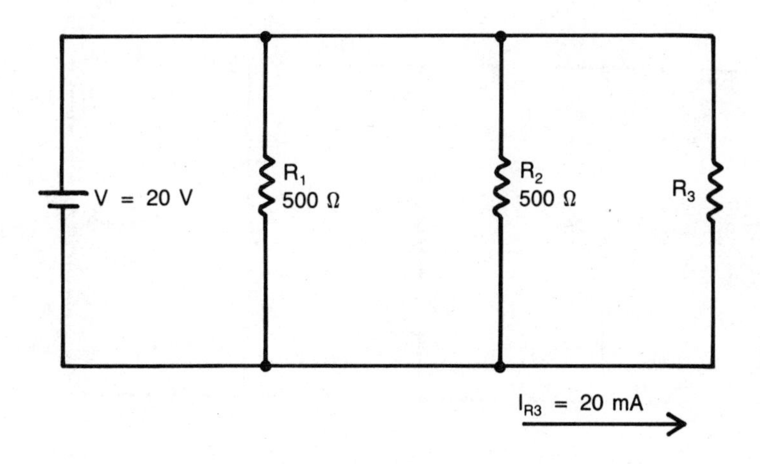

8. Find: R_T, I_A, I_B, I_C

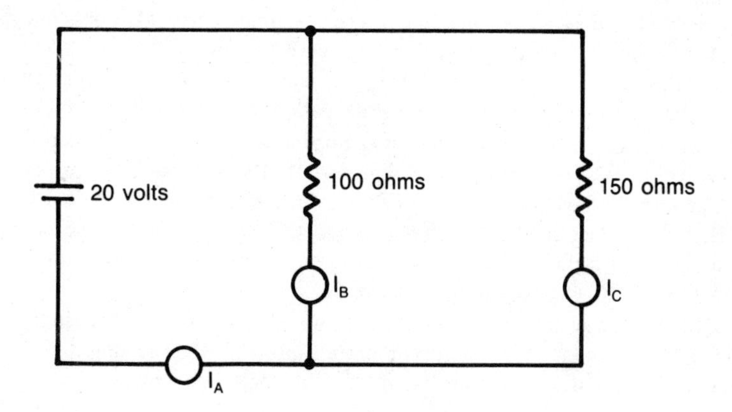

9. Find: R_T, I_A, I_B, I_C, I_D

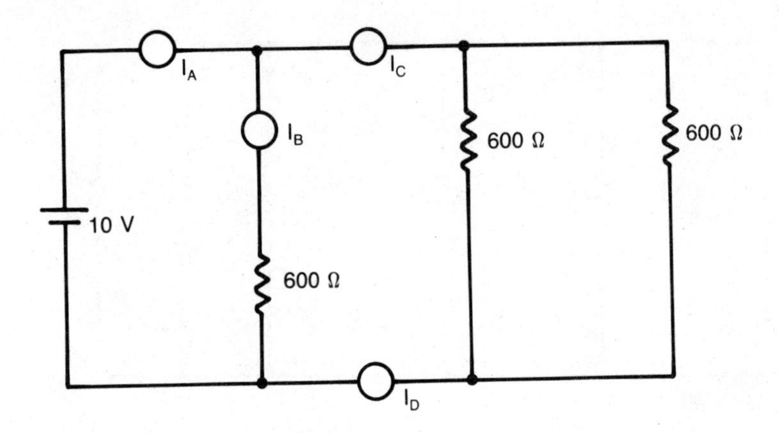

10. Find: R_T, I_A, I_B, I_C, I_D, I_E, I_F, I_G

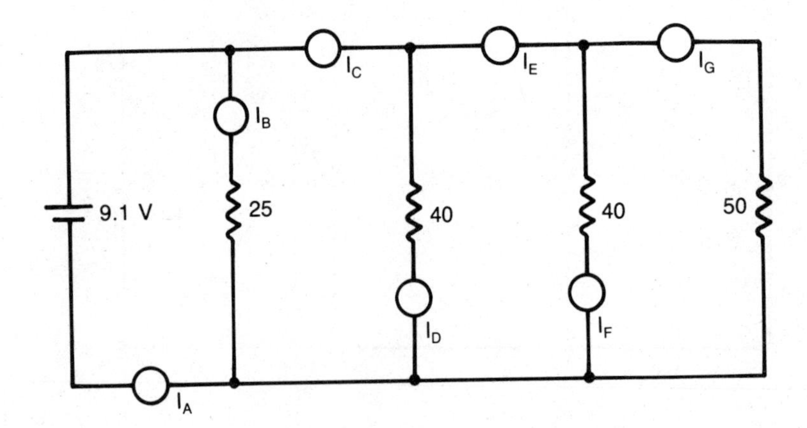

COMPLEX DC CIRCUITS

Complex circuits are a combination of both series and parallel resistors. The most important thing when dealing with complex circuits is to identify which are series and which are parallel branches. Keep in mind, in a series circuit, current has only one path. In a parallel circuit, current has more than one path to flow.

Refer to Fig. 2-7. Notice the current leaving the negative side of the battery is labeled I_T, and it should be since there is only one path for current to flow from the battery. The current then comes to R_4 which is defined as a series resistor, because there is only one path for the current to flow. Therefore, I_T flows through the series resistor, R_4. The current then comes to a place where the circuit has two paths, labeled point b. Here the current divides, flowing to both R_2 and R_3. Therefore, these two resistors are defined as parallel resistors. Then, at point a the current joins again to form I_T to flow through R_1 and return to the battery.

The thought process used here helps to keep straight which resistors are in series and which are in parallel. This is a very important process since solving a circuit requires the use of formulas which are quite different for series and parallel circuits.

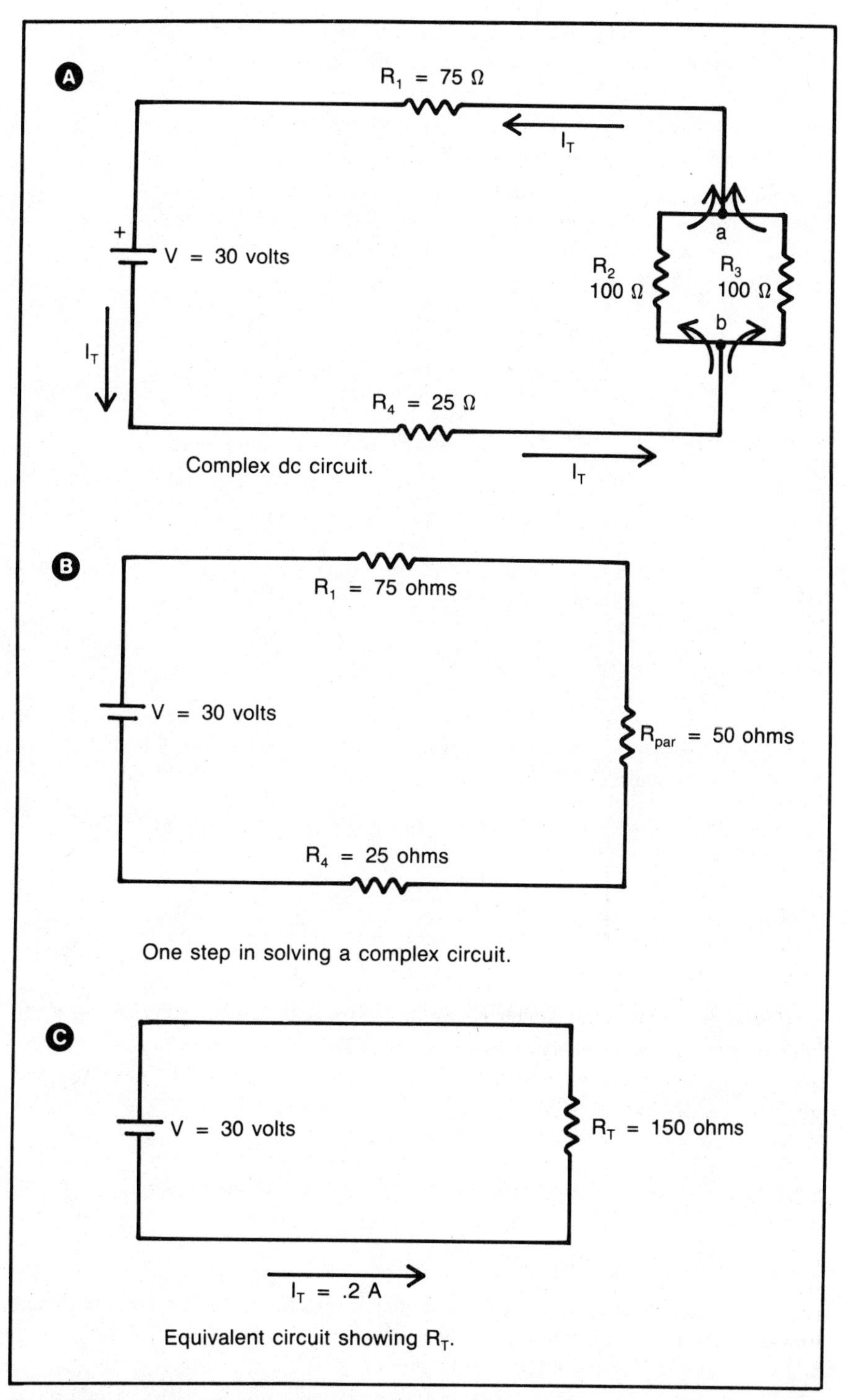

Fig. 2-7. Steps in solving a complex dc circuit: A. Complex dc circuit; B. One step in solving a complex circuit; C. Equivalent circuit showing R_T.

It is generally best to solve the parallel portions of the circuit and then combine them with the series portions. Each circuit is different and will have to be considered on its own.

Solving a Complex Circuit

Refer to Fig. 2-7. It has been identified that R_1 and R_4 are series resistors and R_2 and R_3 are in parallel. The first objective, and usually the hardest, is to solve for R_T. In a complex circuit, this will involve several steps.

Find the equivalent resistance of the parallel branches.

Step 1 $\frac{1}{R_{par}} = \frac{1}{R_2} + \frac{1}{R_3}$ formula to find the parallel equivalent resistance of R_2 and R_3

Step 2 $\frac{1}{R_{par}} = \frac{1}{100} + \frac{1}{100}$ substitute circuit values

Step 3 $\frac{1}{R_{par}} = .01 + .01$ find the decimals equivalents

Step 4 $\frac{1}{R_{par}} = .02$ perform the addition

Step 5 $R_{par} = 50$ ohms

Note Since the two resistors are of equal value, this reciprocal formula could have been replaced by dividing the resistance value by the number of branches with equal resistance.

Refer to Fig. 2-7B. Notice the parallel equivalent, R_{par} is shown as a single resistor in series. Solve for the final R_T.

Step 1 $R_T = R_1 + R_{par} + R_4$ formula
Step 2 $R_T = 75 + 50 + 25$ substitute values
Step 3 $R_T = 150$ ohms

Figure 2-7C shows the equivalent circuit drawn with just the value of R_T. From here we can calculate the total current of the circuit.

Step 1 $I_T = \frac{V}{R_T}$ formula

Step 2 $I_T = \frac{30v}{150\ \Omega}$ substitute value

Step 3 $I_T = .2$ amps perform the division

To find branch currents and voltage drops, it is necessary to first find the voltage drops of the series resistors.

Step 1 $E_{R4} = I_T \times R_4$ formula
Step 2 $E_{R4} = .2$ amps $\times$ 25 ohms substitute values
Step 3 $E_{R4} = 5$ volts voltage drop across R_4

Step 1 $E_{R1} = I_T \times R_1$ formula
Step 2 $E_{R1} = .2$ amps $\times$ 75 ohms substitute values
Step 3 $E_{R1} = 15$ volts voltage drop across R_1

With the voltage drops of the series resistors calculated, the voltage remaining in the circuit is dropped across the parallel circuit.

Step 1 $V_{applied} = E_{R1} + E_{R4} + E_{par}$ formula
Step 2 30 volts = 15 volts + 5 volts + E_{par} substitute values
Step 3 $E_{par} = 10$ volts voltage dropped across both R_2 and R_3

Except for power, the only calculations remaining in this circuit are to find branch currents. This is done using the voltage across the parallel circuit and the resistance of each branch.

Step 1 $I = \frac{E}{R}$ formula

Step 2 $I_{R2} = \frac{10\text{ V}}{100\ \Omega}$ substitute values

Step 3 $I_{R2} = .1$ amps
Step 4 $I_{R3} = I_{R2}$ because this circuit has equal values of resistance in the parallel branches

It is now possible to find the power dissipated by each resistor and also find the total power. Below is a sample power calculation. This resistor was selected because calculating power requires knowing what values to use.

Power calculation for resistor R_2.

Step 1 $P = I \times E$ formula
Step 2 $P = .1$ amps $\times$ 10 volts substitute values
Step 3 $P = 1$ watt power dissipated across R_2

Total power can be calculated by either adding all the powers in the circuit, or using total current.

Step 1 $P_T = I_T \times V$ formula
Step 2 $P_T = .2$ amps $\times$ 30 volts substitute values
Step 3 $P_T = 6$ watts

Solving the Complex Circuit

Figure 2-8A shows the original circuit before any of the resistors have been combined. On this type of circuit, and on many circuits, it is often easiest to start at a point furthest from the power supply. Keep in mind that series and parallel resistors cannot be combined in the same calculations. Saying that resistors are combined means to find the equivalent value of the resistors.

Find the parallel combination of R_5 and R_6. These are two equal resistors in parallel, divide the resistance value by the number of resistors. $\frac{10\text{ kohms}}{2} = 5$ kilohms

Figure 2-8B shows resistor R_4 in series with the combined R_5 and R_6 and this branch is in parallel with R_3. First solve the series combination.

Step 1 $R_4 + R_{5,6}$ formula
Step 2 5 k + 5 k = 10 kilohms this equivalent resistance is in parallel with R_3

Note R_3 and the equivalent resistor are both 10 k. The new equivalent resistance will be 5 k.

Figure 2-8C shows the equivalent resistance from the last calculation, labeled the combined resistance of R_3, R_4, R_5, R_6. The circuit shown in Fig. 2-8C is a series circuit with three resistors and when they are combined, the result will be the final total resistance of the circuit.

Step 1 $R_T = R_1 + R_{3,4,5,6} + R_2$ formula
Step 2 R_T = 2.5 k + 5 k + 1.5 k substitute values
Step 3 R_T = 9 kilohms final circuit total resistance

The next step is always to use the final equivalent circuit to calculate the circuit total current, I_T.

Step 1 $I_T = \frac{V}{R_T}$ formula

Step 2 $I_T = \frac{18 \text{ volts}}{9 \text{ kilohms}}$ substitute values

Step 3 I_T = 2 mA

Using Fig. 2-8C, find the voltage drops in the series equivalent circuit.

Step 1 $E_{R1} = I_T R_1$ formula
Step 2 E_{R1} = 2 mA × 2.5 k substitute values
Step 3 E_{R1} = 5 volts voltage drop across R_1

Step 1 $E_{R3,4,5,6} = I_T R$ formula
Step 2 $E_{R3,4,5,6}$ = 2 mA × 5 k substitute values
Step 3 $E_{R3,4,5,6}$ = 10 volts voltage drop across the parallel circuit

Step 1 $E_{R2} = I_T R_2$ formula
Step 2 E_{R2} = 2 mA × 1.5 k substitute values
Step 3 E_{R2} = 3 volts voltage drop across R_2

Check the voltage drops just calculated to be sure they add to the applied voltage. This is the only time this check can be performed easily.

Step 1 $E_{R1} + E_{R3,4,5,6} + E_{R2} = V$ formula
Step 2 5 + 10 + 3 = 18 volts substitute and add
Step 3 V = 18 volts voltage drops check

Find the current through all the resistors.

R_1 and R_2 are both series resistors, therefore have I_T flowing through them.

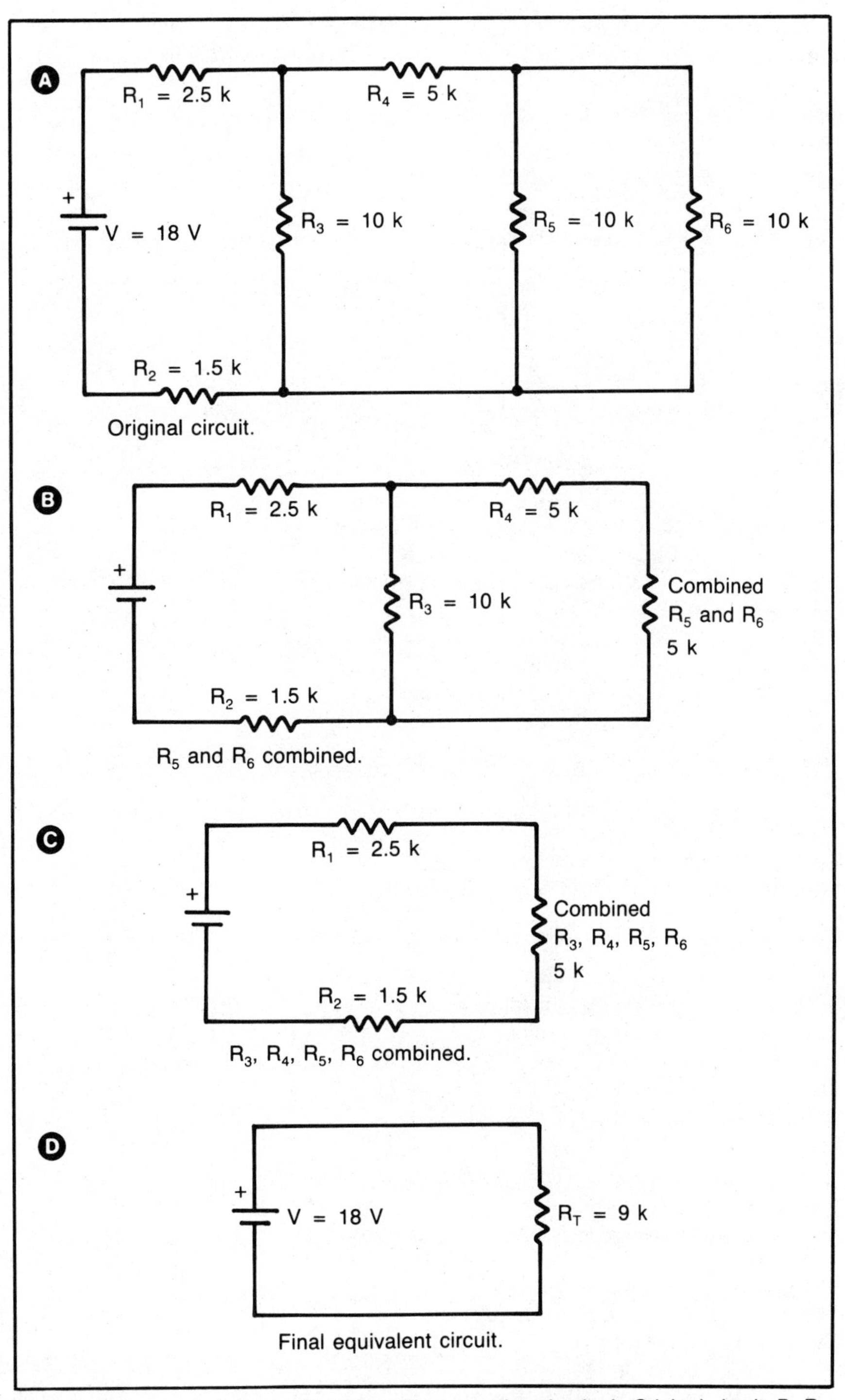

Fig. 2-8. Using equivalent circuits to simplify a complex circuit: A. Original circuit; B. R_5 and R_6 combined; C. R_3, R_4, R_5, R_6 combined; D. Final equivalent circuit.

Find current in R_3. Notice the voltage dropped across $R_{3,4,5,6}$ is all applied directly across R_3.

Step 1 $I_{R3} = \dfrac{E_{R3}}{R_3}$ formula

Step 2 $I_{R3} = \dfrac{10\ V}{10\ k}$ substitute values

Step 3 $I_{R3} = 1$ mA current through R_3

The current in R_3 should be one half the total current since I_T splits to two parallel branches with equal resistance. Refer to Fig. 2-8B. $I_{R4} = I_{R5,6}$ series resistors.

Step 1 $I_{R4,5,6} = \dfrac{E_{R3,4,5,6}}{R_4 + R_{5,6}}$ formula

Step 2 $I_{R4,5,6} = 1$ mA current through R_4 and $R_{5,6}$

This calculated current is the same as through R_3.

The current of $I_{R5,6}$ splits to go to R_5 and R_6 individually. Refer to the original circuit. In order to calculate the current in these resistors, it is first necessary to calculate the voltage drop across the series resistor R_4.

Step 1 $E_{R4} = I_{R4} \times R_4$ formula
Step 2 $E_{R4} = 1\ mA \times 5\ k$ substitute values
Step 3 $E_{R4} = 5$ volts voltage drop across R_4

Voltage dropped across R_5 and R_6, parallel combination will be $E_{R3,4,5,6}$ minus E_{R4}.

Step 1 $E_{R5,6} = E_{R3,4,5,6} - E_{R4}$ formula
Step 2 $E_{R5,6} = 10\ V - 5\ V$ substitute values
Step 3 $E_{R5,6} = 5$ volts voltage across both R_5 and R_6

Note The two will have the same current since they are equal values will equal voltage.

Step 1 $I_{R5} = I_{R6} = \dfrac{E_{R5,6}}{R_5}$ formula

Step 2 $I_{R5,6} = \dfrac{5\ V}{10\ k}$ substituting values

$I_{R5,6} = .5$ mA current in either of the resistors

Summary of voltage and current for each resistor.

$E_{R1} = 5$ volts, $I_{R1} = 2$ mA
$E_{R2} = 3$ volts, $I_{R2} = 2$ mA
$E_{R3} = 10$ volts, $I_{R3} = 1$ mA
$E_{R4} = 5$ volts, $I_{R4} = 1$ mA
$E_{R5} = 5$ volts, $I_{R5} = .5$ mA
$E_{R6} = 5$ volts, $I_{R6} = .5$ mA

Note This circuit is a very good circuit to use as an example on how to work through the different stages of a complex circuit. The values of resistors were selected to make the calculations as easy as possible. When following a sample problem, it is best to have the calculations easy enough to be done in the student's head, when possible.

The purpose of a sample problem is to show the steps and allow the student to follow through the problem while reading the text material.

General Guidelines for Solving a Complex Circuit

Keep in mind the following is intended to be only a general guideline. Each circuit must be worked by its own individuality.

Step 1 Find R_T. When solving complex circuits, it is necessary to find R_T first. This will often involve many steps.

Step 2 Use equivalent circuits for each step of R_T. The equivalent circuits will be a great help when finding currents and voltages.

Step 3 Find I_T. Use the final equivalent circuit showing R_T to find the total current. Remember, total current will flow through any resistor that is in series with the power supply and the rest of the circuit.

Step 4 Calculate the voltage drop for any series resistors. Whatever voltage is not dropped across a series resistor will be available for the parallel circuit combinations.

Step 5 Using the voltage applied to the parallel combination, determine how the current will split to each branch.

Step 6 Use branch currents and resistance values to determine the voltage of each resistor.

Note When working through the steps shown above, keep in mind that the equivalent circuits are quite easy to use to determine the relationships of different resistors, currents and voltages.

Step 7 Make a summary of the voltage drops and currents for each resistor.

Step 8 Find the powers of each resistor using the summary of Step 7.

Practice Problems

Use the schematic diagrams shown to find the unknown quantities.

1. Find: R_T, I_T, E_{R1}, E_{R2}, E_{R3}, I_{R2}, I_{R3}

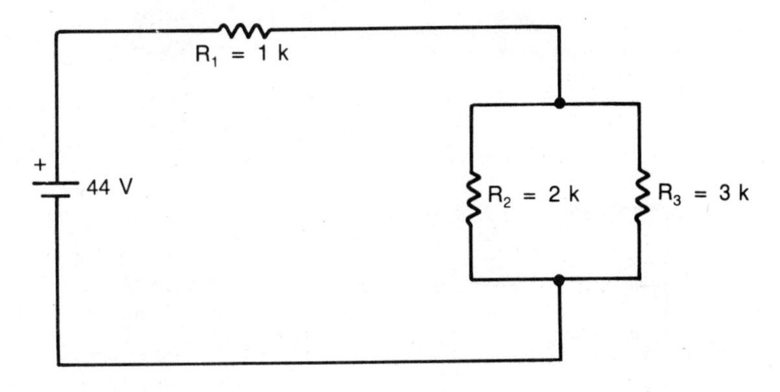

2. Find: R_T, I_T, E_{R1}, E_{R2}, E_{R3}, E_{R4}, I_{R2}, I_{R3}

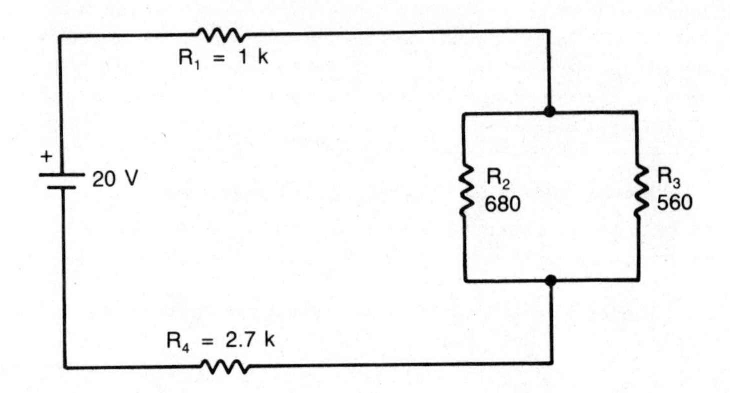

3. Find: R_T, I_T, I_{R2}, I_{R5}, voltage drops for all resistors

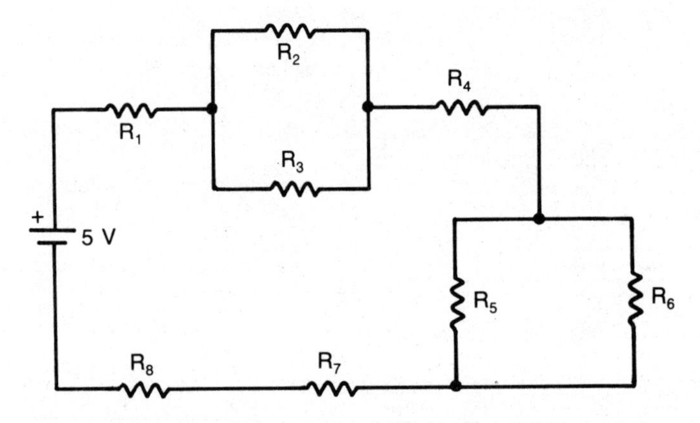

All resistors are 10 ohms.

4. Find: R_T, I_T, current in all resistors and voltage drops across all resistors.

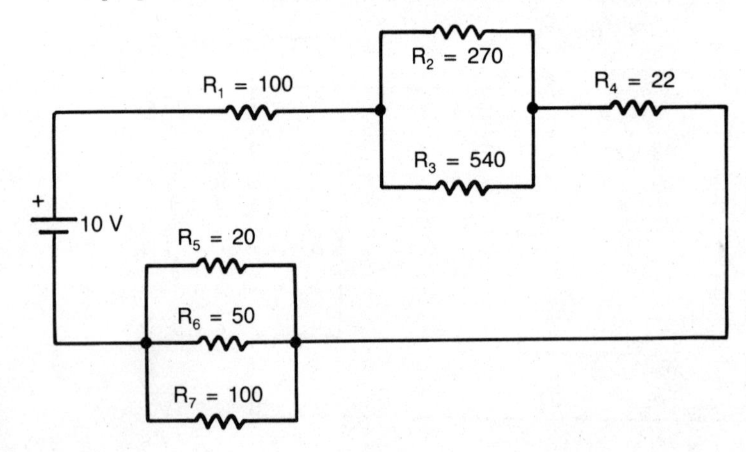

5. Find: R_T, I_T, E_{R1}, E_{R2}, I_{R2}, I_{R3}

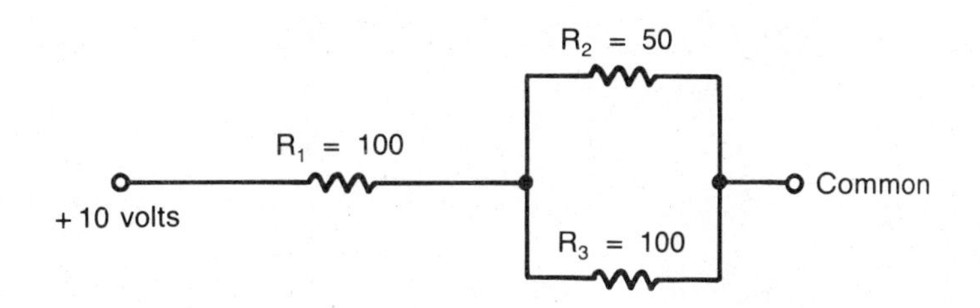

6. Find: R_T

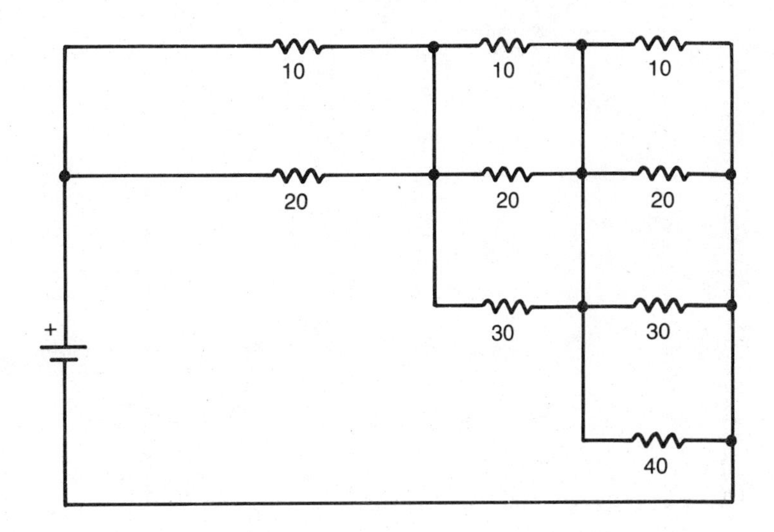

7. Find: R_T

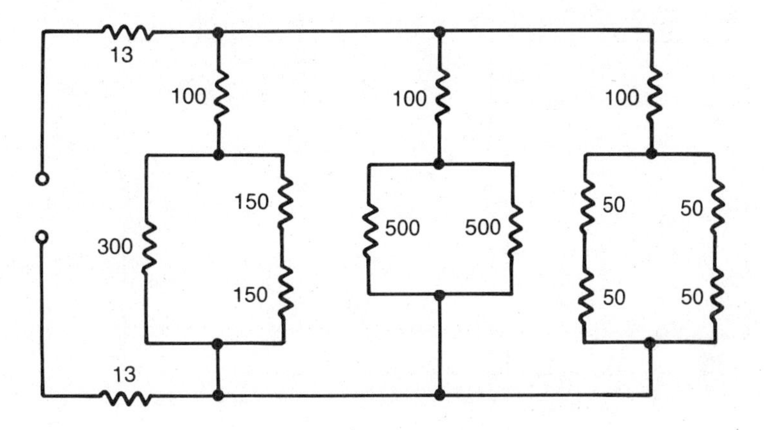

8. Find: R_T

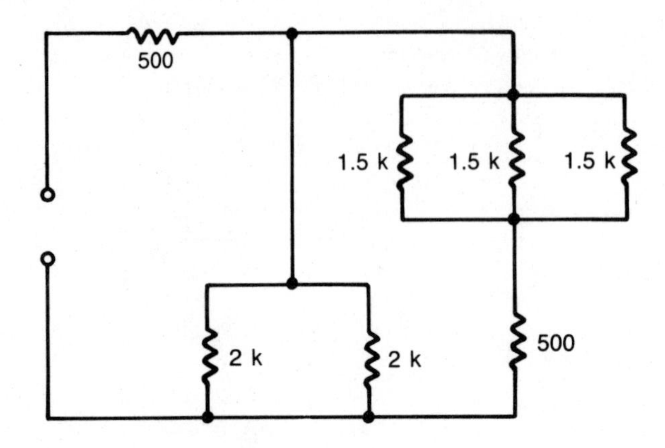

9. Find: R_T, I_T, current in all resistors and voltage drops across all resistors.

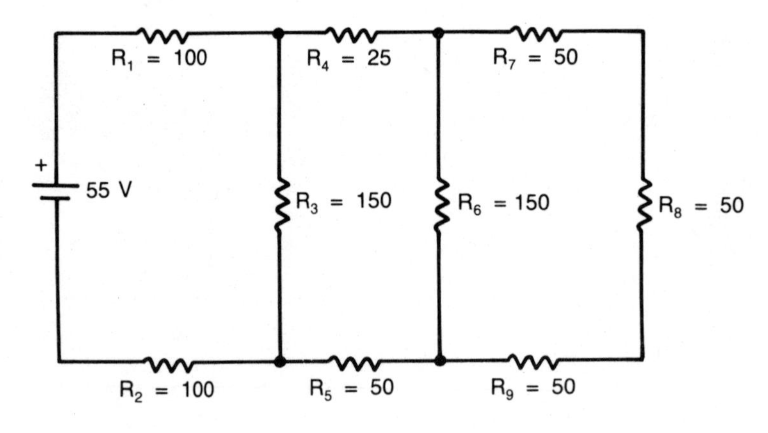

10. Find: R_T, I_T

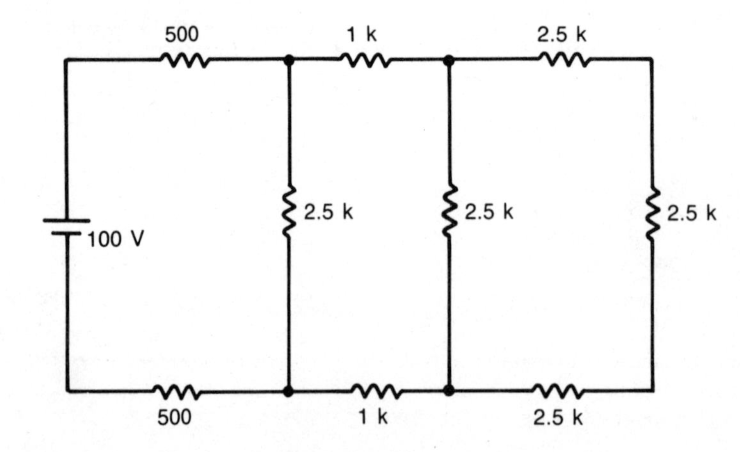

VOLTAGE DIVIDER CIRCUITS

A voltage divider circuit is one application of a series circuit. It is possible using a series circuit to adjust the voltage drops across the resistors, by adjusting the resistance values, and making a circuit with several voltages available from one battery. By what has been described to make a voltage divider circuit, it should be obvious that any circuit can be used as a voltage divider, provided there are at least two series components.

The major consideration for a voltage divider is how each of the voltages available will be used. If they are not used for some other part of a circuit, then there is no sense in simply making a voltage divider circuit. The portion of the circuit that is connected to the voltage divider, is connected across the resistor, in other words, directly in parallel with the voltage divider. Connecting two, or more, resistors in parallel will lower the total resistance. Therefore, when a load is connected to the voltage divider, the resistance will be lowered and the voltage drop in that portion of the voltage divider circuit will also be lowered.

A basic rule when using or designing a voltage divider circuit is that the load resistance should be at least 10 times the resistor it is in parallel with. Another way of saying this is to say the voltage divider should have at least 10 times the current flowing in it, than in the load circuit. The reason the load should have 10 times the resistance, or 1/10 the current, is this can be considered to have so little effect on the voltage divider, it need not be considered when making the calculations. This will limit the uses for the voltage divider circuit, unless the load is to be considered in the circuit.

One very popular application of the voltage divider circuit, when the load does not need to be considered, is in the base circuit of a transistor amplifier. The voltage divider is used to bias the transistor to its normal operating point. The accuracy needed for this application is more than met with the voltage divider circuit.

When calculating the voltages of a voltage divider circuit, there are two ways of performing the necessary calculations. The first method is to use the standard means of solving any series circuit. That is, find the total resistance, find the total current, based on the applied voltage, and then find the voltage drops of the individual resistors in the circuit. The second method is by using the voltage divider formula, which is to first find the total resistance and set up a ratio of the resistor in question, divided by the total resistance, times the applied voltage.

voltage divider formula $V = \frac{R}{R_T} \times V_T$ **Formula 2-8**

Refer to Fig. 2-9. This circuit shows a voltmeter across the 10 ohm resistor. Calculate the voltage at point a is the same as saying the voltage across the resistor, to ground.

Step 1 $V = \frac{R}{R_T} \times V_T$ formula

Step 2 $V = \frac{10}{100} \times 15$ volts substitute values

Step 3 $V = .1 \times 15$ change fraction to decimal

Step 4 $V = 1.5$ volts voltage measured with the voltmeter shown

It is important to note that the voltage across a particular resistor is a ratio of the two resistors. However, the ratio is not one resistor to the other, rather, the ratio is the resistor in question to the total resistance.

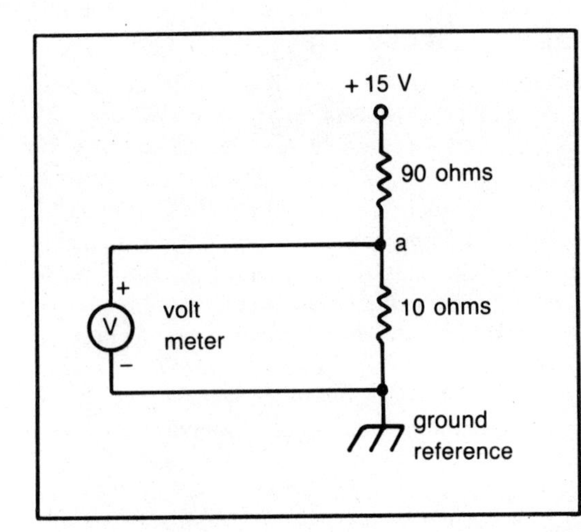

Fig. 2-9. Voltage divider circuit.

Refer to Fig. 2-10. The circuit shown is also a voltage divider circuit. This circuit shows two voltage taps, point a to ground and point b to ground. Whenever a point is shown and it is not specified where the second connection is, always assume the second connection to be on ground.

Calculations

Calculate point a first.

Step 1 $V = \frac{R}{R_T} \times V_T$ formula

Step 2 $V = \frac{40}{90} \times 30$ substitute values, the R is replaced with the voltage from that point to ground

Step 3 $V = 13.3$ V voltage at point a to ground

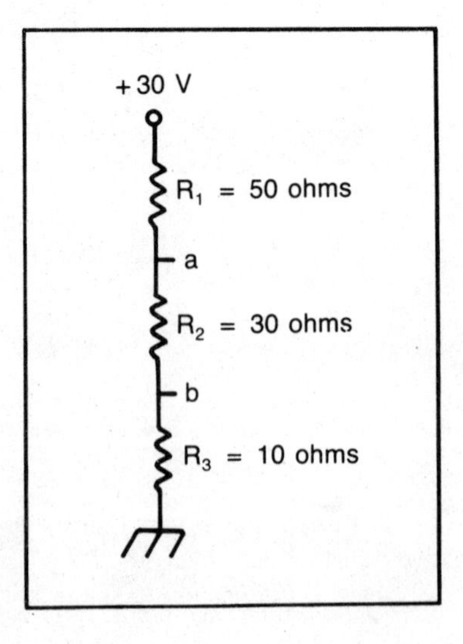

Fig. 2-10. Voltage divider with two voltage taps.

Calculate point b next.

Step 1 $V = \frac{10}{90} \times 30$ substitute values into the voltage divider formula

Step 2 $V = 3.33$ V voltage at point b to ground

Again, refer to Fig. 2-10. It is possible, rather than using the voltage divider formula, to make the calculations for the voltages at the points using Ohm's law with regular dc circuit math.

Calculate total resistance.

Step 1 $R_T = R_1 + R_2 + R_3$ formula
Step 2 $R_T = 50 + 30 + 10$ substitute values
Step 3 $R_T = 90$ ohms

Calculate total current. Remember, current is the same throughout a series circuit.

Step 1 $I_T = \frac{E}{R_T}$ formula

Step 2 $I_T = \frac{30}{90}$ substitute values

Step 3 $I_T = .333$ amps

Calculate the voltage across each resistor.

$E = I_T R$ formula to be used for all voltage drops
Step 1 $E_{R1} = .333 \times 50$ substitute for R_1
Step 2 $E_{R1} = 16.65$ volts voltage drop across R_1

Step 1 $E_{R2} = .333 \times 30$ substitute for R_2
Step 2 $E_{R2} = 9.99$ volts voltage drop across R_2

Step 1 $E_{R3} = .333 \times 10$ substitute for R_3
Step 2 $E_{R3} = 3.33$ volts voltage drop across R_3

Check voltage drops by adding to see if they equal the applied voltage.

Step 1 $E_{R1} + E_{R2} + E_{R3} = V_T$ formula
Step 2 $16.65 + 9.99 + 3.33 = 29.97$ V substitute values, difference is caused by rounding of the numbers

Voltage at point b is equal to the voltage drop across R_3.

Point b = 3.33 volts checks with using the voltage divider formula

Voltage at point a is equal to the voltage drop across $R_3 + R_2$.

Point a = $3.33 + 9.99 = 13.32$ volts checks with using the voltage divider formula.

Refer to Fig. 2-11. This figure shows an arrangement somewhat different from the usual dc circuit. The two power supplies represent the fact that the ground connection is actually a reference point. If all measurements are in reference to the ground connection, then that point is considered zero volts. Therefore, point a is 6 volts above ground, or +6 V. Point c is 3 volts below ground, or −3 V. This results in point b, using a reference of ground, must have a voltage somewhere between +6 V and −3 V. We can, therefore arrive at the conclusion that there is 9 volts difference from points a to c. 9 volts, then, is the total applied voltage to use when calculating the voltage divider.

Calculations

Step 1 $V = \frac{R}{R_T} \times V_T$ formula

Step 2 $V = \frac{60}{140} \times 9$ substitute values

Step 3 $V = 3.8$ volts perform the calculations, this is not the voltage at point b

The voltage calculated above is from point b to point c. Because it is the voltage across the 60 ohm resistor. To find the voltage at point b, to ground, add this value to the −3 V supply.

Step 1 $-3 + 3.8 = .8$ volts at point b to ground

Using the Ohm's law method; total resistance is 140 ohms, calculate current.

Step 1 $I = \frac{E}{R}$ formula

Step 2 $I = \frac{9}{140}$ substitute values

Step 3 $I = 64$ mA

Using the total current, calculate the voltage drops across each resistor.

$E = IR$ formula
Step 1 $E_{R1} = .064 \times 80$ substitute values
Step 2 $E_{R1} = 5.12$ volts voltage drop across R_1

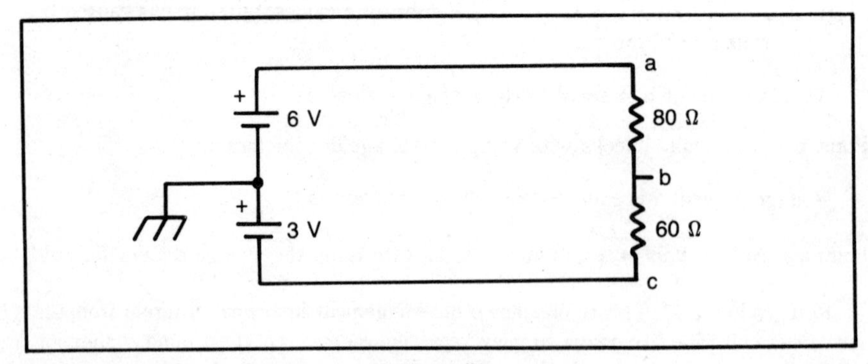

Fig. 2-11. Voltage divider with two power supplies.

Step 1 $E_{R2} = .064 \times 60$ substitute values
Step 2 $E_{R2} = 3.84$ volts voltage drop across R_2

The voltage drop across R_1 is actually from point a to point b. Therefore, point b is less than the voltage at point a by the voltage drop.

point a = 6 V
voltage drop across R_1 = 5.12 volts
voltage at point b = 6 − 5.12 = .88 V

The voltage drop across R_2 is actually from point b to point c. Therefore, point b is higher than point c by the voltage drop across R_2.

voltage drop across R_2 = 3.84 volts
voltage at point c = −3 volts
voltage at point b = 3.84 + −3 = .84 V

The results of calculation 2 = .8 V, calculation 5 = .88 V, calculation 6 = .84 V

The difference between these values is due to rounding of numbers in the calculations.

WHEATSTONE BRIDGE

The Wheatstone bridge is the name of a particular circuit that is used for a very precise measuring instrument. The reason it is being introduced in this chapter is the fact that the Wheatstone bridge is made up of voltage divider circuits and is really nothing more than another dc circuit.

The principle of the bridge circuit is that there are two voltage divider circuits formed. R_1 and R_2 form a voltage divider and R_A and R_B form the other voltage divider. Refer to Fig. 2-12.

Keep in mind that the voltmeter is to be treated like an infinite resistance, or an open circuit. In other words, the right side of the circuit is completely independent of the left side of the circuit. Another point that can be made that is unusual about this circuit is the fact that even if there is a change in the power supply voltage, it will not affect the voltage divider circuits.

This instrument is used for measuring resistance in a very precise manner. R_2 and R_B are fixed resistors of a very precise type. R_1 is part of the instrument and is used for balancing the bridge. R_A is shown as a variable, but is really the unknown resistance being measured. The voltmeter is center zeroed, which means it can swing either way with the center of the meter being zero. The meter is used for balancing the bridge circuit.

There are two possible conditions for the bridge circuit, the first condition is unbalanced and the second condition is balanced. If the bridge is unbalanced, that means there is a difference in the voltage divider ratios between the two sides. This will produce a difference in voltage drops across the resistors and result in the voltmeter showing a voltage. If the bridge is balanced, then the ratio of the two voltage dividers is equal. This will result in an equal voltage drop across the resistors and result in no voltage on the meter.

Sample Wheatstone Bridge

R_1 = 1 k, R_2 = 2 k, R_A = 12 k, R_B = 20 k, V = 10 V

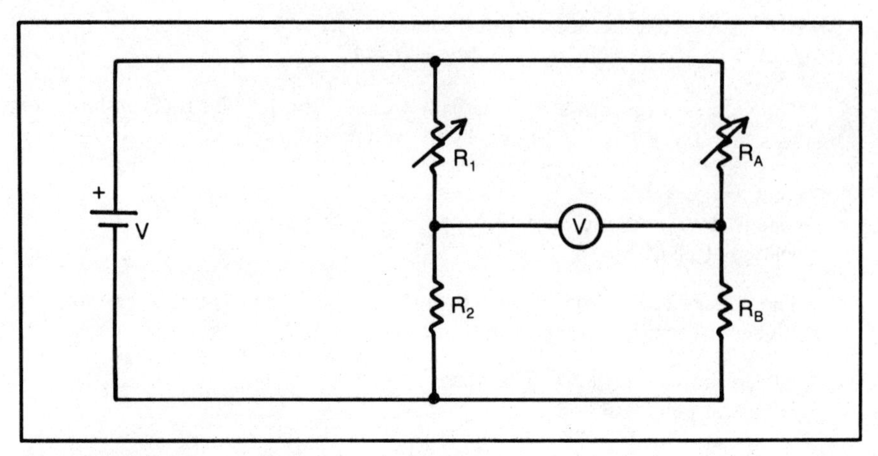

Fig. 2-12. Wheatstone bridge.

Calculate the voltage drop across R_2 and R_B.

$$V = \frac{R}{R_T} \times V_T \text{ formula}$$

Step 1 $V_{R2} = \frac{2\text{ k}}{3\text{ k}} \times 10\text{ V}$ substitute values

Step 2 $V_{R2} = 6.67\text{ V}$ voltage across R_2

Step 1 $V_{RB} = \frac{20\text{ k}}{32\text{ k}} \times 10\text{ V}$ substitute values

Step 2 $V_{RB} = 6.25\text{ V}$ voltage across R_B

Notice there is a difference in the two voltages. V_{R2} is .42 V (6.67 − 6.25) higher than V_{RB}. This means the voltmeter will have this reading, with a positive on the left and negative on the right. This is considered an unbalanced bridge. To balance the bridge is simply a matter of having the two voltage divider ratios equal. Notice the voltage divider ratio portion of the circuit is only the R/R_T. This leads to a new formula for Wheatstone bridges:

balancing a Wheatstone bridge $\frac{R_1}{R_2} = \frac{R_A}{R_B}$ **Formula 2-9**

Balancing Using the Sample Values

$R_1 = 1\text{ k}$, $R_2 = 2\text{ k}$, $R_A = \text{unknown}$, $R_B = 20\text{ k}$

Step 1 $\frac{1\text{ k}}{2\text{ k}} = \frac{R_A}{20\text{ k}}$ substituting values in the Wheatstone formula

Step 2 $.5 = \frac{R_A}{20\text{ k}}$ change the left side to decimal

Step 3 $.5 \times 20\text{ k} = R_A$ follow the rules of algebra
Step 4 $R_A = 10\text{ k}$ value of R_A to balance the bridge

MAXIMUM TRANSFER OF POWER

The maximum transfer of power occurs when the value of the load is equal to the internal resistance of the power supply.

Refer to Fig. 2-13 for a sample circuit of maximum transfer of power. Notice that this is a very simple circuit. It is easier this way to prove the point, however, any circuit can be used.

Calculations

The maximum transfer of power states the load must be equal to the internal resistance of the power supply, represented by r_i. This means R_{load} must be equal to 200 ohms.

Calculate power with the 200 ohm load.

Step 1 $P = \frac{E^2}{R}$ formula

Step 2 $P = \frac{15^2}{400}$ substitute values, total circuit resistance

Step 3 $P = .563$ watts maximum power, 1/2 to the load
Step 4 $P_{load} = .282$ watts power to the load

To prove the power in the previous calculation is the maximum power that can be delivered to the load from a 15 volt power supply with 200 ohms internal resistance, a value of load resistor above 200 ohms and one below will be tried.

Use a larger resistor. It is probably easier to calculate load power by first finding circuit current. Let's try 500 ohms for the load.

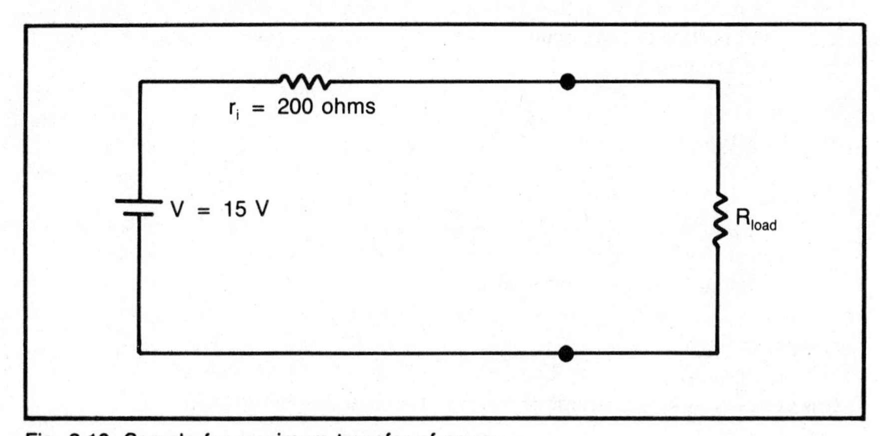

Fig. 2-13. Sample for maximum transfer of power.

Step 1 $I = \frac{E}{R}$ formula for current

Step 2 $I = \frac{15}{700}$ substitute values, total resistance

Step 3 I = .021 amps

Calculate load power.

Step 1 P = I^2R formula
Step 2 P = $.021^2 \times 500$ substitute values, load resistor
Step 3 P = .221 watts power to the 500 ohm load is lower

Use a smaller resistor. Try 75 ohms.

Step 1 $I = \frac{15}{275}$ substitute values, total resistance

Step 2 I = .055 amps

Calculate load power.

Step 1 P = I^2R formula
Step 2 P = $.055^2 \times 75$ substitute values, load resistor
Step 3 P = .227 watts power to a 75 ohm load is lower

CHAPTER SUMMARY

Direct current circuit math is one of the most important single subjects in the study of electronics. It sets the basic rules for all of electricity and electronic circuits. The following is a list of the key points in this chapter.

- Stated by Ohm's law: voltage = amps × ohms
- In any dc circuit there are four quantities that can be calculated; voltage, current, resistance, and power.
- I is directly related to P.
- E is directly related to I.
- E is directly related to R.
- E is directly related to P.
- R is indirectly related to I.
- R is indirectly related to P.
- In a series circuit, there is only one current path. Current is the same throughout a series circuit.
- Voltage drops at each resistance directly related to the size of the resistance.
- The sum of the voltage drops in a series circuit must equal the supply voltage.
- The sum of the powers dissipated in a series circuit must be equal to the total power.
- Electron current flow is from negative to positive. Conventional current flow is from positive to negative.
- In a series circuit, the largest resistance will drop the most voltage.
- In a series circuit, the largest resistance will dissipate the most power.

- A short circuit has a resistance of zero. This condition has unlimited current, zero voltage drop and zero power.
- An open circuit has a resistance of infinity. This condition causes zero current, applied voltage dropped at open, zero power.
- In a parallel circuit, the current divides to the individual branches indirectly related to the size of the resistors.
- Total current in a parallel circuit is equal to the sum of the individual branch currents.
- Voltage is the same throughout a parallel circuit.
- The sum of the powers dissipated in a parallel circuit is equal to the total power.
- The total resistance of a parallel circuit is smaller than the smallest resistance of any branch.
- To find the total resistance of equal resistances in parallel, divide the resistance of one branch by the number of branches with equal resistance.
- In a parallel circuit, if branches have equal resistance, the current through each branch will be equal.
- The total current divided by the number of branches with equal resistances will give the individual branch currents.

Summary of Formulas

Ohm's law $E = I \times R$ **Formula 2-1**

power formula $P = I \times E$ **Formula 2-2**

modified power formula $P = I^2R$ **Formula 2-3**

modified power formula $P = \frac{E^2}{R}$ **Formula 2-4**

total of resistors in series $R_T = R_1 + R_2 + R_3 + \ldots$ **Formula 2-5**

total of resistors in parallel reciprocal formula $\frac{1}{R_T} = \frac{1}{R_1} + \frac{1}{R_2} + \frac{1}{R_3} + \ldots$

Formula 2-6

two resistors in parallel shortcut formula $R_T = \frac{R_1 \times R_2}{R_1 + R_2}$ **Formula 2-7**

voltage divider formula $V = \frac{R}{R_T} \times V_T$ **Formula 2-8**

balancing a Wheatstone bridge $\frac{R_1}{R_2} = \frac{R_A}{R_B}$ **Formula 2-9**

Practice Problems

Use the schematic diagrams for each problem to find the unknown values. It may be easier with some problems to redraw the schematic to determine which resistors are in series and which are in parallel.

1. Find: R_T, I_T, E_{R1}, E_{R2}

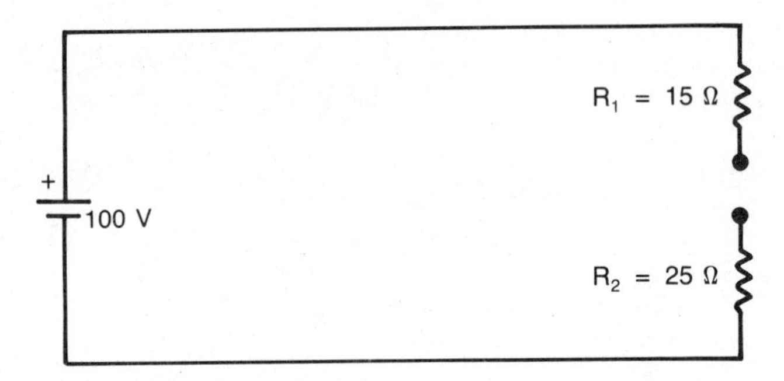

2. Find: R_T, I_T, E_{R1}, E_{R2}, E_{R3}

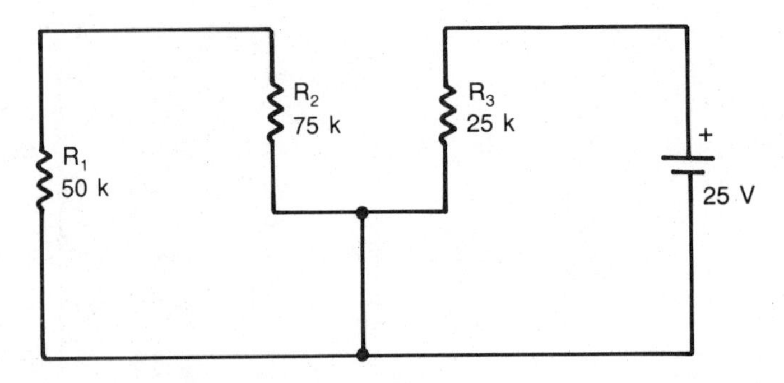

3. Find: R_T, I_T, V_a, V_b

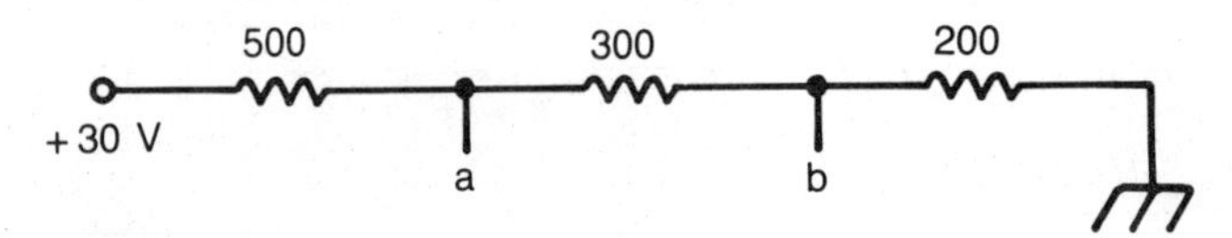

4. Find: R_T, I_T, I_{R1}, I_{R2}

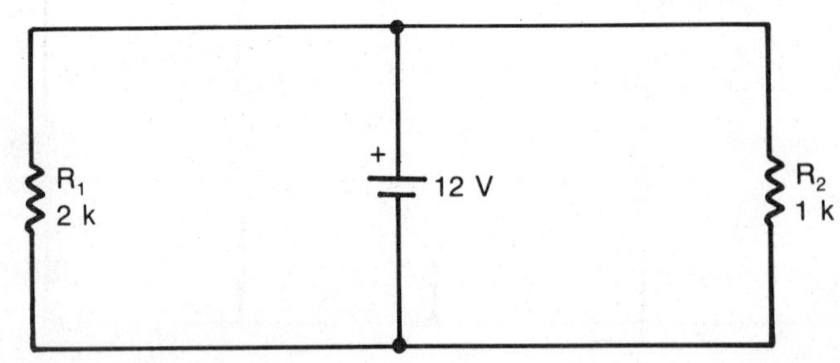

5. Find: R_T, I_T, I_{R1}, I_{R2}, I_{R3}

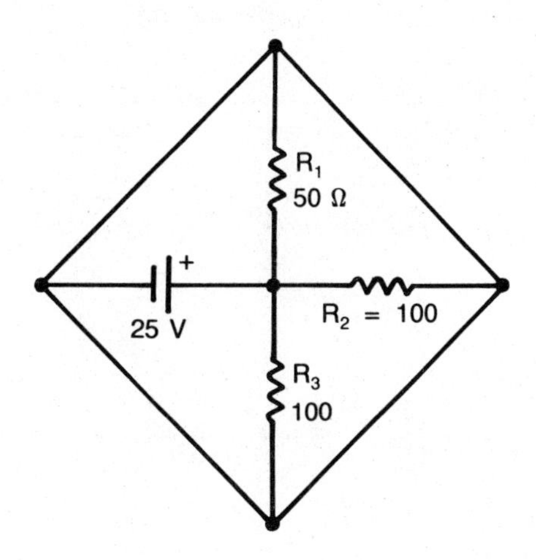

6. Find: R_T, I_T, P_T

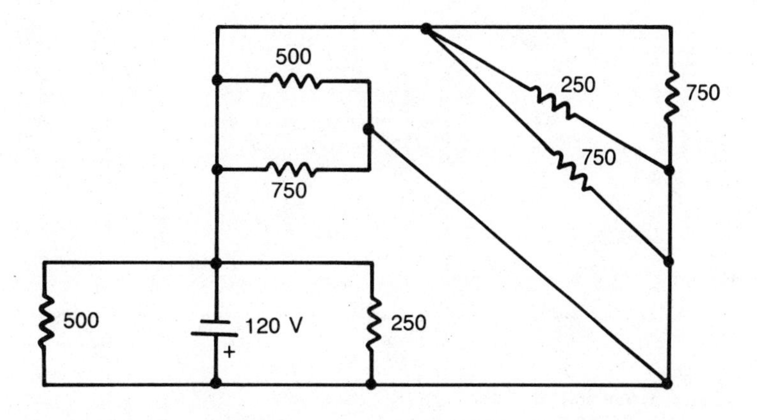

7. Find: R_T, I_T, P_T, R_1, R_2, R_3, R_4

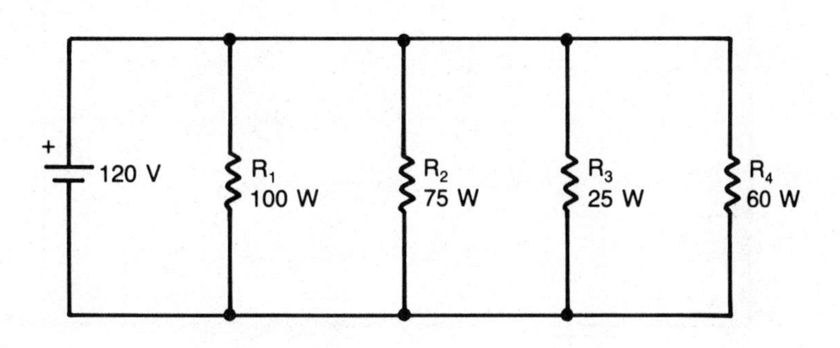

8. Find: R_T, I_T, I_{R1}, I_{R2}, I_{R3}, E_{R1}, E_{R2}, E_{R3}
What percentage of total current flows through R_2?

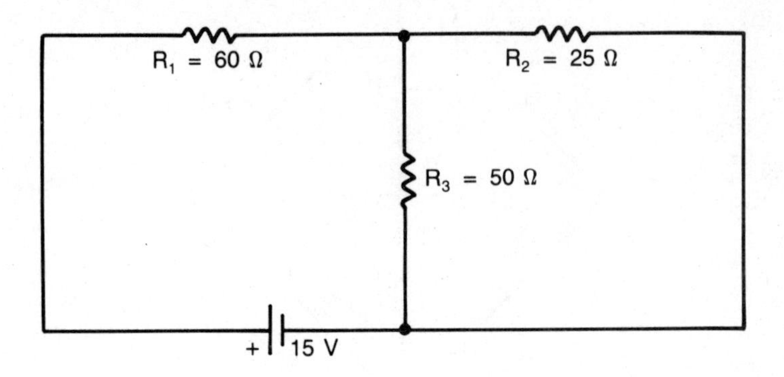

9. Find: R_T, I_T, I_{R1}, I_{R2}, I_{R3}, I_{R4}, I_{R5}, I_{R6}, E_{R1}, E_{R2}, E_{R3}, E_{R4}, E_{R5}, E_{R6}

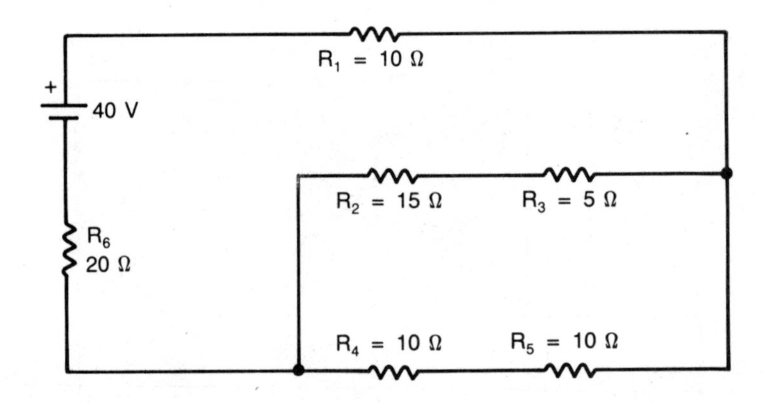

10. Find: Current in the meter.

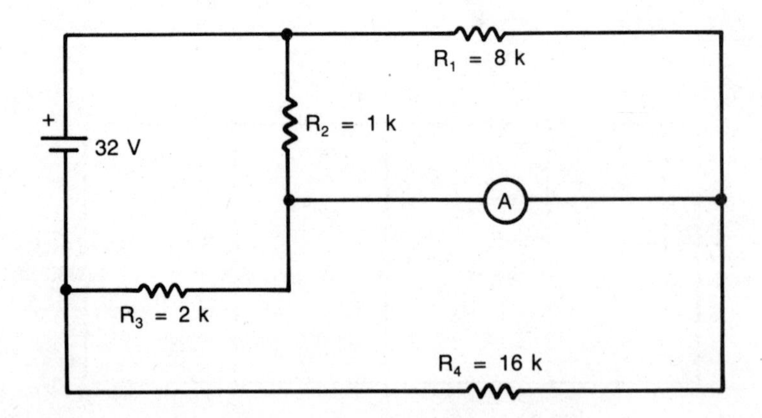

Chapter 3

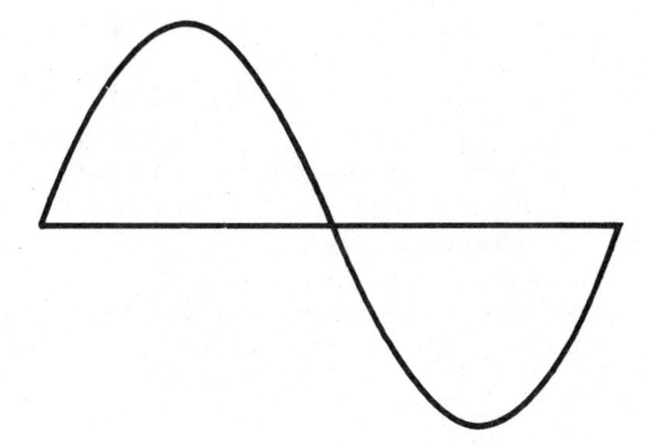

The Sine Wave

The sine wave is the most important waveform in all electronic circuits and is often referred to as the primary frequency. The basic components of inductors and capacitors have ac resistance, called reactance, based on the sine wave.

One common example of the use of a sine wave is household electricity. Household electricity used in the United States is a 120-volt, 60-cycle sine wave.

PRODUCING A SINE WAVE

There are basically two ways to produce a sine wave. One way is with an oscillator and the other way is with a generator.

An oscillator is an electronic circuit that changes a dc supply voltage into the sine wave. The circuit contains a parallel resonant circuit and an amplifier. Further discussion is beyond the scope of this text. An oscillator is usually used to produce high frequencies.

A generator is the method used to produce the sine wave of household electricity. Because the generator is an easy way to see the formation of a sine wave, this method will be discussed.

The Alternating Current (ac) Generator

The ac generator produces electricity due to the fact that a wire is rotated through a magnetic field. The wire moving in the magnetic field causes a voltage to be induced, with the strongest voltage happening when the wire is perpendicular (at right angles) to the magnetic lines of force. Another way of saying this is, cutting across the magnetic lines produces the maximum voltage. When the wire is going in the same direction as the magnetic lines of force, there is no voltage induced. All the positions of the wire, between the maximum and minimum (zero) voltage points, produce a voltage proportional to the position.

As the wire is rotated through a complete circle, the wire will have a position associated with the opposite magnetic pole. This will produce electricity having an opposite polarity.

Figure 3-1 shows a loop of wire rotating in a magnetic field produced by a two-pole generator. Position a represents 0 degrees, the starting point. At 0 degrees the wire is in direct line with the magnetic lines of force and therefore, there is no voltage produced. Position b represents one-quarter turn, or 90 degrees of rotation. At 90 degrees the wire is cutting the lines of force at a maximum point, producing maximum voltage. This is the positive peak of the sine wave. Position c the wire has rotated through one half of its full cycle, or 180 degrees of rotation. At this point the wire is again in line with the magnetic lines of force, except the wire is heading in the opposite direction. There is zero voltage produced. Position d shows the wire cutting the maximum lines of force, except produced by the opposite pole. This is three-quarters of a cycle, or 270

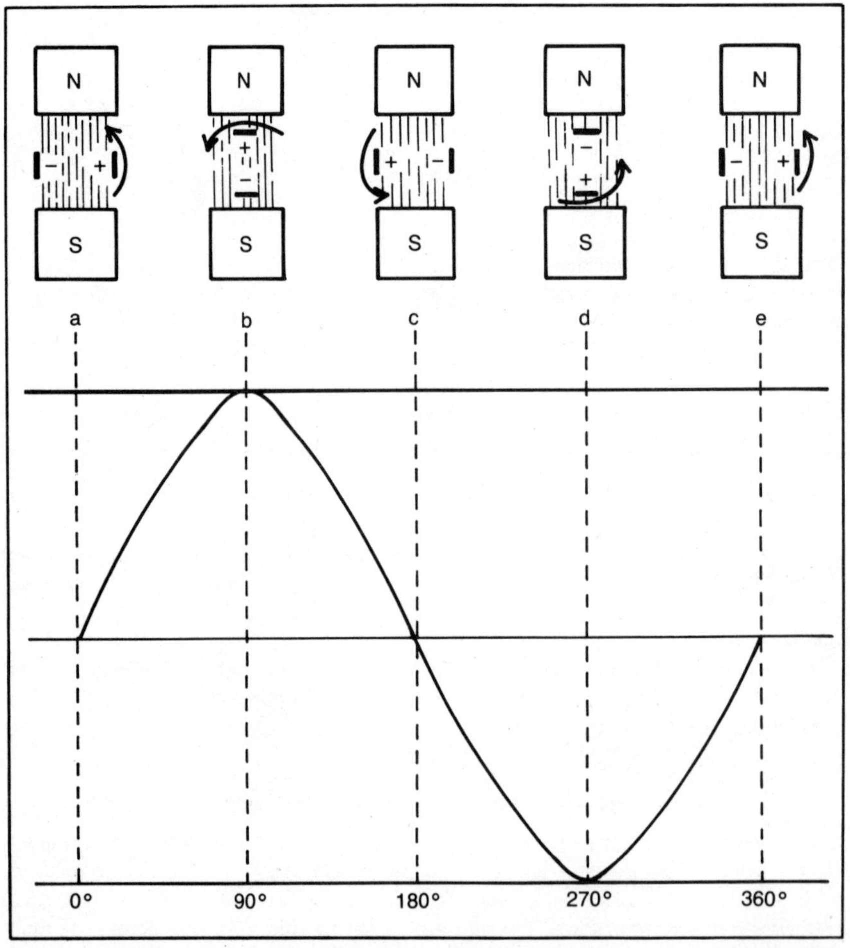

Fig. 3-1. One cycle of a sine wave produced by rotating a loop of wire in a two pole generator.

degrees, producing voltage in the opposite direction, or a negative voltage at the maximum, or peak of the sine wave. Then in position e the wire has returned to the original starting point. At this point, it again produces zero volts and the waveform is headed in the positive direction. This is one complete cycle or 360 degrees.

If this wire were to be rotated in this generator at a rate of 60 complete cycles every second, the sine wave would be reproduced at a rate of 60 cycles per second.

PLOTTING A SINE WAVE

The value of a sine wave can be plotted, or calculated with the use of a formula.

instantaneous voltage of a sine wave $V = V_m \sin \Theta$ **Formula 3-1**

The formula states the instantaneous voltage (V) is equal to the maximum voltage the sine wave reaches (V_m, also called the peak value) times the sine of the angle at the particular point in time the calculation is made. Note sin is the abbreviation of sine.

Figure 3-2 shows one cycle of a sine wave plotted using the formula for instantaneous voltage. Refer to Table 3-1 for the results of the calculations. The calculations were made, for this example, every 15 degrees. The voltage selected for the maximum voltage is 100 volts. This voltage was chosen because it makes it very easy to change it to percent of full voltage.

Sample Calculation for Table 3-1, Fig. 3-2

Step 1 $V = V_m \sin \Theta$ formula
Step 2 $V = 100 \sin 45$ substitute values
Step 3 sin 45 degrees = .707 find the sin of 45 degrees on a calculator
Step 4 $V = 100 \times .707$ substitute values
Step 5 V = 70.7 volts multiply (this point is the RMS value)

When dealing with sine waves, there are several new words and definitions to be learned. Refer to Figs. 3-2 and 3-3.

cycle A complete cycle is when the waveform repeats itself. Figures 3-2 and 3-3A show one cycle, Fig. 3-3B is two cycles, Fig. 3-3C is three cycles, and Fig. 3-3D is four cycles.

Table 3-1. $V = V_m \sin \Theta$.

Degrees	Voltage	Degrees	Voltage
0	0	180	0
15	25.9	195	-25.9
30	50.0	210	-50.0
45	70.7	225	-70.7
60	86.6	240	-86.6
75	96.6	255	-96.6
90	100	270	-100
105	96.6	285	-96.6
120	86.6	300	-86.6
135	70.7	315	-70.7
150	50.0	330	-50.0
165	25.9	345	-25.9
180	0	360	0

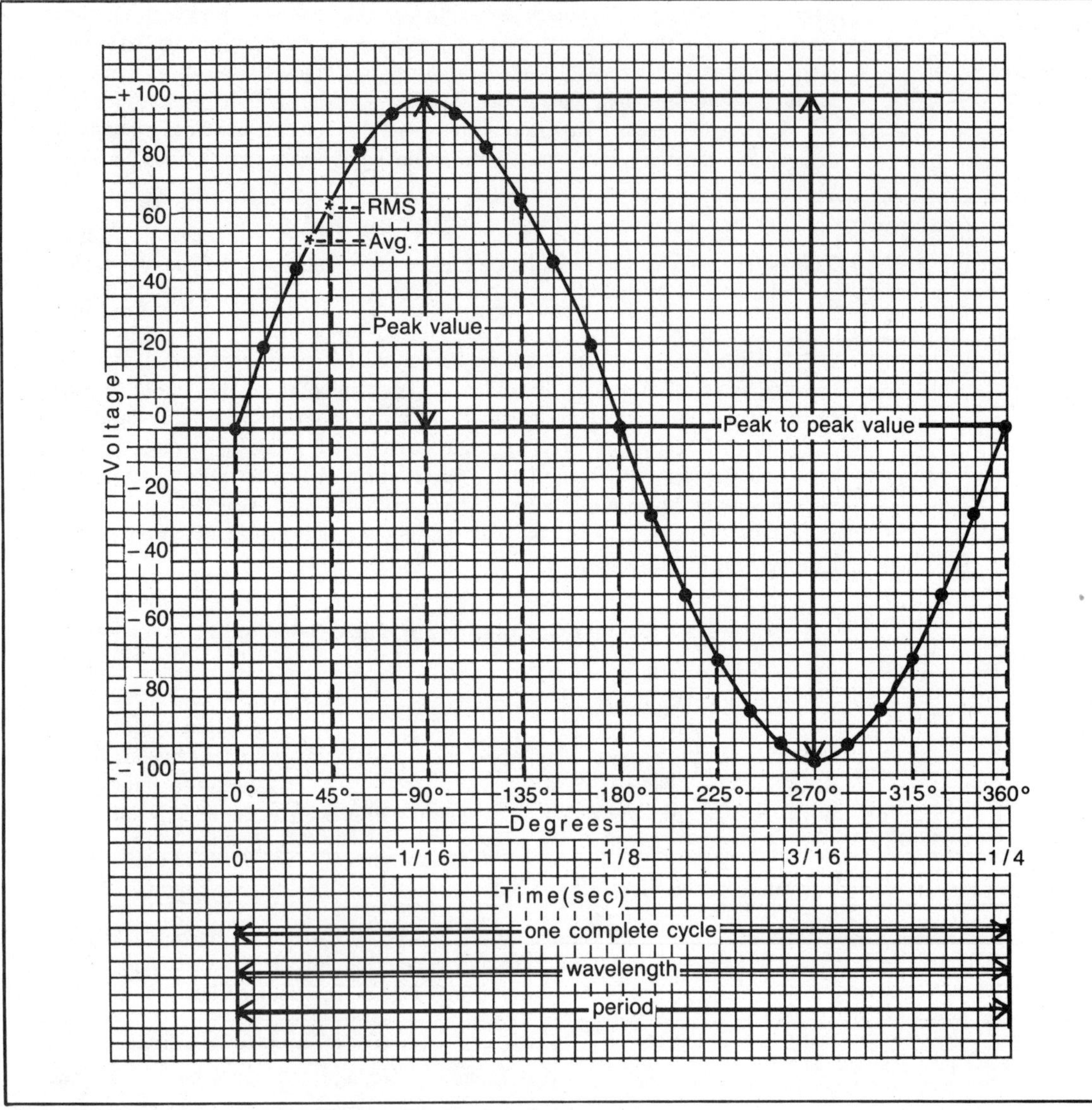

Fig. 3-2. Plotting one cycle of a sine wave using the formula; $V = V_m \sin \Theta$.

period The length of time it takes for the wave to make one cycle. Figure 3-2 has a period of 1/4 second, Fig. 3-3A is 1/2 second, Fig. 3-3B is 1/4 second, Fig. 3-3C is 1/6 second, and Fig. 3-3D is 1/8 second.

wavelength The distance a wave will move in air. The symbol for wavelength is the Greek letter lambda λ.

This calculation results in lambda in centimeters.

$$\lambda = \frac{3 \times 10^{10} \text{ cm/s}}{\text{freq (Hz)}}$$

Formula 3-2

The wavelength of radio waves is an important consideration when determining the length of an antenna.

frequency The number of cycles in one second. The unit of measure for frequency is cycle per second (cps) or hertz (Hz). The formula to calculate frequency is:

$$f = \frac{1}{T}$$ **Formula 3-3**

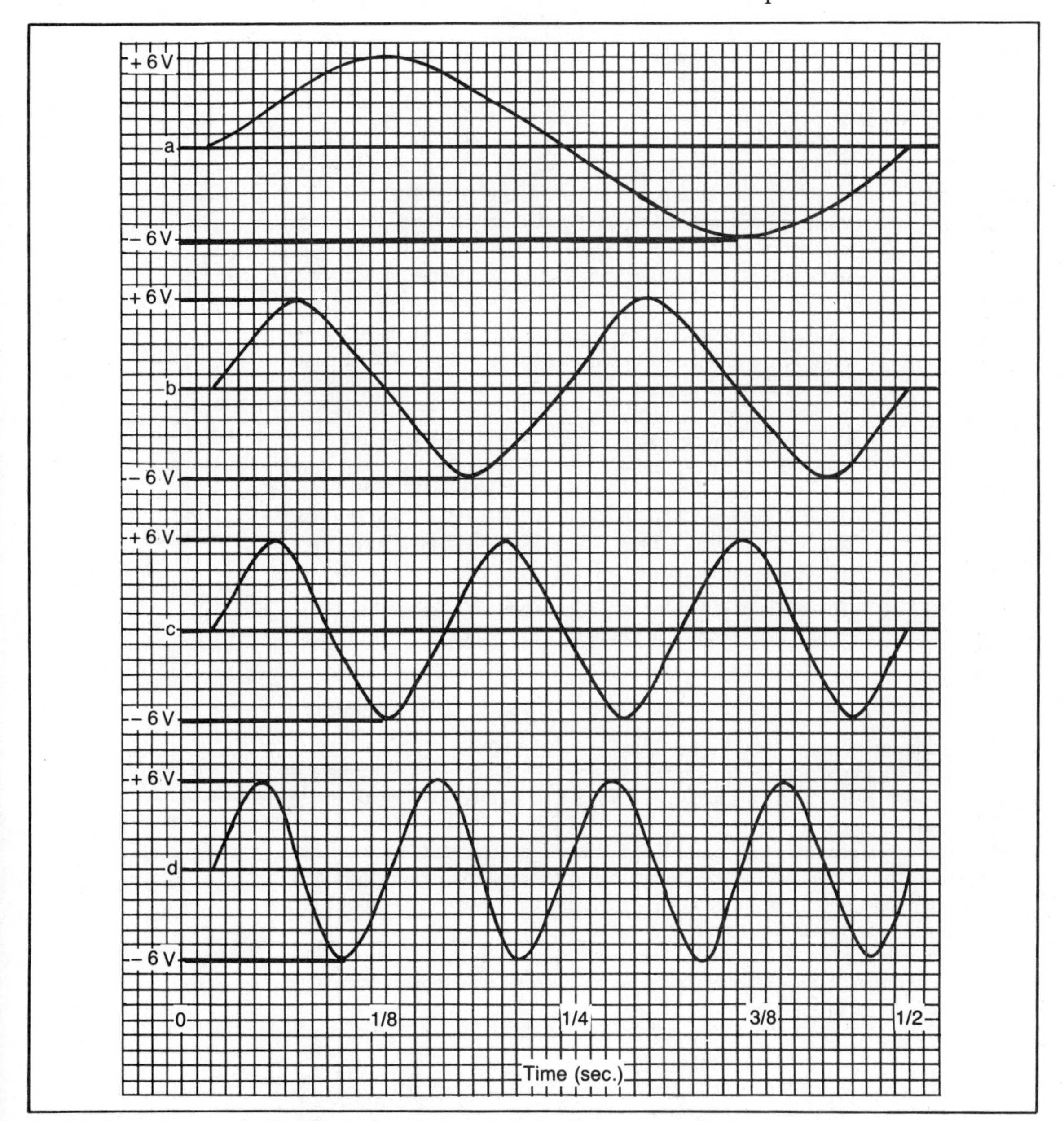

Fig. 3-3. Four sine waves with different frequencies, all having the same amplitude.

Frequency is the inverse of time.
Figure 3-2 has a frequency of 4 Hz, Fig. 3-3A is 2 Hz, Fig. 3-3B is 4 Hz, Fig. 3-3C is 6 Hz, and Fig. 3-3D is 8 Hz.

peak value The maximum value a sine wave reaches as measured from the zero reference line. Peak value can be either positive or negative. Figure 3-2 has a peak value of 100 V, 3-3 (all four waveforms, A-D) have a peak value of 6 volts.

peak to peak The value of the sine wave from the negative peak to the positive peak. It is a value equal to 2 times the peak. Figure 3-2 has a p to p value of 200 volts, and all four in Fig. 3-3 have a p to p of 12 volts.

average The average value of one-half the sine wave has a value equal to .636 × peak. This occurs at a point equal to 39.5 degrees. This takes into account the fact that the sine wave is constantly changing. Notice that the average value of both halves of the sine would be 0. That is because the wave goes as much negative as it does positive.

RMS RMS stands for Root Mean Square. It is also called the effective value or the dc equivalent value. RMS is equal to .707 × peak and occurs at 45 degrees. The RMS value is the amount of ac voltage needed to produce the same amount of heat as an equal amount of dc voltage. Most all meters read the RMS value. The household electricity in the United States is said to be 120 volts (approximately), 60 hertz. The stated 120 volts is the RMS value, the peak value would be 170 volts peak, the peak to peak value would be 340 volts p to p. Notice, the frequency has no effect on voltage measurements.

amplitude The height of the waveform. This is a word developed from the use of oscilloscopes. It can mean either peak or peak to peak.

CONVERTING VALUES IN RMS, PEAK, PEAK TO PEAK, AND AVERAGE

From the definitions given for the different values of a sine wave, there are four names that are all related to the voltage or current waveform. With any given sine wave it is possible to describe it in any of the four different names and still not change the value of the sine wave.

Therefore it is necessary to have a means of converting from one form to another and vice-versa. Table 3-2 shows the possible ways of converting from one to the other.

To use the table, find the given value in the left hand column and, looking to the right, perform the arithmetic indicated.

The table is formed by modifying the relationships given in the definitions. The following list of formulas is the relationships given by the basic definitions. Notice they are all given in terms of the peak value.

Peak to Peak = peak × 2 **Formula 3-4**
Average = peak × .636 **Formula 3-5**
RMS = peak × .707 **Formula 3-6**

Table 3-2. Conversion Factors.

Given value	Peak	Peak to peak	RMS	Avg.
Peak	---	× 2	× .707	× .636
Peak to peak	× 2	---	$\times \frac{.707}{2}$	$\times \frac{.636}{2}$
RMS	× 1.414 or $\times \frac{1}{.707}$	Peak × 2	---	× .9 or Peak × .636
Average	× 1.57 or $\times \frac{1}{.636}$	Peak × 2	× 1.11 or Peak × .707	---

Conversion Problems

With each of the following, use a peak value of 20.

Find peak to peak

Step 1 p to p = 2 × peak — conversion factor
Step 2 p to p = 2 × 20 — substitute
Step 3 p to p = 40 — solve

Find RMS

Step 1 RMS = .707 × peak — conversion factor
Step 2 RMS = .707 × 20 — substitute
Step 3 RMS = 14.14 — solve

Find average

Step 1 average = .636 × peak — conversion factor
Step 2 average = .636 × 20 — substitute
Step 3 average = 12.72 — solve

Find the peak value for each.

peak to peak = 280;

Step 1 peak = p to p $\times \frac{1}{2}$ — conversion factor
Step 2 peak = 280 × .5 — substitute
Step 3 peak = 140 — solve

RMS = 156;

Step 1 peak = RMS $\times \frac{1}{.707}$ — conversion factor
Step 2 peak = 156 × 1.414 — substitute
Step 3 peak = 220 — solve

avg = 63.6;

Step 1 peak = average × $\frac{11}{.636}$ conversion factor

Step 2 avg = 63.6 × 1.57 substitute

Step 3 avg = 100 solve

Practice Problems

Convert each of the following.

1. 30 volts peak, convert to: a) RMS b) p to p c) avg
2. 100 volts peak, convert to: a) p to p b) avg c) RMS
3. 80 volts p to p, convert to: a) RMS b) peak c) avg
4. 260 volts p to p, convert to: a) peak b) avg c) RMS
5. 70 volts RMS, convert to: a) peak b) p to p c) avg
6. 12 volts RMS, convert to: a) avg b) p to p c) peak
7. 16 volts avg, convert to: a) RMS b) peak c) p to p
8. 120 volts avg, convert to: a) peak b) p to p c) RMS
9. 85 volts RMS, convert to: a) peak b) p to p c) avg
10. 70 volts peak, convert to: a) p to p b) avg c) RMS

In problems 11 through 15, find: a) p to p b) peak c) RMS d) avg

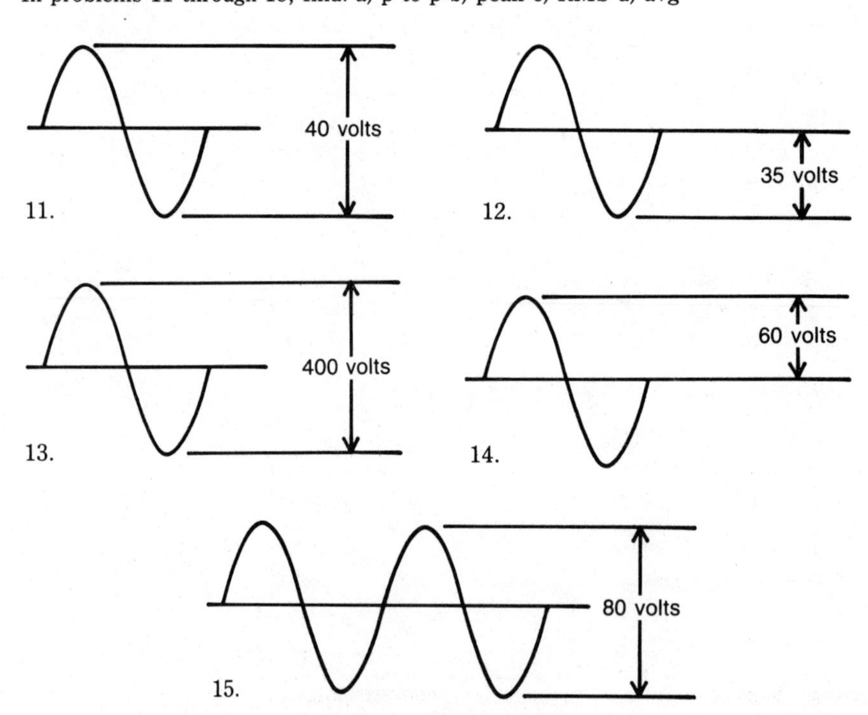

CONVERTING VALUES IN FREQUENCY, PERIOD, AND WAVELENGTH

When dealing with a sine wave, the amplitude is important for measuring the voltage. The frequency and period are also important characteristics of the sine wave. When using an oscilloscope, the amplitude is the vertical portion of the waveform and the time (frequency and period) are the horizontal portion of the waveform.

Wavelength is a measure of how long one cycle is when it is traveling through the air. Therefore, wavelength has a unit of measure in distance.

Frequency and Period

Although frequency has a unit of measure in hertz, its alternate unit of measure is cycles per second, which means frequency is very closely related to time. Period is the measure of time of one cycle. The higher the frequency, that is, the more cycles per second, the shorter period of time to complete one cycle. It stands to reason then, that the relationship of frequency to time is an inverse relationship.

$$\text{frequency} = \frac{1}{\text{time}} \qquad \text{time} = \frac{1}{\text{frequency}}$$

Frequency and Time Problems

The period of a sine wave is 1 ms. What is the frequency?

Step 1 $\text{frequency} = \dfrac{1}{\text{time}}$ formula

Step 2 $\text{frequency} = \dfrac{1}{1\text{ ms}}$ substitute values

Step 3 frequency = 1000 Hz solve

What is the period (time) of the 60 hertz household line voltage?

Step 1 $\text{time} = \dfrac{1}{\text{frequency}}$ formula

Step 2 $\text{time} = \dfrac{1}{60}$ substitute

Step 3 time (or period) = 16.7 ms solve

Wavelength λ

When dealing with radio waves traveling through the air, it is necessary to know the wavelength of the frequency being transmitted. Radio waves are considered to travel at a velocity equal to the speed of light which is 186,000 mi/s or 3×10^{10} cm/s.

$$\text{wavelength} = \frac{\text{velocity (speed of light for radio waves)}}{\text{frequency (frequency of the transmitted signal)}}$$

$$\text{wavelength } \lambda = \frac{3 \times 10^{10}\text{ cm/s}}{\text{frequency (Hz)}}$$

Wavelength Problems

What is the wavelength of a radio station's signal when the station frequency is 630 kHz? (AM broadcast)

Step 1 $\lambda = \dfrac{3 \times 10^{10}\ \text{cm/s}}{\text{frequency}}$ formula

Step 2 $\lambda = \dfrac{3 \times 10^{10}\ \text{cm/s}}{630{,}000\ \text{Hz}}$ substitute values

Step 3 $\lambda = 47{,}619\ \text{cm}$ solve

The wavelength of the same 630 kHz signal in feet is?
2.54 cm = 1 inch and 12 inches = 1 foot

Step 1 $\lambda = \dfrac{47{,}619\ \text{cm}}{2.54\ \text{cm/in}} = 18{,}747\ \text{inches}$ divide

Step 2 $\lambda = \dfrac{18{,}747\ \text{in}}{12\ \text{in/foot}} = 1562\ \text{feet}$ divide

Practice Problems

Calculate the period of the following frequencies:

1. 10 Hz
2. 100 Hz
3. 1 kHz
4. 10 kHz
5. 100 kHz
6. 1 MHz
7. 10 MHz
8. 100 MHz
9. 1 GHz
10. 10 GHz

Calculate the frequencies whose period is:

11. 50 ms
12. 5 ms
13. .5 ms
14. 50 μs
15. .5 s

Calculate the wavelength in cm of the following frequencies:

16. 500 kHz
17. 5 MHz
18. 50 MHz
19. 5 GHz
20. 500 GHz

USING AN OSCILLOSCOPE TO MEASURE A SINE WAVE

The oscilloscope is probably the most useful instrument a technician has at his disposal for measuring *any* kind of waveform. Learning to make measurements on the oscilloscope with a sine wave is a very logical starting point.

The scope is capable of two types of measurement, voltage and time. Other things can be found from these, but usually require an intermediate step with calculations. For example, frequency can be found by measuring the time of one complete cycle and calculating the frequency based on the time measurement.

The front of the scope is often filled with many knobs, switches, and connectors. The controls can usually be placed into one of three categories: 1. voltage, amplitude, the up and down of the waveform, 2. time, back and forth, sideways, 3. trigger, or provide a stable trace on the screen. The two categories of prime interest here are the volts and time.

Measuring Voltage on the Oscilloscope

The primary control on the scope for use when controlling the amplitude is called volts per division. The oscilloscope screen is divided into blocks or squares called divisions. Each of the divisions have four small marks to divide the divisions into subdivisions (most scope screens have the subdivisions marked only on the center axis) which represent part of a division, for example: .2, .4, .6, .8, of a division.

Most scopes have the volts per division control calibrated in steps such as: .01, .05, .1, .5, 1, 5. When preparing the scope display to measure voltage, it is best to adjust the volts per division for a waveform that makes the screen as full as possible (from top to bottom) without going beyond the limits of the screen. Whenever the waveform has an amplitude that is too small, it will be very difficult to read the screen and as a result, the accuracy will suffer.

Figure 3-4 shows three examples of adjusting the oscilloscope, all for the same sine wave. Notice the difference in the amplitude of the waveforms as the volts per division

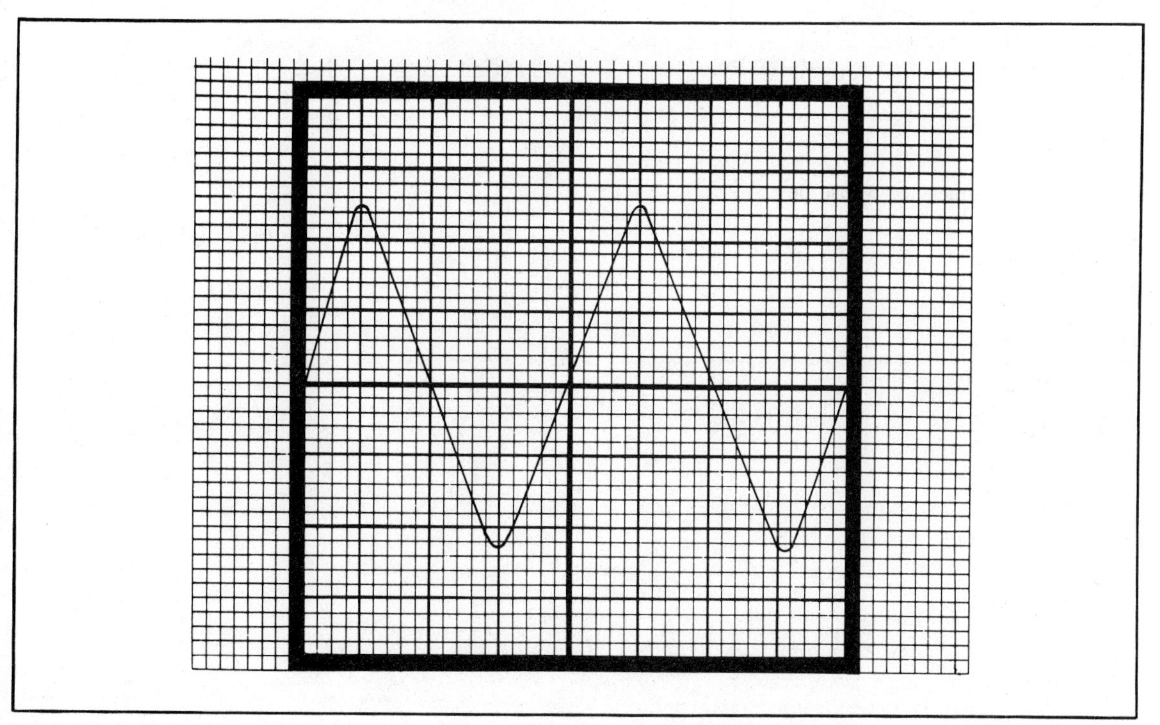

Fig. 3-4A. Oscilloscope display with volts per division and time per division correctly adjusted. .1 volts per division/20 ms time per division.

is adjusted. Figure 3-4A shows the correct way to adjust the scope to make measurements of the voltage and the frequency.

Refer to Fig. 3-4A. The easiest way to measure the voltage is to start off by adjusting the position of the waveform to a location that is convenient to work with. It should be noted that since the voltage measurement is to be a peak to peak measurement, it is not necessary to have the center of the sine wave on the center line of the screen. The drawings shown here do have the center of the sine wave on the center line of the scope screen.

To make the voltage measurement, start at the peak of the wave (either positive or negative peak) and count the number of divisions and subdivisions to the opposite peak. In Fig. 3-4A, starting at the negative peak, the divisions are read as .5 divisions. That is to say, the negative peak is located half way between two major divisions, or .5 to the first major division. Now, count the major divisions; four divisions are counted, plus another .5 divisions. This results in a total of 4 divisions plus two .5 divisions or a total of 5 divisions.

The next step is to read the control setting for volts per division. The setting for Fig. 3-4A is .1 volts per division. Multiply the number of divisions by the setting and the peak to peak voltage is 5 divisions × .1 volts per division = .5 volts p to p.

Measuring Frequency on the Oscilloscope

The frequency is measured similar to the voltage. The difference being that the frequency is a function of time and is therefore measured on the horizontal. To obtain

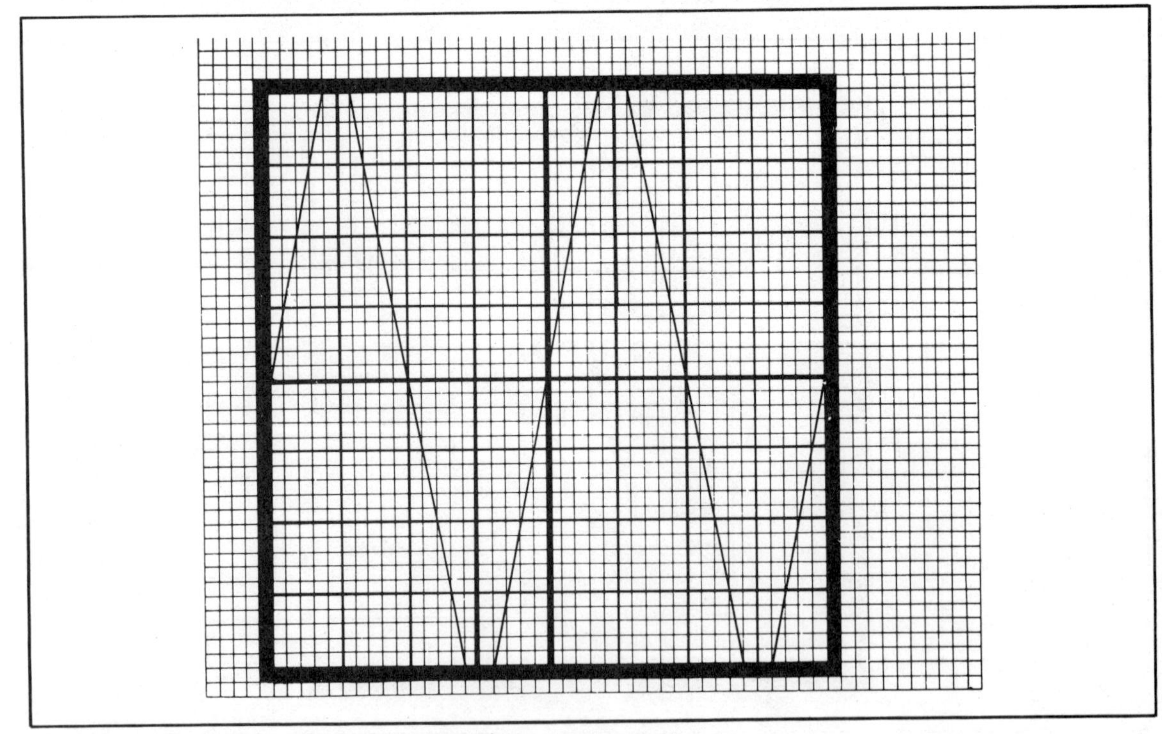

Fig. 3-4B. Volts per division expanded too much. .05 volts per division.

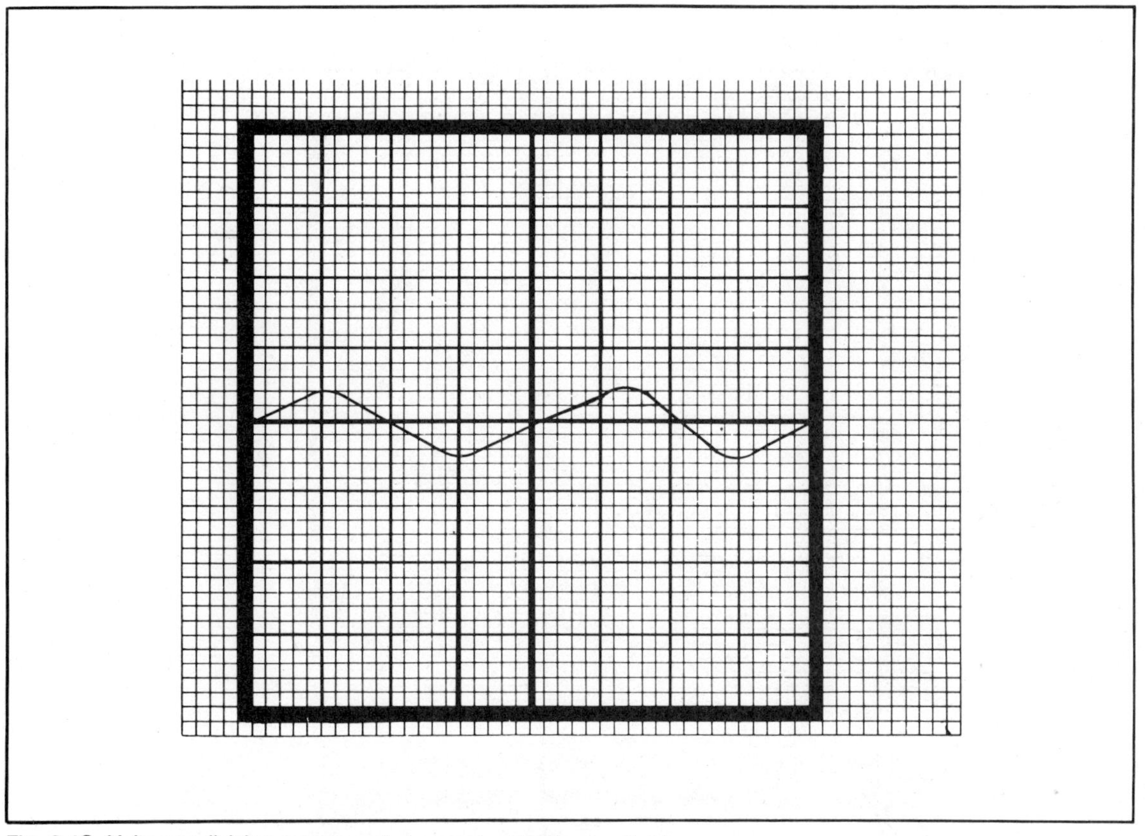

Fig. 3-4C. Volts per division not expanded enough. .5 volts per division.

the most accurate results, the best place to measure the time required for one complete wave is at the exact center of the waveform. The reason this is the most accurate place to measure is because at this point the line crosses the center axis at the most straight up and down place on the waveform. It is best then, to center the waveform exactly on the scope's center axis line.

To measure the time of one cycle, called the period, start where the wave crosses the center line at zero, going in the positive direction. Count the number of major divisions to the point where the wave again crosses the center line, in the positive direction. The sine wave of Fig. 3-4A shows a period of 4 divisions. Multiply the number of divisions by the time per division to arrive at the period.

Step 1 4 divisions × 20 ms per division = 80 ms
Step 2 period = 80 ms
Step 3 frequency = 1/t = 1/80 ms = 12.5 hertz

Practice Problems

Each of the following drawings represents an oscilloscope display of a sine wave. Determine the peak to peak voltage, period and frequency.

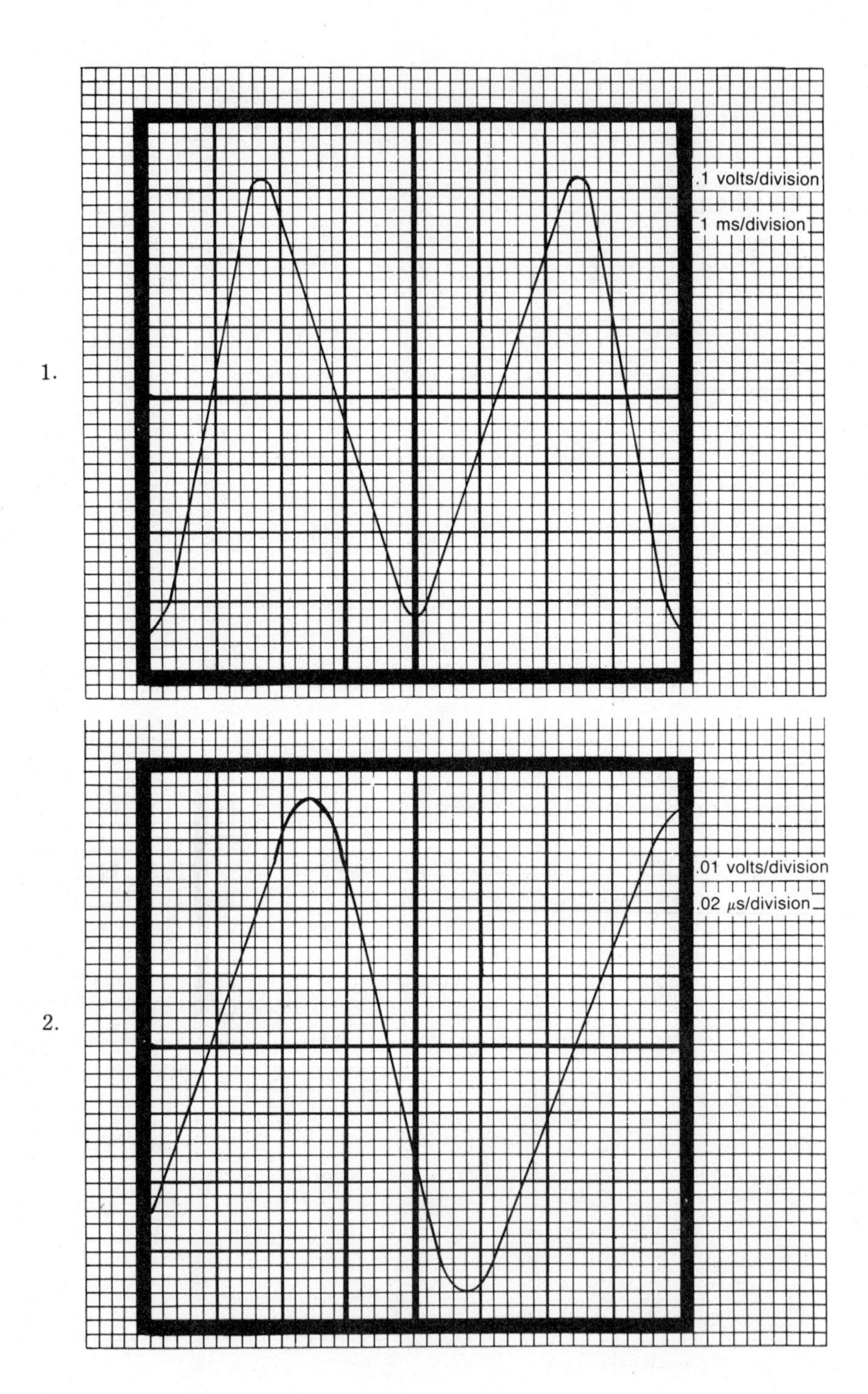
1.
.1 volts/division
1 ms/division
2.
.01 volts/division
.02 μs/division

3.

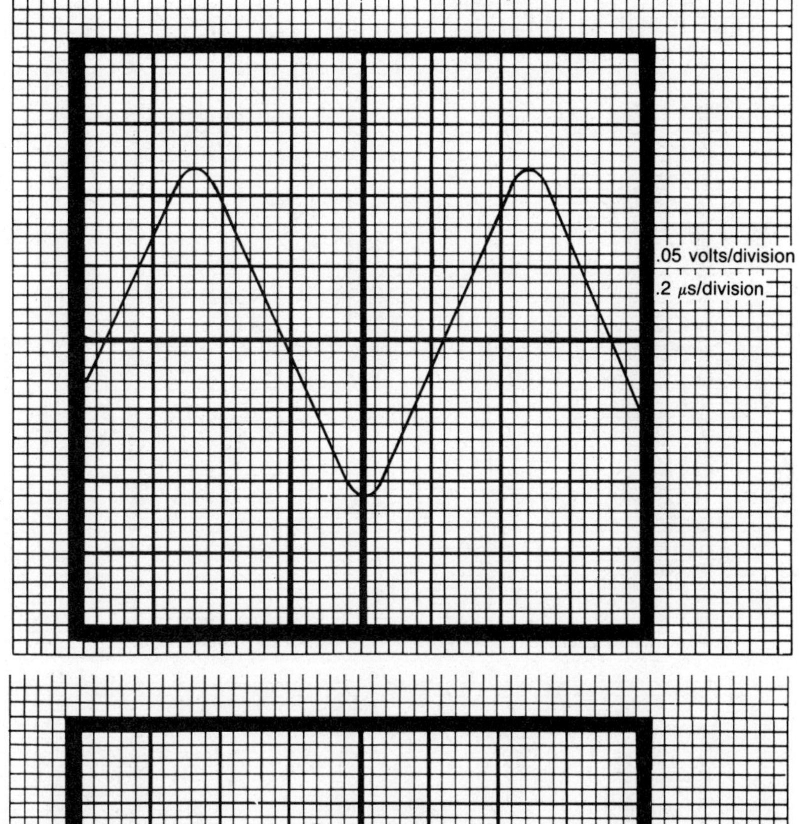
.05 volts/division
.2 μs/division

4.

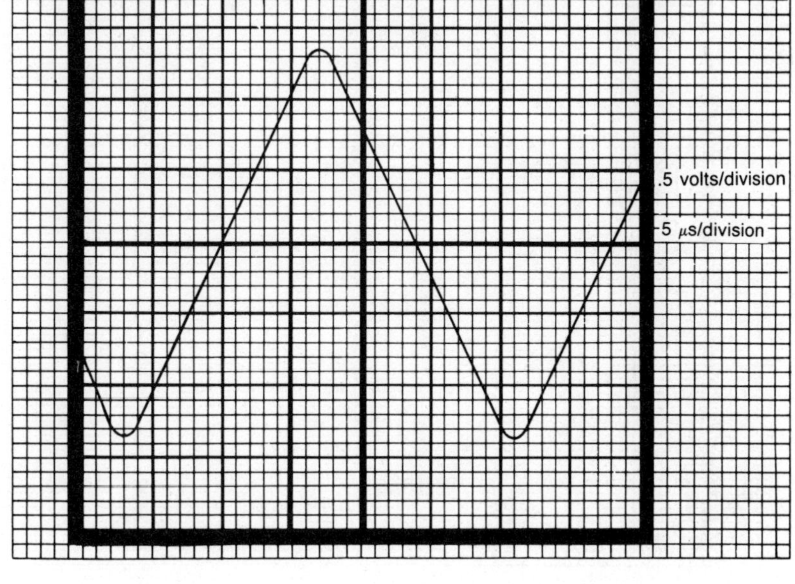
.5 volts/division
5 μs/division

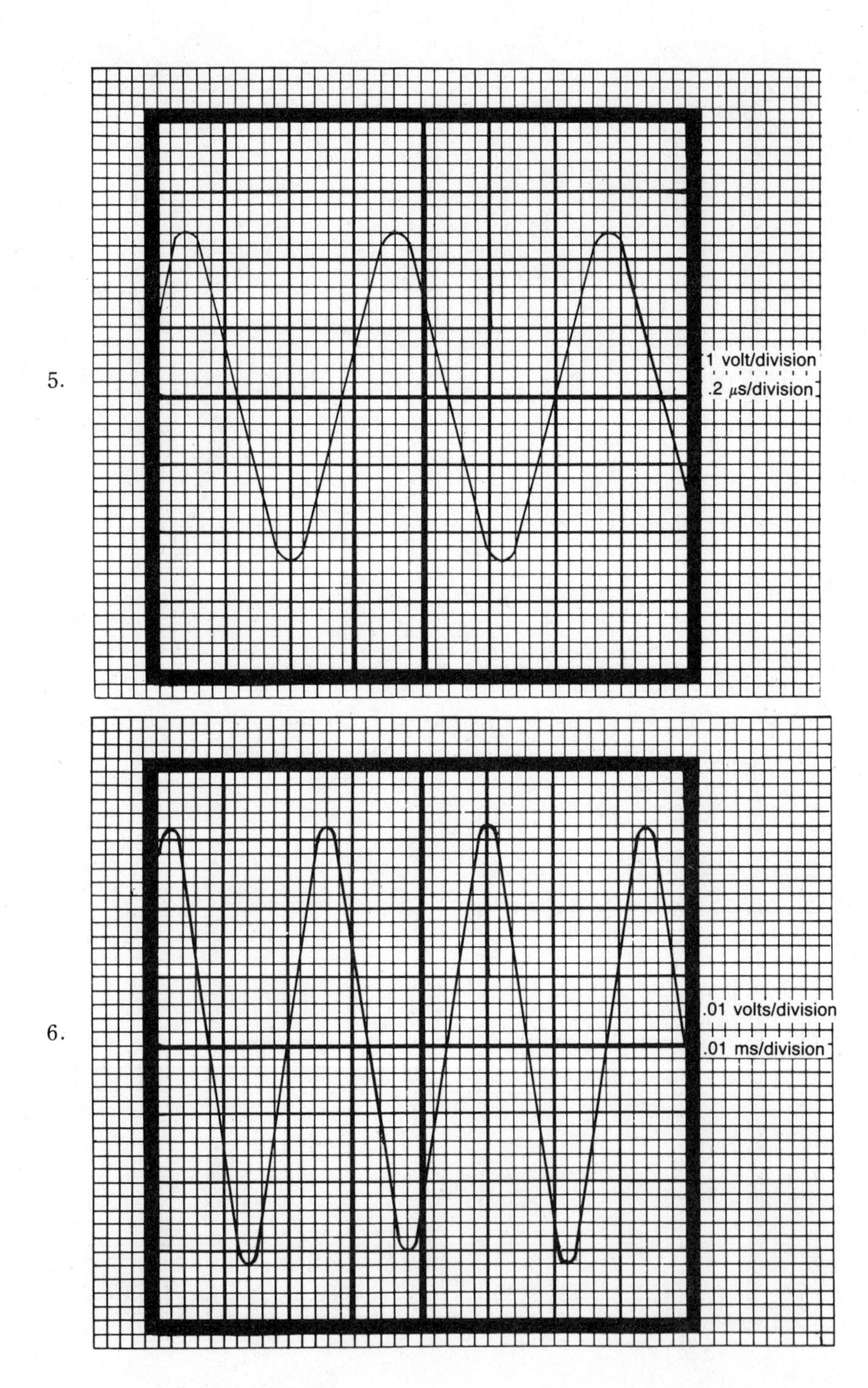
5.
1 volt/division
.2 μs/division
6.
.01 volts/division
.01 ms/division

7.

8.

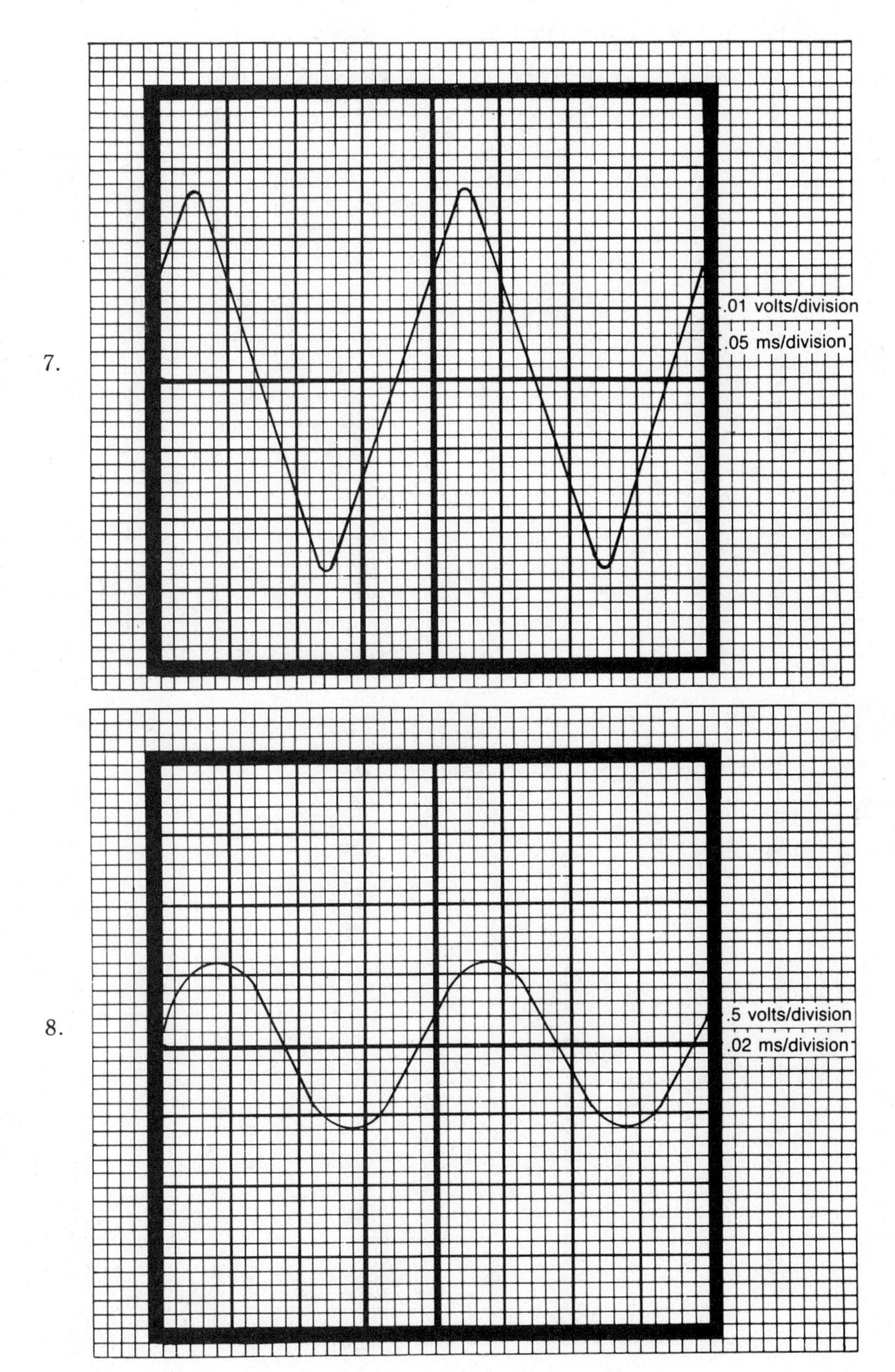
.01 volts/division
.05 ms/division
.5 volts/division
.02 ms/division

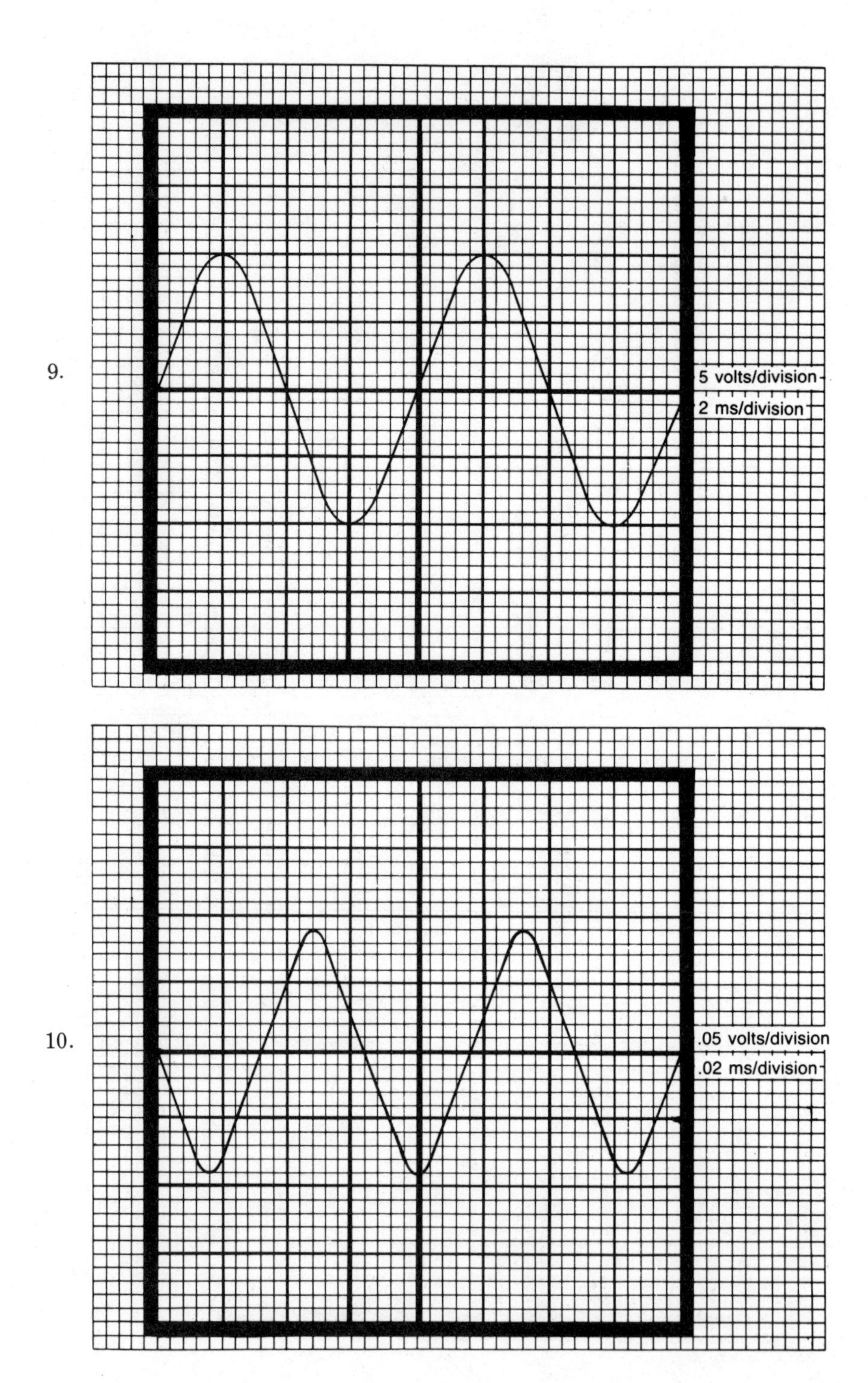
9.
5 volts/division
2 ms/division
10.
.05 volts/division
.02 ms/division

CHAPTER SUMMARY

The sine wave is a continuously varying voltage, that will go from zero, through a maximum positive voltage, through zero to a maximum negative voltage and return to zero again in a time frame reference of 360 degrees.

A sine wave voltage can be measured in terms of peak to peak, peak, RMS, or average. There are conversion factors to convert the different voltages from one form to another.

The time frame used by a sine wave can be expressed in one of five different ways: degrees, time, frequency, period, and wavelength. These all have conversion factors, except degrees.

The oscilloscope, discussed in this chapter is seen as one of the most valuable instruments for use when measuring waveforms, whether they are sine waves or any other waveform.

Summary of Formulas

instantaneous voltage of a sine wave $V = V_m \sin \Theta$ **Formula 3-1**

wavelength in centimeters $\lambda = \dfrac{3 \times 10^{10} \text{ cm/s}}{\text{freq (hz)}}$ **Formula 3-2**

frequency $f = \dfrac{1}{T}$ **Formula 3-3**

peak to peak $\text{Peak to Peak} = \text{peak} \times 2$ **Formula 3-4**

Average $\text{Average} = \text{peak} \times .636$ **Formula 3-5**

RMS $\text{RMS} = \text{peak} \times .707$ **Formula 3-6**

Chapter 4

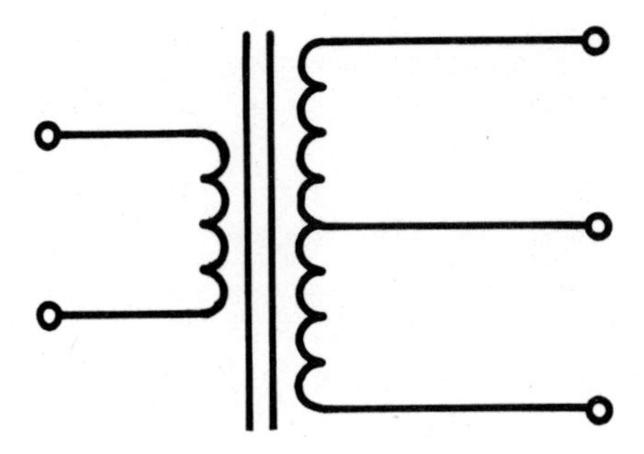

Transformers

Transformers are found in almost every piece of electronic equipment. A transformer is usually used in the power supply section of a circuit, therefore, the study of transformers is fairly significant.

A transformer is basically two coils of wire wound together in such a way that there is as near perfect magnetic coupling as possible between the two coils of wire.

The mathematics of transformers deals with the turns ratio, voltage ratio, current ratio, power relationships, and efficiency.

TURNS RATIO, VOLTAGE RATIO, CURRENT RATIO

The turns ratio is the mathematical ratio that describes the relationship of the number of turns in the primary of the transformer (input) to the number of turns in the secondary (output).

The voltage ratio is based on the turns ratio and is used to calculate the amount of voltage found at the secondary when a certain voltage is applied to the primary.

The current ratio is an inverse relationship with the turns ratio and is used to calculate the current of the secondary when current is applied to the primary.

Turns Ratio

The turns ratio is expressed as the number of turns in the primary to the number of turns in the secondary. The capital letter N is used to represent the number of turns.

transformer turns ratio $\frac{N_p}{N_s}$ **Formula 4-1**

For example, if a particular transformer has 600 turns in the primary and 60 turns in the secondary, the ratio is:

$$\frac{600}{60} = \frac{10}{1} \text{ or 10:1 primary to secondary}$$

The turns ratio results in pure numbers, with no units. Also, the turns ratio does not necessarily state the actual number of turns of wire, it is only a ratio for use with calculations.

Voltage Ratio

The voltage ratio is the most common way of stating the capability of a transformer. The voltage ratio is equal to, and directly proportional to, the turns ratio. The reason voltage is often stated as the rating of a transformer is the fact that it can be related to how the transformer will be used.

$$\text{voltage ratio } \frac{N_p}{N_s} = \frac{V_p}{V_s}$$ **Formula 4-2**

An example of using the voltage ratio; a particular transformer has a turns ratio of 10:1 and 120 volts RMS is applied to the primary. What is the secondary voltage?

Step 1 $\frac{N_p}{N_s} = \frac{V_p}{V_s}$ formula

Step 2 $\frac{10}{1} = \frac{120}{V_s}$ substitute values

Step 3 $V_s = \frac{120 \times 1}{10}$ transpose equation

Step 4 $V_s = 12$ volts RMS secondary voltage

Current Ratio

Current ratio is helpful when determining the size of the fuse to use that will protect a circuit. The current ratio is the inverse of the turns, or voltage ratio. This means a current in the secondary of a step up (step up or step down always refers to the voltage) transformer will be higher than the current of the primary.

$$\text{current ratio } \frac{N_p}{N_s} = \frac{I_s}{I_p}$$ **Formula 4-3**

For an example of using the current ratio; a step-down transformer has a turns ratio of 6:1 and 120 volts is applied to the primary. If 2 amps flows in the secondary, what is the primary current?

Step 1 $\frac{N_p}{N_s} = \frac{I_s}{I_p}$ formula (voltage given will not be used)

Step 2 $\frac{6}{1} = \frac{2}{I_p}$ substitute values

Step 3 $I_p = \frac{2 \times 1}{6}$ transpose equation

Step 4 I_p = .333 amp

Voltage and current ratios calculated by the methods shown here assume a perfect transformer. It should be noted that actual transformers have losses. The voltage ratio is affected by the coefficient of coupling. That means the ability of the transformer winding to couple the magnetic field from the primary to the secondary. The current used by the secondary in relation to the current developed in the primary also is effected by the various losses in a transformer. Any calculations made in this section will assume perfect transformers. Calculating transformer efficiency will be discussed later.

When transformer ratios are given in terms other than 1:something, the problem is solved using proportions.

Example

A transformer has a turns ratio of 5:2 with a primary voltage of 120. Find the secondary voltage.

Step 1 $\frac{N_p}{N_s} = \frac{V_p}{V_s}$ formula

Step 2 $\frac{5}{2} = \frac{120}{V_s}$ substitute values

Step 3 $V_s = \frac{120 \times 2}{5}$ transpose equations

Step 4 V_s = 48

Practice Problems

Complete Table 4-1.

	Turns ratio (pri:sec)	Primary volts (RMS)	Secondary volts (RMS)	Primary current (amps)	Secondary current (amps)
1	10:1	120	- - -	.1	- - -
2	6:1	- - -	20	- - -	.6
3	1:5	30	- - -	- - -	.2
4	- - -	24	96	- - -	1
5	- - -	- - -	27	.2	.6
6	1:7	17	- - -	- - -	.03
7	2:5	10	- - -	.3	- - -
8	3:4	- - -	160	- - -	1.3
9	4.5:2	- - -	53.3	.5	- - -
10	7:6	28	- - -	- - -	.21

POWER RELATIONSHIPS IN A TRANSFORMER

If a transformer is considered to have no losses (losses do exist in an actual transformer) then, there is a 1:1 relationship of power in the primary to power in the secondary. That is to say, power in the primary is equal to power in the secondary. If the primary is considered as the input and the secondary is considered as the output, $P_{in} = P_{out}$. As an example, some of the practice problems from the previous section are shown below.

Step 1 Pri volts = 120, Sec volts = 12
Pri current = .1, Sec current = 1
Step 2 $P = I \times E$ formula
Step 3 $P_{pri} = .1 \times 120$ substitute primary values
Step 4 $P_{pri} = 12$ watts primary power
Step 5 $P_{sec} = 1 \times 12$ substituting values for secondary
Step 6 $P_{sec} = 12$ watts secondary power

As this example demonstrates, the power in the primary = the power in the secondary when there are no transformer losses. In fact, it is from this power relationship that the inverse current relationship can be demonstrated.

Step 1 $P_{pri} = P_{sec}$ original relationship
Step 2 $I_pE_p = I_sE_s$ use the power formula to substitute
Step 3 $\frac{E_p}{E_s} = \frac{I_s}{I_p}$ transpose the equation to have the form of the voltage to current ratio of a transformer

TRANSFORMER EFFICIENCY

Since nothing can operate without losses, that is to say, nothing is 100 percent efficient, it stands to reason that transformers cannot be 100 percent efficient.

Transformer windings cannot have perfect coupling, the core material cannot be a perfect conductor of magnetic lines of force, the wire itself is not a perfect conductor of electricity. In other electrical subject areas, the losses are usually small enough to be ignored. However, in transformers, if there is a significant amount of power transfer there can be significant amount of losses.

Transformer losses usually result in heat, which is a direct loss of power. Efficiency is stated in terms of percent, the closer to 100 percent, the better the system, or the smallest amount of losses.

Efficiency of any system is calculated by

$$\text{Eff} = \frac{P_{out}}{P_{in}} \times 100\%$$ **Formula 4-4**

The formula states that power out is always less than or equal to power in, never larger.

Example

$P_{out} = 300$ watts, $P_{in} = 400$ watts

Step 1 $\text{Eff} = \frac{P_{out}}{P_{in}} \times 100\%$ formula

Step 2 $\text{Eff} = \dfrac{300}{400} \times 100\%$ substitute values

Step 3 $\text{Eff} = 75\%$

Along with transformer efficiency is the application of voltage and current in a normal, working transformer. When a transformer has no load connected to the secondary, no current is being drawn from either the secondary or the primary. With no load, the transformer output voltage is at its maximum value. This is the equivalent of 100 percent efficient. Since there is no current being drawn, there is no power.

As the transformer load current increases, the current flowing in the windings will cause I^2R drops in the wire to increase, resulting in a power loss. Also, current increase in the secondary causes a current increase in the primary which results in I^2R losses in the primary, reducing its power. Current flowing in the windings also increase the core losses, further decreasing the efficiency.

When a transformer is in actual operation, the main concern is of the output voltage. This whole discussion on transformer efficiency was to prove why a transformer output voltage will decrease as more current is drawn from the secondary.

The following key point should be noted.

- The output voltage of a transformer decreases as the load current is increased.

TESTING A TRANSFORMER WITH AN OHMMETER

A transformer can be thought of as two coils of wire; the primary and the secondary. Some transformers have a multi-tap secondary, which means there is more than one secondary winding.

Consider the primary first. Remember, an ohmmeter must be used in a circuit with no voltage applied. Some transformers will have a primary with a tap for connecting to a higher voltage. Consider using the ohmmeter on the primary without a tap first.

The ohmmeter reads the dc resistance of the winding, which is nothing more than a continuous piece of wire. If the coil is normal, it will read a dc resistance of between 1 and 200 Ω, depending on the coil. The most likely trouble that will happen to a transformer winding is an open coil. With a condition of an open, the ohmmeter will read infinite resistance. If the primary winding has a tap, it will have three wires coming out. The tap is usually at a point of approximately one-half the total resistance of the full coil.

The secondary is usually made with taps from the same coil. Often, the secondary will have several completely different coils. The individual coils will have to be tested with the ohmmeter on an individual basis. If the secondary is different windings, a faulty winding does not affect any other winding.

If a primary winding is open, there will be no voltage induced in the secondary.

CHAPTER SUMMARY

Transformers have a magnetic coupling between the primary winding and the secondary winding. An ac voltage applied to the primary winding will induce a voltage in the secondary winding, resulting in a transfer of power from the primary to the secondary.

An actual transformer will have an efficiency of less than 100 percent due to the losses in the transformer construction.

The following keypoint should be noted.

- The output voltage of a transformer decreases as the load increases.

Summary of Formulas

transformer turns ratio $\frac{N_P}{N_S}$ **Formula 4-1**

voltage ratio $\frac{N_P}{N_S} = \frac{V_P}{V_S}$ **Formula 4-2**

current ratio $\frac{N_P}{N_S} = \frac{I_S}{I_P}$ **Formula 4-3**

efficiency in a transformer $\text{Eff} = \frac{P_{out}}{P_{in}} \times 100\%$ **Formula 4-4**

Chapter 5

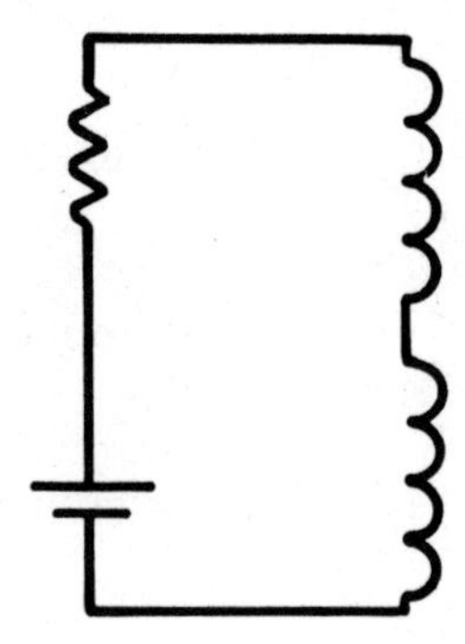

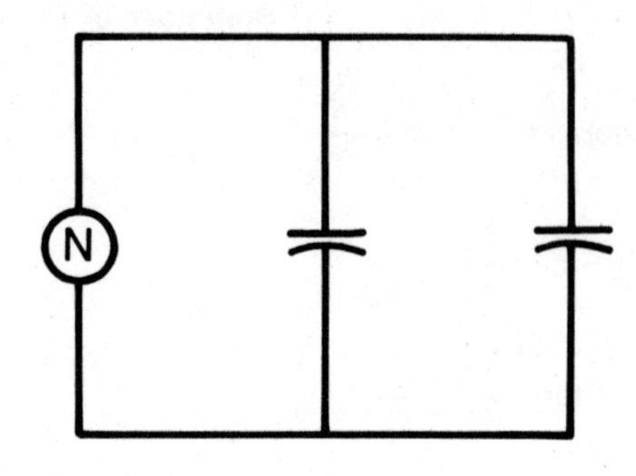

Inductors and Capacitors

Along with resistors, inductors and capacitors are the three most basic components in the study of electricity and electronics. Inductors and capacitors are important because of their capability of storing electrical energy, called a charge.

These components have very useful properties in both dc and ac circuits. Inductance is the principle by which a transformer works. The coil in the ignition system of an automobile is an inductor which stores energy to fire the spark plug. Capacitors make filters for use with a power supply, fire the electronic flash unit on a camera and are also used to form timing circuits.

PROPERTIES OF INDUCTANCE

Whenever electrical current is flowing in a wire there will be a magnetic field developed around the wire. Inductance is the ability of a conductor to use the magnetic field developed to induce a voltage in itself. The only time a coil can develop voltage within itself is when there is a changing current flow in the conductor. This changing current does not need to be an ac current flow, dc will have the same effect, if the current flow is changing. The voltage that is developed in an inductor is opposing the applied voltage and is called back emf.

- Back emf is developed in a coil of wire only if there is changing current in the coil.

When current first starts to flow, it must build the magnetic field in the inductor. This process of trying to build the magnetic field causes the back emf to oppose this changing current. Then, when the current is decreasing the magnetic field that has been charged will discharge in the same direction as the current flow, adding to the current, which has the effect of opposing the current from decreasing.

- Inductance is the characteristic that opposes any change in current.

Like resistors, it is possible to connect inductors in series or parallel in order to obtain a resultant, total inductance, similar to the idea of a total resistance.

Although inductors could be used for various applications in dc circuits, inductors serve an important function when used with ac circuits, especially when used with a sine wave. Inductors can be thought of when used as a transformer, but a single coil of wire can be used to form an electric motor. Also, the property of a magnetic field being developed around any coil of wire can also be used to make an electro-magnet.

In order to demonstrate the strength an electro-magnet can have when a large enough coil is used with a large enough current (remember it is current flow that develops the magnetic field), consider this example; large electro-magnets are connected to the end of cable and are used to lift very large pieces of metal, such as cars. This would be a dc voltage applied to the coil and the coil would be rather large. As long as the current is flowing, the magnetic field will hold solid, the moment power to the coil is removed, the magnetic field will collapse.

The example shown above is on a large scale, but it does serve to demonstrate the ability of the magnetic field that can be developed in a coil of wire. In electronics, however, the size of the inductance is usually quite small.

The unit for inductance is the henry, symbol H. A medium sized transformer, or a filter choke, which is a coil used in power supply filters, will have inductance in the range of 1 or 2 henries. An inductor used as part of a low frequency resonant circuit will be in the neighborhood of about 100 mH. A high frequency coil would be in the range of 1 or 2 μH. The larger the coil, the larger the henries. A small coil will have a small number of henries.

INDUCTORS IN SERIES

More than one coil can be connected in series to produce different values of inductance. It is possible, however, to have the two coils physically located near each other to cause the magnetic field of one coil to interact with the field of the other coil. When this happens, it produces some transformer action, which is called mutual inductance. Unless the mutual inductance is a transformer, mutual inductance is undesirable.

The equation for calculating two or more inductors in series assumes no mutual inductance.

$$L_T = L_1 + L_2 + L_3 \ldots \text{etc.}$$ **Formula 5-1**

Figure 5-1 shows the schematic diagram of two coils in series.

Mutual Inductance

As in the case of transformers, mutual inductance is helpful in producing an end result that is not available from any one single coil. This is found in the secondary of the transformer when it is intended for coils to be connected with either series aiding or series opposing field to have a phasing dot on the schematic diagram. Figure 5-2 shows coils wound on a core material (This is not a schematic diagram) showing the direction of the windings. The large dots just above and to one side of the coil are the phasing dots. Another way of thinking about the application of the phasing dots is to consider the

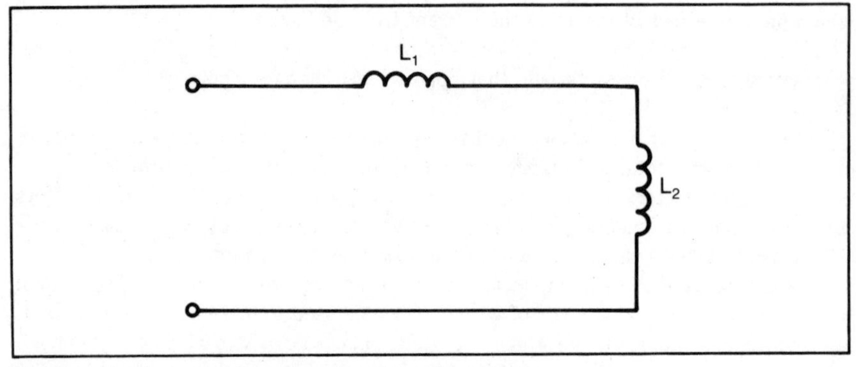

Fig. 5-1. Two inductors in series. No mutual inductance.

sine wave with its positive half cycle and negative half cycle. When one phasing dot is positive, the phasing dot on the neighboring coil will also be positive. After drawing in the instantaneous polarity of the coils, think of them in terms of series aiding or series opposing batteries.

The formula for calculating total inductance when the coils are in series with mutual inductance;

Series Aiding Mutual Inductance	$L_T = L_1 + L_2 + 2L_M$	**Formula 5-2**
Series Opposing Mutual Inductance	$L_T = L_1 + L_2 - 2L_M$	**Formula 5-3**

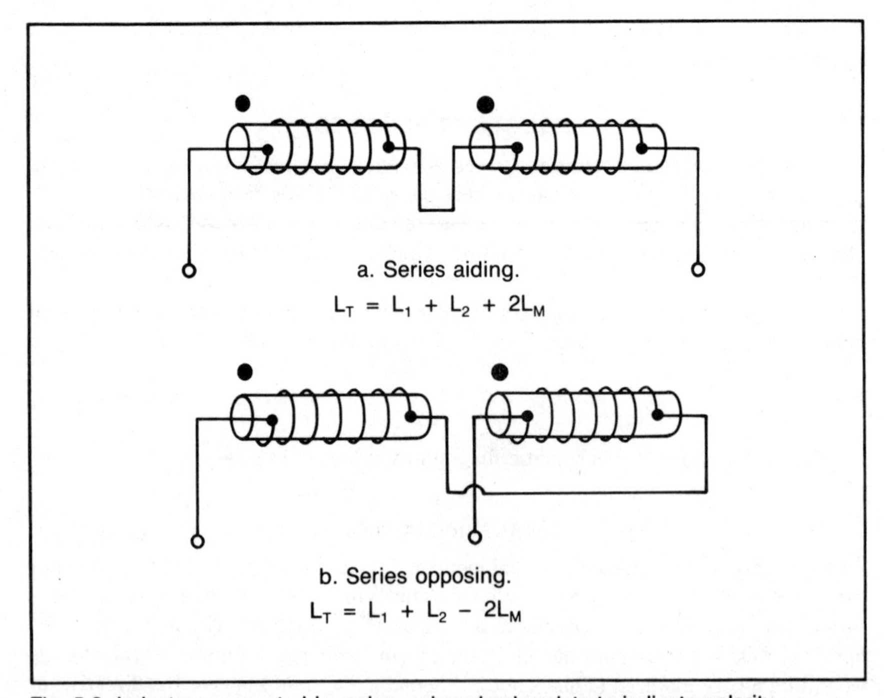

Fig. 5-2. Inductors connected in series, using phasing dots to indicate polarity.

INDUCTORS IN PARALLEL

Inductors connected in parallel have a total inductance calculated in the same manner as resistors in parallel, using the reciprocal formula. The short cut formulas used in resistance can be used with inductors in parallel.

Parallel Inductors $$\frac{1}{L_T} = \frac{1}{L_1} + \frac{1}{L_2} + \cdots$$ **Formula 5-4**

It is possible to have inductors connected in parallel with mutual inductance. However, due to the inverse relationships, the calculations become quite complicated and the predictions of the end result make a parallel connection almost never used in actual practice.

Example Problems

What is the equivalent inductance of three inductors connected in series with no mutual inductance? The values of the inductors are; 25 mH, 50 mH, 75 mH.

Step 1 $L_T = L_1 + L_2 + L_3$ formula
Step 2 L_T = 25 mH + 50 mH + 75 mH substitute values
Step 3 L_T = 150 mH total equivalent inductance

Two 100 mH coils are connected in series aiding with a 10 mH mutual inductance. What is the total inductance?

Step 1 $L_T = L_1 + L_2 + 2L_M$ formula
Step 2 L_T = 100 mH + 100 mH + 2×10 mH substitute values
Step 3 L_T = 200 mH + 20 mH combine inductors and multiply the mutual inductance portion
Step 4 L_T = 200 mH total equivalent inductance

A 500 μH inductor is connected in series opposing with a 400 μH inductor with 50 μH mutual inductance. What is the effective inductance?

Step 1 $L_T = L_1 + L_2 - 2L_M$ formula
Step 2 L_T = 500 μH + 400 μH − 2×50 μH substitute values
Step 3 L_T = 800 μH total equivalent inductance

Practice Problems

Find the total inductance of each of the following combinations. Assume $L_M = 0$ when no value is given.

1. Series circuit; 500 μH, 700 μH
2. Series circuit; 10 mH, 50 mH, 30 mH
3. Series circuit; 100 mH, .2 H, .15 H, L_M = 10 mH, aiding
4. Series circuit; 1 H, 1.5 H, 1200 mH, L_M = 25 mH, opposing
5. Parallel Circuit; 100 mH, 200 mH

6.

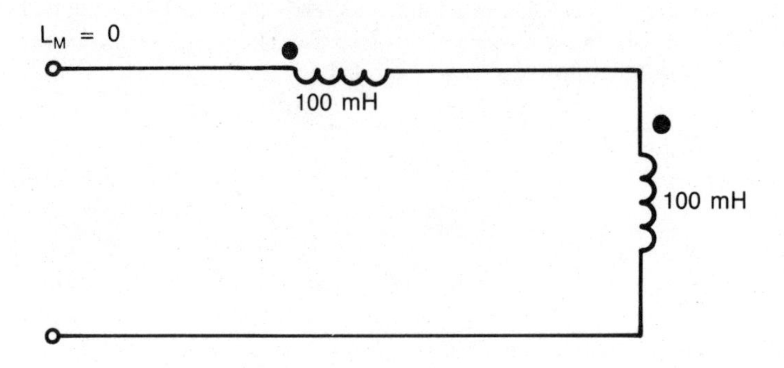

$L_M = 0$
100 mH
100 mH

7.

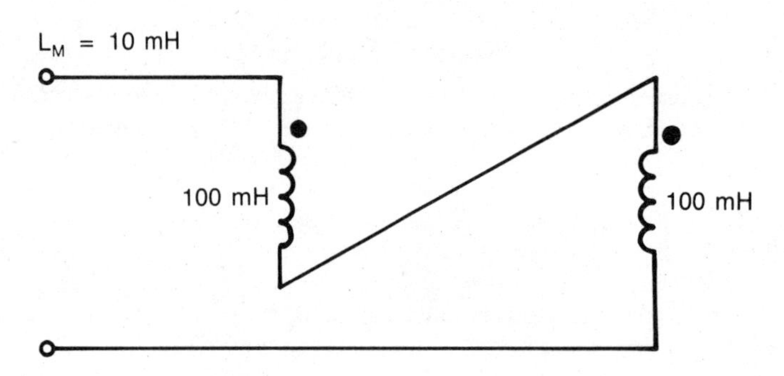

$L_M = 10$ mH
100 mH
100 mH

8.

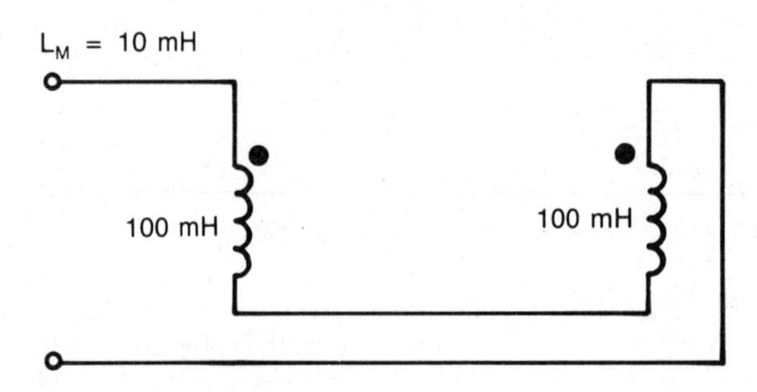

$L_M = 10$ mH
100 mH
100 mH

9.

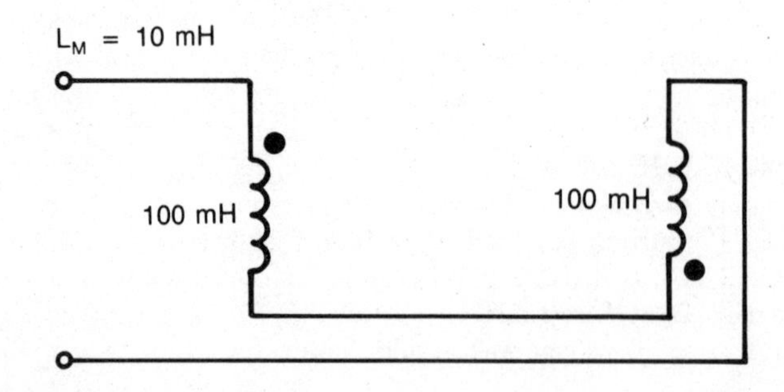

$L_M = 10$ mH
100 mH
100 mH

10.

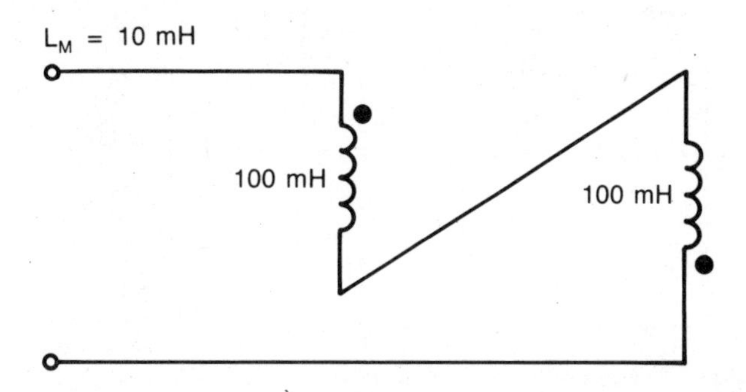

PROPERTIES OF CAPACITANCE

Capacitors are a component in very frequent usage, however, many electronics students find it difficult to fully grasp how the capacitor works. In order to form a capacitor there are two factors that must be satisfied. The first being two separate conductors, the second factor being that the conductors are separated by an insulator, called a dielectric. The following description of how a charge is stored is over-simplified but should help the student to a better understanding of the capacitor.

When a dc voltage is applied to a capacitor, the capacitor will take on a charge with a voltage equal to the applied voltage. The capacitor requires a period of time to charge. This will be discussed in the next chapter on time constants, however for this discussion, enough time is allowed for a full charge.

Refer to Fig. 5-3. In Fig. 5-3A, the capacitor has no charge across it. The switch is in the **off** position, no current will flow in the circuit because there is not a complete current path. In Fig. 5-3B, the switch is closed to the **a** position. The capacitor will start charging. Electrons will leave the negative side of the battery, travel to the capacitor to the plate marked negative. This plate will allow the electrons to gather. The opposite plate, marked positive, will allow electrons to leave the plate and travel to the battery. For every electron that leaves the negative side of the battery and stores on the negative plate of the capacitor, there will be an equal number that will vacate the positive side of the capacitor and return to the battery. The battery considers this situation to be a closed circuit. The current will continue to flow until the capacitor has stored enough electrons on the negative terminal and removed enough electrons from the positive to have a difference in potential between the two plates equal to the battery, in this circuit it's 10 volts. When the capacitor has stored the 10 volts, current will stop flowing in the circuit, even though the switch remains closed.

In Fig. 5-3C, the switch is returned to the **off** position. The capacitor had a full charge and this charge will remain across the capacitor since there is no path for current to flow to allow the electrons to balance the difference in potential between the two plates. In actual practice, the dielectric, the insulator between the plates, is not a perfect insulator, no matter what it is made of. Since it is not a perfect insulator, some electrons can travel through the insulator and eventually discharge the capacitor. This is called leakage, and is generally so low as not to cause any problems.

In Fig. 5-3D, the switch is moved to the **b** position. In this position, the capacitor has a current path. Note, the battery is removed from the circuit and current through the resistor is in the opposite direction than it was in Fig. 5-3B. Current is now supplied by the capacitor. Its starting voltage was 10 volts, however, as electrons leave the nega-

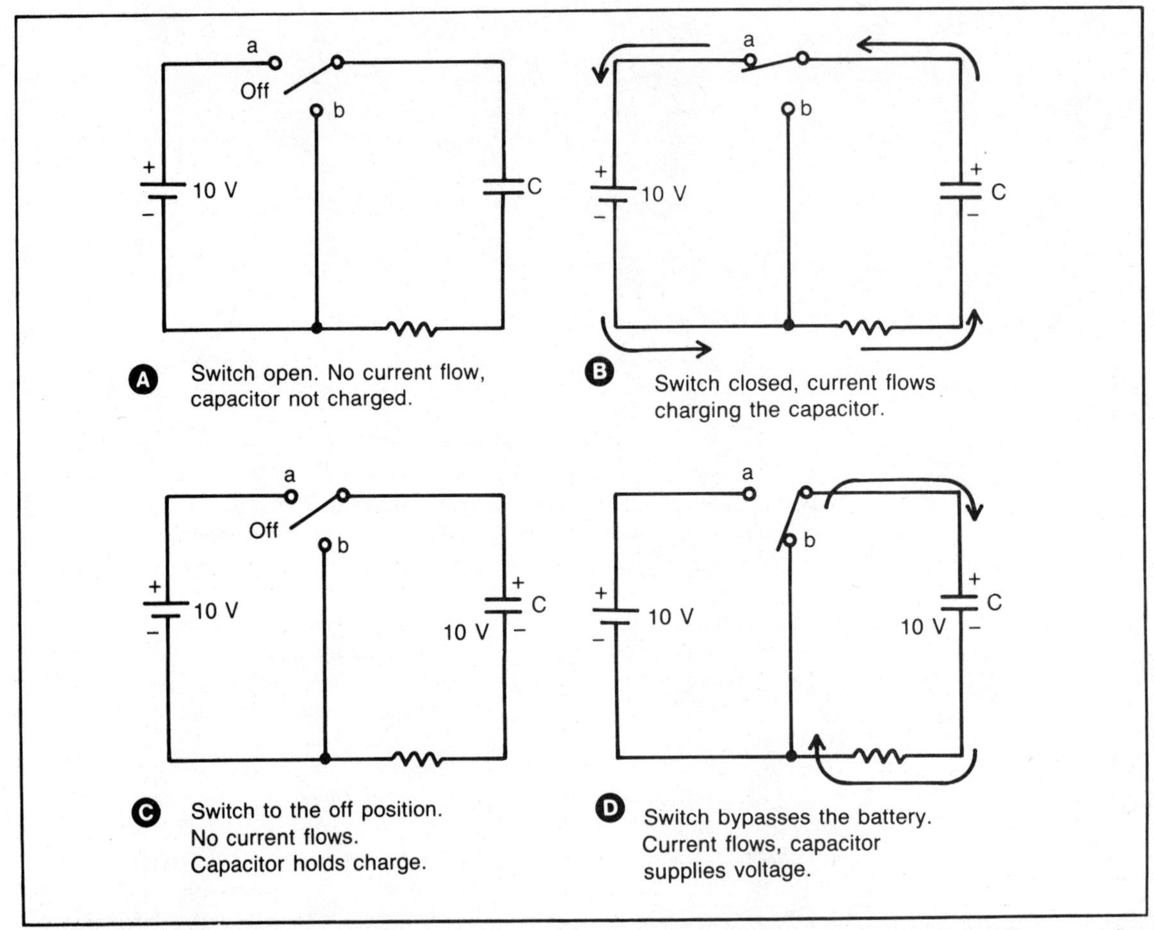

Fig. 5-3. A dc circuit to show how a capacitor charges and stores voltage. A. Switch open. No current flow, capacitor not charged. B. Switch closed, current flow charging the capacitor. C. Switch to the off position. No current flows. Capacitor holds charge. D. Switch bypasses the battery. Current flows, capacitor supplies voltage.

tive terminal and return to the positive plate, the potential difference across the plates is lowered, neutralized. Current will continue to flow, although the force drops off, until the capacitor is fully discharged.

Because a capacitor stores a voltage, it is the property of a capacitor to try to maintain a constant voltage in the circuit. A circuit voltage cannot change quickly because it must wait for the capacitor to charge. Notice, also that current can flow in a dc circuit only while the capacitor is charging (or discharging).

- The property of a capacitor is to oppose any **change** in **voltage**.
- When a capacitor is in a dc circuit, current flows **only** when the capacitor is charging (or discharging).

The unit of capacitance is the farad. The farad is a measure of the ability of a capacitor to store a charge. The two factors that affect a capacitor's ability to store a charge are the plate area and the dielectric thickness. The larger the plate area, the more charge can be stored. The closer the plates (thinner dielectric) the stronger is the electric field developed resulting in a larger charge.

The unit farad is much too large of a unit to deal with in an actual circuit. The most common unit size is the micro-farad (μF, 10^{-6}) and the next most common is the pico-farad (pF, 10^{-12}). Capacitors are almost always measured in terms of one of these two unit sizes.

CAPACITORS IN PARALLEL

When capacitors are connected in parallel, the resultant, effective capacitance is the addition of the individual capacitors.

Parallel capacitance $C_T = C_1 + C_2 + C_3$ **Formula 5-5**

In Fig. 5-4, two capacitors are shown in parallel. By connecting the capacitors in parallel, the effective plate area is increased, thus increasing the capacitance value. Since it is a parallel connection, voltage is the same across both capacitors, therefore, they will charge to the same voltage. In the event the capacitance values were different, the stored voltage would be the same but the stored number of electrons would be different. The smaller capacitor would store a smaller number of electrons, which simply means the voltage that is stored will discharge quicker.

When capacitors are connected in parallel (or series) to make a new equivalent capacitance, the capacitors should be of the same type. For example; two electrolytic, two disc, two mica, etc. The reason to try to match the types is the current storage capacity. A comparison to this might be connecting a large automotive battery in series with a small flashlight battery. The small battery could be destroyed by the current capacity of the larger battery.

CAPACITORS IN SERIES

When capacitors are connected in series, the effect is to increase the thickness of the dielectric, the insulator between the plates, resulting in a lowered value of capacitance. The reciprocal formula is used to calculate the effective capacitance.

Series Capacitance $\frac{1}{C_T} = \frac{1}{C_1} + \frac{1}{C_2} \ldots$ etc. **Formula 5-6**

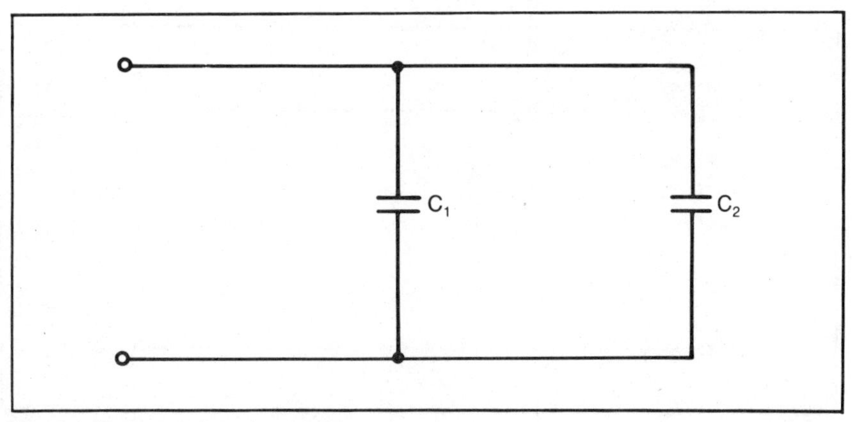

Fig. 5-4. Two capacitors in parallel.

Any of the short cut formulas apply to the reciprocal formula, regardless if it is used for parallel resistors, parallel inductors, or series capacitors.

Often, it is difficult to visualize the current path in a circuit with series capacitors. For every electron that leaves the negative terminal of the battery and travels to the negative plate of the first capacitor, an equal number of electrons will depart the positive plate of the series capacitor, leaving a charge on the first capacitor. The electrons that left the plate of the positive terminal of the first capacitor are collected on the negative terminal of the series capacitor, then an equal number of electrons will leave the positive terminal of that capacitor and return to the power supply. Because an equal number of electrons that left the battery from the negative terminal returns to the positive terminal, there is a complete current path. Even though it appears that the wire between the two capacitors receives no current, it is important to note that in a series circuit, current is the same throughout the series circuit. Figure 5-5 shows two capacitors in series.

CAPACITIVE VOLTAGE DIVIDERS

One very useful function of capacitors in series is the capacitive voltage divider. Voltage drops in any voltage divider is determined by the ratio of the components, not the applied voltage or the current flow.

voltage divider formula for finding the voltage across C_1

$$V_{C1} = \frac{C_2}{C_1 + C_2} \times V$$

Formula 5-7

voltage divider formula for finding the voltage across C_2

$$V_{C2} = \frac{C_1}{C_1 + C_2} \times V$$

Formula 5-8

The two formulas shown above are simply a modification of the resistive voltage divider formula. As with any voltage divider, it is important to notice the ratio is based on the ratio of the individual component values to the total value, found by adding the components values. This ratio is then multiplied by the applied voltage.

Figure 5-6 shows a capacitive voltage divider with two equal components, capacitive values. The voltage drops across each should be the same.

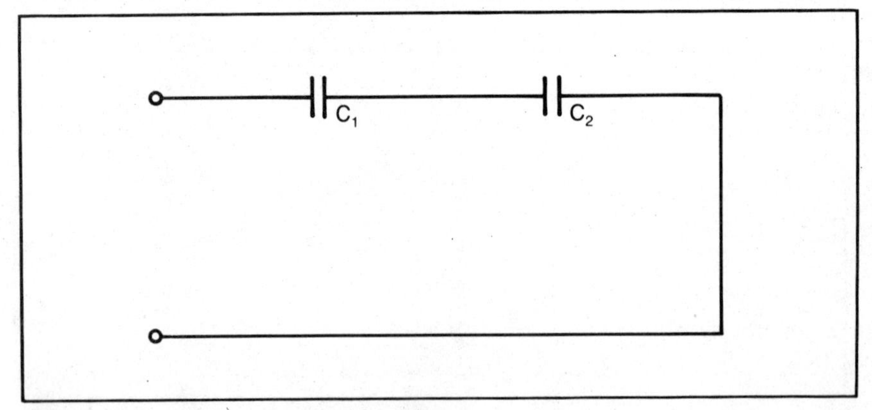

Fig. 5-5. Two capacitors connected in series.

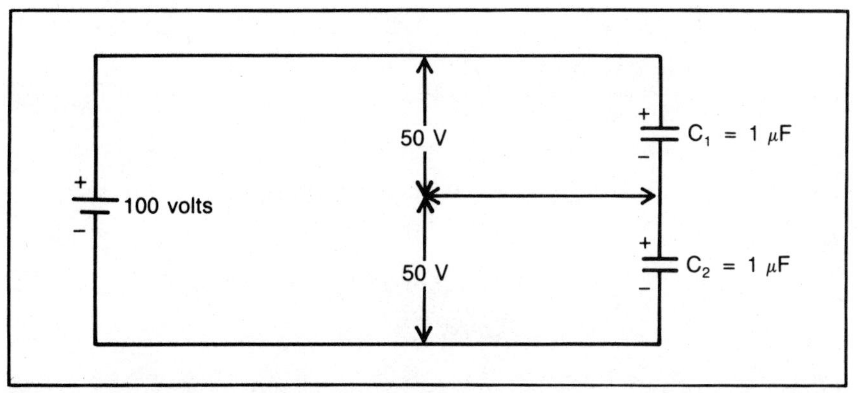

Fig. 5-6. Capacitive voltage divider with equal values of capacitance, equal voltage drops.

Substitute values and prove.

Step 1 $V_{C1} = \dfrac{C_2}{C_1 + C_2} \times V$ formula

Step 2 $V_{C1} = \dfrac{1\ \mu F}{1\ \mu F + 1\ \mu F} \times 100\ V$ substitute values

Step 3 $V_{C1} = \dfrac{1}{2} \times 100$ simplify ratio fraction

So if $V_{C1} = 50$ V (voltage across C_1), then $V_{C2} = 50$ V

Figure 5-7 shows a capacitive voltage divider with unequal capacitance values. It stands to reason that there should be unequal voltage drops, but the big question when trying to set up the ratios is which capacitor will get the highest voltage.

- In a capacitive voltage divider, the greater value capacitor gets the least voltage.
- In a capacitive voltage divider, the least value capacitor gets the greatest voltage.

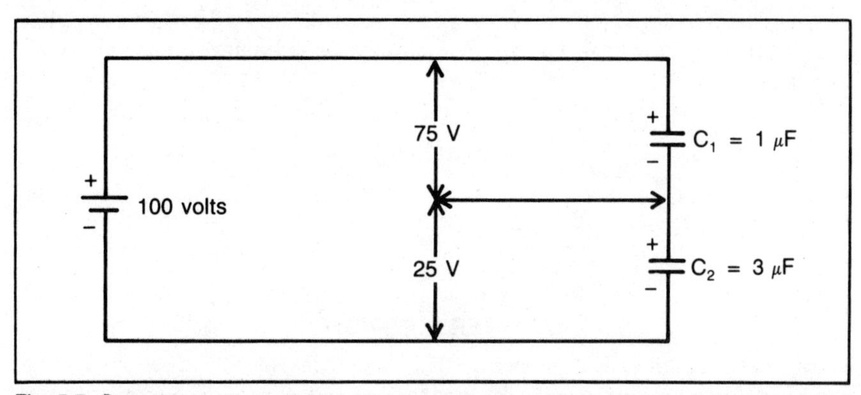

Fig. 5-7. Capacitive voltage divider with unequal values of capacitance. Larger capacitor, smaller voltage.

Find V_{C1} for Fig. 5-7.

Step 1 $V_{C1} = \dfrac{C_2}{C_1 + C_2} \times V$ formula

Step 2 $V_{C1} = \dfrac{3\ \mu F}{1\ \mu F + 3\ \mu F} \times 100\ V$ substitute values

Step 3 $V_{C1} = \dfrac{3}{4} \times 100$ simplify ratio fraction

Step 4 $V_{C1} = 75\ V$ voltage across C_1

Find V_{C2} for Fig. 5-7.

Step 1 $V_{C2} = \dfrac{C_1}{C_1 + C_2} \times V$ formula

Step 2 $V_{C2} = \dfrac{1\ \mu F}{1\ \mu F + 3\ \mu F} \times 100$ substitute values

Step 3 $V_{C2} = \dfrac{1}{4} \times 100$ simplify ratio fraction

Step 4 $V_{C2} = 25\ V$ voltage across C_2

Practice Problems

Questions 1 through 5, find the total capacitance.

1. Parallel circuit; 2 μF, 3 μF
2. Parallel circuit; 5 pF, .001 μF, 10 pF
3. Parallel circuit; .01 μF, .001 μF, .22 μF
4. Parallel circuit; 47 μF, 47 μF, 47 μF
5. Parallel circuit; 100 μF, 100 μF, 100 μF, 100 μF, 100 μF

Questions 6 through 10 are all series circuit voltage dividers. Given is the capacitor values and the supply voltage. Find the total capacitance and voltage drops across each capacitor.

6. $C_1 = 1\ \mu F$, $C_2 = 2\ \mu F$, $V = 10\ V$
7. $C_1 = 5\ pF$, $C_2 = 10\ pF$, $V = 10\ V$
8. $C_1 = .01\ \mu F$, $C_2 = .1\ \mu F$, $V = 5\ V$
9. $C_1 = 10\ \mu F$, $C_2 = 10\ \mu F$, $C_3 = 10\ \mu F$, $V = 15\ V$
10. $C_1 = .1\ \mu F$, $C_2 = .2\ \mu F$, $V = 20\ V$

CHAPTER SUMMARY

Inductors and capacitors have properties that make them valuable in certain applications. This chapter was intended to discuss the hows and whys of these properties, not to discuss the applications. In future chapters, some of the applications will be discussed.

Although inductors and capacitors are often thought of as being useful only to ac circuits, it has been demonstrated in this chapter, dc circuits provide an understandable discussion in how they work.

- Back emf is developed in a coil of wire only if there is changing current in the coil.
- Inductance is the characteristic that opposes any change in current.
- The property of a capacitor is to oppose any change in voltage.
- When a capacitor is in a dc circuit, current flows only when the capacitor is charging (or discharging).
- In a capacitive voltage divider, the higher value capacitor gets the smallest voltage drop.
- In a capacitive voltage divider, the lower value capacitor gets the largest voltage drop.

Summary of Formulas

series inductors $L_T = L_1 + L_2 + L_3 \ldots$ etc. **Formula 5-1**

series aiding mutual inductance $L_T = L_1 + L_2 + 2L_M$ **Formula 5-2**

series opposing mutual inductance $L_T = L_1 + L_2 + 2L_M$ **Formula 5-3**

parallel inductors $\frac{1}{L_T} = \frac{1}{L_1} + \frac{1}{L_2} + \frac{1}{L_3} \ldots$ **Formula 5-4**

parallel capacitors $C_T = C_1 + C_2 + C_3$ **Formula 5-5**

series capacitors $\frac{1}{C_T} = \frac{1}{C_1} + \frac{1}{C_2} + \frac{1}{C_3} \ldots$ **Formula 5-6**

voltage divider formula for the voltage across C_1 $V_{C1} = \frac{C_2}{C_1 + C_2} \times V$ **Formula 5-7**

voltage divider formula for the voltage across C_2 $V_{C2} = \frac{C_1}{C_1 + C_2} \times V$ **Formula 5-8**

Chapter 6

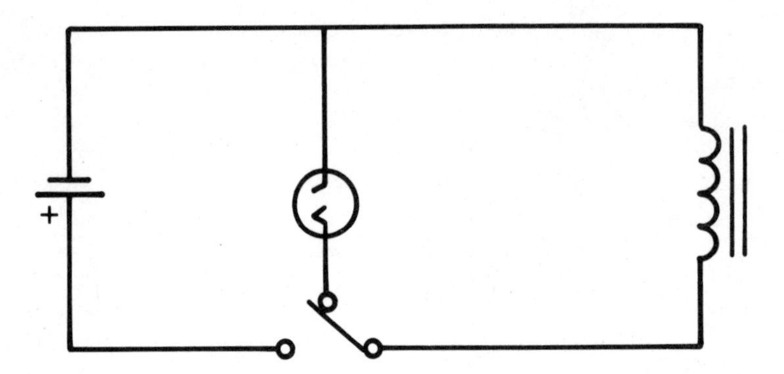

Time Constants

When a voltage is applied to an inductor or a capacitor, the result is predictable. In fact, during the first few moments that the voltage is applied, or removed, the building, or collapsing of the charge will always follow a curve with the same shape. This is called the **universal time constant curve**.

The time involved in the charging or discharging of an inductor or capacitor is called a time constant. The time constant finds a great many applications in electronic circuitry.

UNIVERSAL TIME CONSTANT CURVE

The time constant is a mathematical formula used to calculate the time required to reach a percentage of full charge or discharge. Time constant formulas will be shown later.

- A time constant is defined as the length of time required to reach 63 percent of full charge or discharge.
- Five time constants is defined as the time required to reach full charge or discharge.

The universal time constant curve is actually made up of two curves (refer to Fig. 6-1). The bottom axis of the curve is the number of time constants, with 5 time constants indicating a full charge or discharge. The vertical axis on the left side is labeled in percentage of full charge. When using the discharge curve, the percentage of full charge actually indicates the charge still remaining after a period of discharge.

The purpose of the universal time constant curve is to predict the voltage of either charge or discharge for any time constant between 0 and 5 time constants even for decimal time constants. The curve provides a solution by graphical analysis, which although it is somewhat inaccurate, is often accurate enough and actually shows the charging/discharging action.

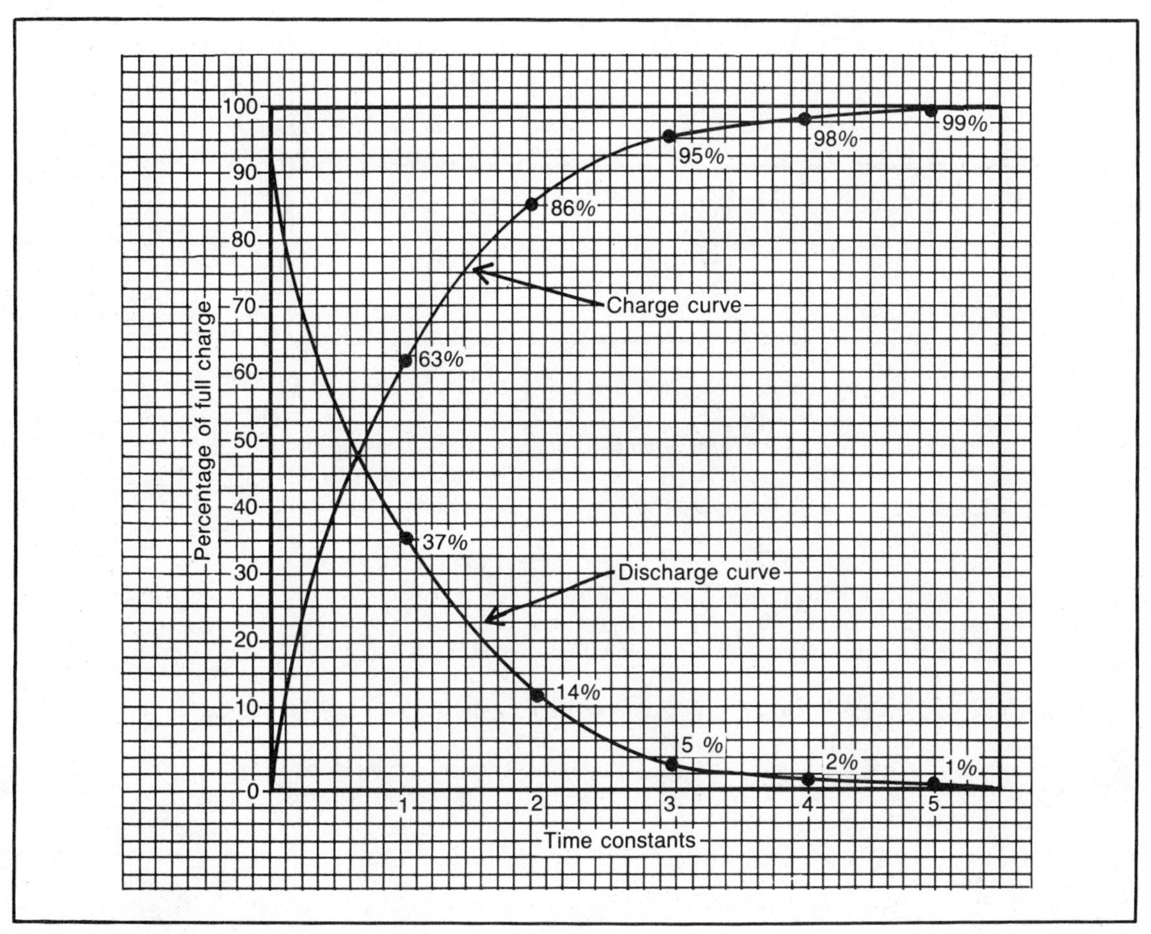

Fig. 6-1. Universal time constant curve.

Plotting the Charge Curve

There are two ways to plot the curves. The first method is to use the 63 percent as the definition of a time constant. This is an approximation. The second method of plotting the curve is through the use of the time constant equation.

The following is the calculations to plot the charge curve using 63 percent as one time constant. The results are shown in Table 6-1.

Starting from a time of 0, the charging starts toward 100 percent of the full charge.

Step 1 At the first time constant, 63 percent of the full charge has been completed, leaving 37 percent.

Step 2 After the second time constant 63 percent of the 37 percent remaining from step 1 will be completed. ($.63 \times .37 = .23$). This means 23 percent of the full charge is completed during this step. Adding to what has been done so far, $63\% + 23\% = 86\%$ completed. 14 percent remains of the full charge.

Step 3 The third time constant will charge 63 percent of the remaining portion. ($.63 \times .14 = .09$). This means 9 percent of the full charge is completed during

Table 6-1. Plotting the Universal Time Constant Curve.

	Charging		Discharging	
Time constant	63%	$\% = 1 - e^{-T}$	63%	$\% = e^{-T}$
0	0	0	100%	100%
1	63	63.2	37	36.8
2	86	86.5	14	13.5
3	95	95.0	5	5.0
4	98	98.2	2	1.8
5	99	99.3	1	.67

this step. Adding to what has been done so far, 63% + 23% + 9% = 95% completed, 5 percent remains.

Step 4 The fourth time constant will charge 63 percent of the remaining portion. (.63 × .05 = .03). This step provides 3 percent of the full charge. Adding to what has been done so far; 63% + 23% + 9% + 3% = 98% completed, 2 percent remains.

Step 5 The fifth time constant will charge 63 percent of the remaining portion. (.63 × .02 = .01). This step provides 1 percent of the full charge. Adding to what has been done so far; 63% + 23% + 9% + 3% + 1% = 99% completed, 1 percent (actually less because of rounding) remains to be fully charged.

It should be noted, if each time constant only does 63 percent of what remains, the unit will never reach a full charge. This makes sense since an actual component, regardless of how good it is will have some losses. Therefore, 5 time constants is considered to be as close to a full charge as is possible. Any further attempt at charging will only overcome the losses.

The next set of calculations is based on the formula for calculating the instantaneous charge.

Instantaneous charge $\%\ \text{charge} = 1 - e^{-T} \times 100\%$ **Formula 6-1**

e is the natural log
T is the time constant

The formula 6-1 is called the instantaneous charge formula because it gives the exact value anywhere along the charge curve. It does not have to be used at the whole number time constants, but can be used at any point, for example; 2.25 T. The capital letter T is used to indicate the time constant. The formula uses a $-T$. When the calculator is used to calculate the e^{-T} a decimal value will result which will be the equivalent of the percentage of the charge remaining. By subtracting this from 1, the value will then be the decimal of the percentage charged during that time. Multiplying by 100 converts the decimal to a percentage.

The results of the following calculations are in Table 6-1. The calculations shown are for the instantaneous charge, calculated at the whole number time constants.

$\%\ \text{charge} = 1 - e^{-T} \times 100\%$ **Formula 6-1**

Note e^{-T} is shown on the calculator as e^x simply replace the e^x with the proper value of $-T$

Calculate the first time constant.

Step 1 % = 1 − e^{-1} × 100% substitute the time constant.
Step 2 % = 1 − .368 × 100% perform the e^x
Step 3 % = 63.2% of full charge at the first time constant

Calculate the second time constant.

Step 1 % = 1 − e^{-2} × 100% substitute the time constant
Step 2 % = 1 − .135 × 100% perform the e^x
Step 3 % = 86.5% of full charge at the second time constant

Calculate the third time constant.

Step 1 % = 1 − e^{-3} × 100% substitute the time constant
Step 2 % = 1 − .0498 × 100% perform the e^x
Step 3 % = 95% of full charge at the third time constant

Calculate the fourth time constant.

Step 1 % = 1 − e^{-4} × 100% substitute the time constant
Step 2 % = 1 − .0183 × 100% perform the e^x
Step 3 % = 98.2% of full charge at the fourth time constant

Calculate the fifth time constant.

Step 1 % = 1 − e^{-5} × 100% substitute the time constant
Step 2 % = 1 − .0067 × 100% perform the e^x
Step 3 % = 99.3% of full charge at the fifth time constant

Comparing the results of the calculations in Table 6-1, it should be recognized that even though the equation using the natural log, e^{-T}, is a more accurate calculation, the approximation method of 63 percent is close. Therefore, the definition of one time constant being 63 percent is a good approximation.

Plotting the Discharge Curve

In plotting the discharge curve portion of the universal time constant curve there are two ways to find the location of the whole number time constants on the curve. The first method is to use the time constant equals 63 percent. The second method is to use the following formula.

Instantaneous discharge % of full charge = e^{-T} × 100% **Formula 6-2**
e is the natural log
T is the time constant

Formula 6-2 will give the exact instantaneous value of the discharge for any point along the curve. Notice it is stated in terms of the percent of full charge.

Calculating by the method of the 63 percent time constant;

Time zero of the discharge curve is located at the maximum charge point, 100 percent.

Step 1 Starting from 100 percent, discharge the first time constant brings the curve to 100 − 63% = 37% of the full charge.

Step 2 Starting the second time constant from the point left off after the first time constant, discharge 63 percent of what is remaining. (.63 × .37 = .23). The second time constant discharges 23 percent which means the point of the discharge curve is now 100 − 63 = 37. 37 − 23% = 14% of full charge remaining.

Step 3 Time constant 3 will discharge 63 percent of the 14 percent remaining. (.63 × .14 = .09) The third time constant discharges 9 percent. 14% − 9% = 5% of full charge remains.

Step 4 The fourth time constant will discharge the remaining portion 63 percent. (.63 × .05 = .03). The fourth time constant discharges 3 percent of the total. 5% − 3% = 2% remaining of the original full charge.

Step 5 The fifth time constant will discharge the remaining portion 63 percent. (.63 × .02 = .01). The fifth time constant discharges 1 percent of the total. 2% − 1% = 1% remaining of the full charge after 5 time constants.

Formula 6-2 will be used to calculate the plotting of the discharge curve in order to determine the exact values of each point since the 63 percent method is only an approximation. The results of these calculations are found in Table 6-1.

Calculate the first time constant.

Step 1 % of full charge = e^{-1} × 100% substitute the time constant
Step 2 % = .368 × 100% perform the e^x
Step 3 % = 36.8% remains of full charge after 1 time constant

Calculate the second time constant.

Step 1 % = e^{-2} × 100% substitute the time constant
Step 2 % = .135 × 100% perform the e^x
Step 3 % = 13.5% remains after 2 time constants

Calculate the third time constant.

Step 1 % = e^{-3} × 100% Substitute the time constant
Step 2 % = .0498 × 100% perform the e^x
Step 3 % = 5.0% remains after 3 time constants

Calculate the fourth time constant.

Step 1 % = e^{-4} × 100% Substitute the time constant
Step 2 % = .0183 × 100% perform the e^x
Step 3 % = 1.8% remains after 4 time constants

Calculate the fifth time constant.

Step 1 % = e^{-5} × 100% Substitute the time constant
Step 2 % = .0067 × 100% perform the e^x
Step 3 % = .67% remains of full charge after 5 time constants.

The calculations shown above are all to plot the universal time constant curve. It is a valid curve for any circuit, regardless of the component values. However, the standard means of calculating time constants involves the application of actual component values for a particular circuit.

L/R TIME CONSTANT

Whenever an inductor has electricity applied to it, there will be a magnetic field developed around the coil. The magnetic field does not charge instantly, but rather takes a period of time based on the value of the inductor and the amount of resistance that is in the circuit. The same also applies when the electricity is removed, the inductor will discharge through the path provided for discharging. The length of time involved will depend on the value of the inductor and the resistance in the circuit.

Even though the actual length of time is different for each different circuit, one time constant is still 63 percent of full charge and it takes five time constants to either charge or discharge.

The formula for calculating the time constant of an inductive circuit is:

$$T = \frac{L}{R}$$ **Formula 6-3**

Time constant for an inductor with series resistance. Time in seconds if resistance in ohms and inductance in henries.

Refer to Fig. 6-2. Whenever analyzing a time constant circuit, trace the current path for charge and discharge.

The charge path of the inductor is from the battery's negative terminal, through the inductor, through the resistor and returning to the positive side of the battery through the switch in the top position. When charged, the inductor will have an effective polarity as shown with the plus at the top.

The discharge path is always opposite the charge path, as far as the inductor is concerned. The discharge path starts at the inductor, on the bottom side, through the center path to the bottom side of the switch, through the resistor and returning to the inductor.

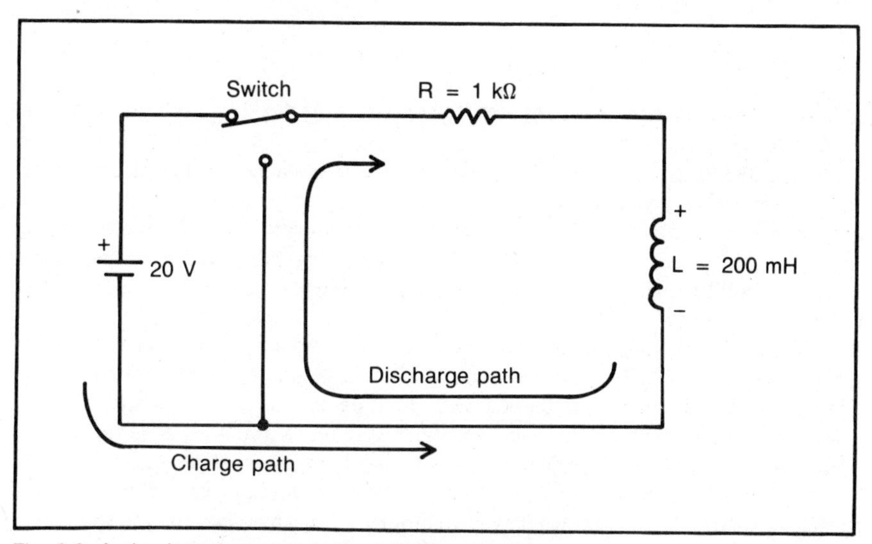

Fig. 6-2. A circuit to demonstrate the L/R time constant.

The calculations for the charge and discharge time constants for this circuit are exactly the same because the values of inductance and resistance are the same for both charge and discharge.

Step 1 $T = \frac{L}{R}$ formula

Step 2 $T = \frac{.2\ H}{1000\ ohms}$ substitute values, first change units to henries and ohms

Step 3 T = .0002 seconds
T = .2 ms value of one time constant

Step 4 One time constant equals .2 ms, therefore full charge is reached 5 × .2 ms (5 time constants) = 1 ms for full charge. The discharge time constant is exactly the same for this circuit.

The analysis of Fig. 6-2 is not complete with just calculating the value of time constant. The property of an inductor is to build a magnetic field by having current flow through the windings. The inductor opposes any change in current and this accounts for the time constant. Since the magnetic field cannot build or collapse instantly if the power is instantly removed from the inductor, the magnetic field will try to maintain the same current through its discharge path.

In Fig. 6-2, the inductor is represented as having zero ohms coil resistance, with the resistor being a separate component. Current in the charge path, therefore, is calculated by using Ohm's law;

Step 1 $I = \frac{E}{R}$ formula

Step 2 $I = \frac{20\ V}{1000\ ohms}$ substitute values

Step 3 I = 20 mA current after full charge

The current calculated here is after full charge is reached because prior to that the inductor is opposing.

Twenty mA of current is flowing after full charge is reached, as long as the battery is connected. When the switch is changed to the discharge position, the inductor will try to maintain the same current. The voltage across the inductor will switch polarity, allowing the current to discharge the inductor in the reverse direction. If the resistor is considered the load, the current through the resistor will develop a voltage drop which can be calculated.

The voltage developed across the resistor at the very start of discharge will be equal to the battery voltage because the current is the same and the same resistor is used for charge and discharge. As the magnetic field decays, as calculated with the time constants, the current will also decrease at the same rate and because the current decreases, the voltage developed across the resistor will also decrease.

Figure 6-3 is an inductive circuit that has different resistances in the charge and discharge paths. This circuit will have two different time constants and the voltage developed across the discharge resistor should be higher. Calculate time constants first.

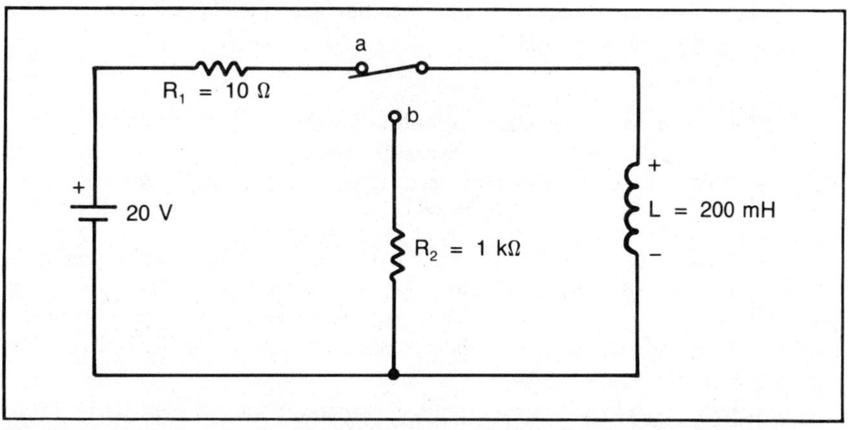

Fig. 6-3. An inductive circuit with different time constants for charge and discharge.

Step 1 $T = \frac{L}{R}$ formula

Step 2 $T = \frac{.2\ H}{10\ \Omega}$ substitute the values for charge

Step 3 T = .02 seconds
20 ms time constant for charge

Step 4 5 × T = 100 ms for full charge

Step 5 $T = \frac{.2\ H}{1000\ \Omega}$ substitute values for discharge

Step 6 T = .0002 seconds
.2 ms time constant for discharge

Step 7 5 × T = 1 ms full discharge

For the calculations involved here assume the switch is in the proper position for a period of time much greater than is required for 5 time constants. In other words, assume the inductor will reach full charge and discharge. Next, calculate the charge current.

Step 1 $I = \frac{E}{R}$ formula

Step 2 $I = \frac{20\ V}{10\ ohms}$ substitute values of charge path

Step 3 I = 2 amps current after full charge

When the switch is moved to position b the battery will be disconnected and the inductor will be allowed to discharge through resistor R_2. Not only does the discharge path have a shorter time constant, but notice what happens with the voltage developed across the resistor at the instant the switch changes positions. The current flowing at the moment the switch changes will continue to flow due to the back emf in the inductor caused by the magnetic field.

Step 1 $E = I \times R$ formula
Step 2 $E = 2 \text{ amps} \times 1000 \text{ ohms}$ substitute full charge current with the discharge resistance
Step 3 $E = 2{,}000 \text{ volts}$ voltage developed across the resistor at the moment of switch change

The 2 thousand volts developed across the resistor is an instantaneous value at the very start of discharge. If it is desired to determine the developed voltage at any other point in time, the universal time constant curve can be used to determine the instantaneous current during discharge and the instantaneous voltage can then be calculated.

HIGH VOLTAGE PRODUCED BY OPENING AN R-L CIRCUIT

Refer to Fig. 6-4. This circuit represents the ignition system of an automobile. The switch is the points that open and close to allow the coil to charge and then discharge through the spark plug. R_1 represents the dc resistance of the coil windings. R_2 represents the resistance of the spark plug gap. This resistance value is purely fictitious since the spark plug is actually an open circuit. Some value is needed to be given in the discharge path and 250 k ohms represents an open circuit.

The important calculation here is the voltage developed across the spark plug when the switch allows the inductor to discharge.

Current at full charge must first be calculated and then, using the charge current, calculate the voltage across the spark plug at the first moment of discharge.

Step 1 $I = \frac{E}{R}$ formula for charge current

Step 2 $I = \frac{12 \text{ V}}{100 \text{ ohms}}$ substitute values of charge path

Step 3 $I = .12 \text{ A}$ full charge current
Step 4 $E = I \times R$ formula for voltage during discharge

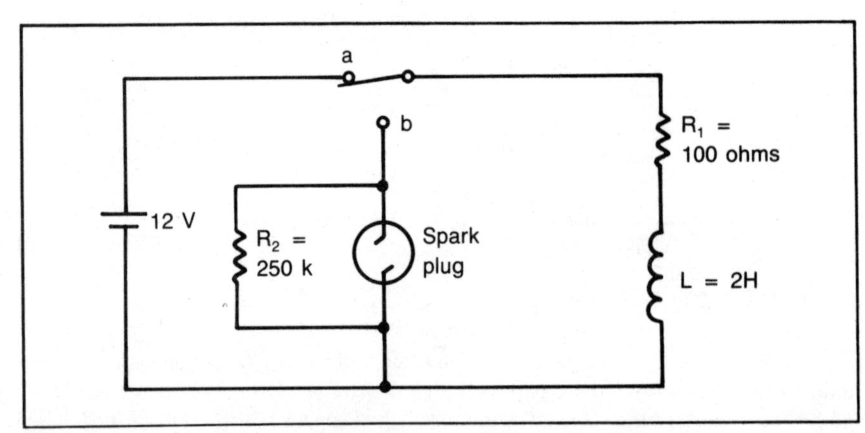

Fig. 6-4. High-voltage developed by opening an L/R circuit.

Step 5 E = .12 A × 250,000 ohms substitute values of discharge path
Step 6 E = 30,000 V voltage developed at the instant of the start of discharge

The calculations of the voltage developed across a spark plug are somewhat simplified here, but the principle of operation is accurate. It takes an extremely large voltage to cause the spark to jump the gap of a spark plug. the modern day electronic ignition systems work much the same with the major difference being the points (the switch) has been replaced with an electronic switch that does not wear out.

When large motors are used in a circuit, it is necessary to provide protection for the switch. The electric motor is a very large inductor. When the circuit is turned off, there must be a discharge path provided or the switch will become the discharge path and a spark will result across the contacts of the switch, which usually causes considerable damage to the switch.

RC TIME CONSTANT

Whenever the term time constant is mentioned, the first thing that should come to mind is the capacitive circuit. The reason for this is the fact that there are many more applications for the capacitive time constant circuit than there are for the inductive circuit. This does not in any way diminish the practical inductive circuits, it simply means capacitors are more common.

The property of a capacitor is to store a charge that can be said to be the same as storing a voltage. Because a capacitor stores a charge of voltage, it will oppose any change in voltage. What this means is that the capacitor must try to store a charge equal to the change in voltage and this cannot be done instantly. The length of time involved in the storage of the charge is called the time constant of the R-C circuit.

The formula for calculating the time constant of a capacitive circuit is:

$$T = R \times C$$ **Formula 6-4**

Time constant for a capacitor with its series resistance. Time in seconds, resistance in ohms and capacitance in farads.

Remember, whenever using any formula, be sure to keep in mind that the formula will have the right final units if the correct units are used in the calculations. Notice, in the R-C time constant formula, the capacitance is in farads. Actual capacitor values are seldom, if ever, in farads but rather in microfarads of picofarads.

Figure 6-5 shows a circuit that demonstrates the charge and discharge paths for a sample R-C time constant. The charge path is from the negative side of the battery, to the bottom, negative plate of the capacitor. For every electron that leaves the negative terminal of the battery and collects on the negative terminal of the capacitor, an equal number of electrons leave the positive, top, plate of the capacitor, travel through the 100 ohms resistor and return to the positive terminal of the battery. This causes the capacitor to charge by having a storage of electrons on the negative plate and a shortage of electrons on the positive plate. The storage of electrons creates an electric field and is a difference in potential, which is the definition of voltage. Therefore, the capacitor is storing voltage.

The discharge path is when the switch is placed in the b position. The battery is then disconnected and the capacitor is allowed to neutralize the difference in potential of stored electrons. The discharge path is from the negative plate of the capacitor, through the center path, to the b contact of the switch, through the resistor and replacing the missing electrons on the positive plate.

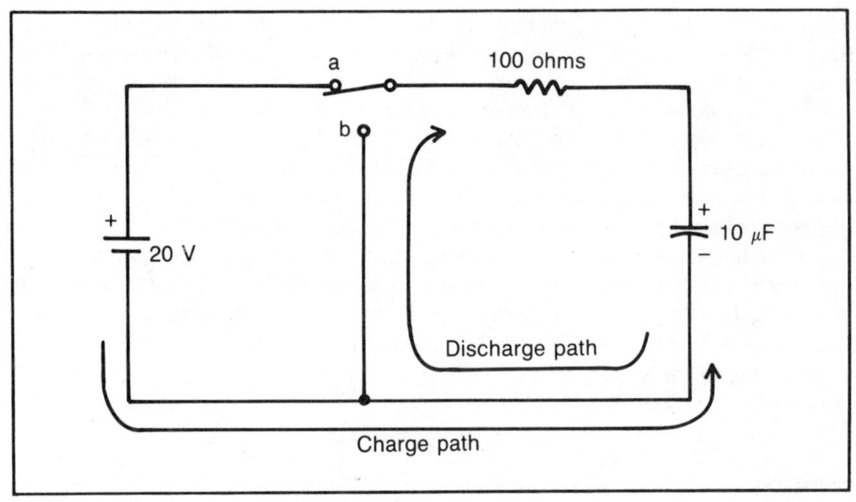

Fig. 6-5. A circuit to demonstrate the R-C time constant.

When calculating the time constant for this R-C circuit, both the charge path and discharge path have the same values of capacitance and resistance. Therefore, the time constant for both charge and discharge are the same.

Step 1 T = R × C formula for time constant
Step 2 T = 100 ohms × 10μF substitute values
Step 3 T = 1000 μs time constant for both
T = ms charge and discharge
Step 4 5 × 1 ms = 5 ms for a full charge

When the capacitor reaches full charge after 5 time constants, the voltage will be 20 volts across the capacitor. Once the capacitor is charged to full value, current can no longer flow in the circuit because the battery and the capacitor will have no difference in potential. If the battery remains connected after full charge is reached, even though it is said to have zero current flow, an actual capacitor will have a very slight leakage through the dielectric and the battery will have a very slight leakage current to replace any lost charge. Another point to make at full charge is the fact that since no current flows (leakage current is so small, it is usually considered zero) then there cannot be any voltage developed across the resistor.

When the switch is placed in position b the capacitor will be allowed to discharge. At the first moment of discharge, the voltage developed across the resistor will be equal to the capacitor voltage, since it is a simple series circuit. At the first moment of discharge, the voltage is at its highest value and the current through the resistor, known as the instantaneous discharge current, can be calculated using Ohm's law.

When examining the inductor and L/R time constants, a different discharge was used than the charge path to generate a very high voltage. The capacitor circuit and R-C time constants can be used to also have a different discharge path. The difference will be that the capacitor will produce a high current.

Figure 6-6 shows a possible schematic diagram for an electronic flash unit in a camera. Light bulbs have a light intensity based on the wattage. There are three different ways to develop the needed wattage; raise the voltage, raise the current, or a combination of both. A camera's electronic flash unit is very light and portable because it uses very small batteries. The batteries are often the AAA type. This type of battery has 1.5 volts per battery and a very limited current capacity. It would be impossible to use a battery of this type to ignite a very high intensity flash bulb very many times.

Figure 6-6 shows the charge path for the capacitor through a 100 ohm resistor, to charge the 250 μF capacitor. This gives a time constant for charge of 25 ms. The reason the resistor is needed is to limit the charge current. When the switch is moved to position b the capacitor will discharge through the light bulb, shown as having 1 ohm resistance. This gives a discharge time constant of 250 μs. Keep in mind the flash on a camera is very fast.

When the capacitor is fully charged, it will have a voltage of 3 volts across it. At the time of discharge, the 3 volts stored in the capacitor will be seen across the 1 ohm flash bulb. Instantaneous discharge current will then be 3 amps. Flash bulbs are made in such a way as to produce a high intensity light with this current.

CHAPTER SUMMARY

The universal time constant curve is a very useful tool to graphically show the charge or discharge of an inductive or capacitive circuit. Note that 63 percent of full charge or discharge represents 1 time constant. It takes 5 time constants to reach full charge or full discharge.

When an inductor is placed in series with a resistor, there will be a length of time for the inductor to reach full charge or discharge, this is the time constant. An inductor needs to build the magnetic field by current flowing in the turns. An inductor opposes any change in current. At full charge, the inductor will have current limited by the circuit resistance.

At the first moment the inductor is caused to discharge, the current will be equal to the last charge current. This current will flow through whatever discharge path is provided. If the discharge path has a very high resistance, there will be a very high voltage developed at the first moment of discharge.

When a capacitor is placed in series with a resistor, there will be a length of time required for the capacitor to charge or discharge, this is the time constant. A capacitor needs to store electrons, and therefore a voltage. At full charge, the dc current will stop flowing, except for a very small leakage current. A capacitor opposes any change in voltage. At full charge the capacitor will store the applied voltage.

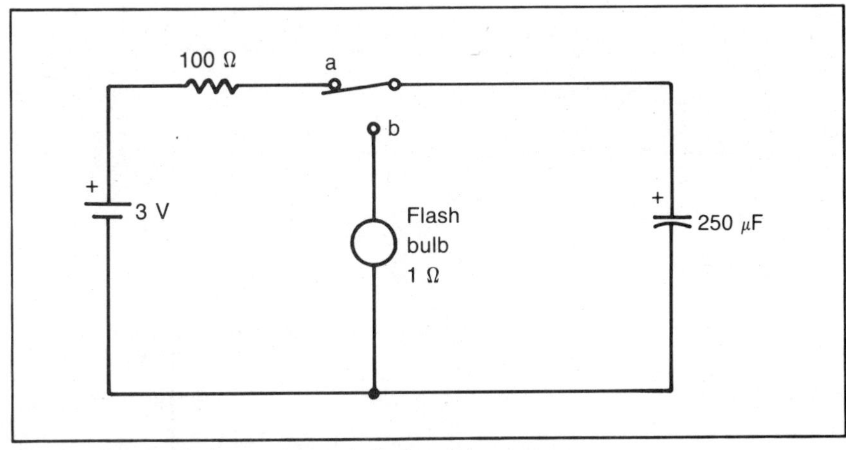

Fig. 6-6. Schematic for an electronic flash unit in a camera.

At the first moment the capacitor is discharged, the voltage applied to the discharge path will equal the last voltage stored in the capacitor. If the discharge path has a very low resistance, there will be a very high current during discharge through the resistor.

Summary of Formulas

instantaneous charge % of full charge $= 1 - e^{-T} \times 100\%$ **Formula 6-1**

instantaneous discharge % of full charge remaining $= e^{-T} \times 100\%$

Formula 6-2

time constant for an inductor in series with a resistor $T = \frac{L}{R}$ **Formula 6-3**

time constant for a capacitor in series with a resistor $T = R \times C$ **Formula 6-4**

Practice Problems

Use the schematic diagrams of constant currents shown to find charge time constant, full charge voltage, discharge time constant and the current developed at first instant of charge.

1.

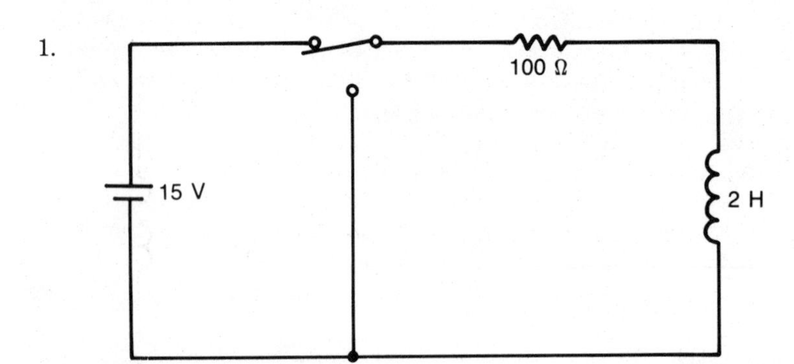

2.

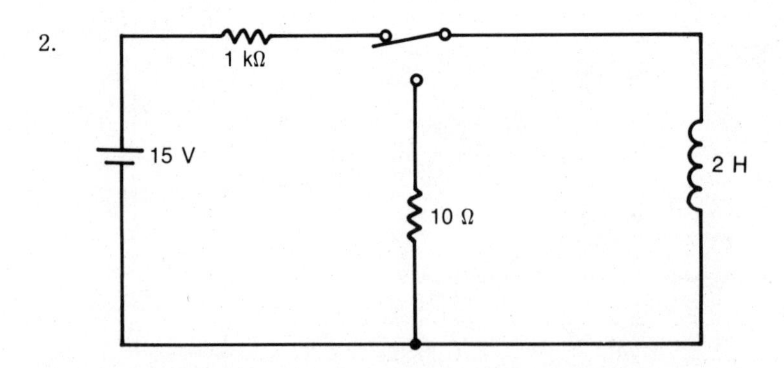

3.

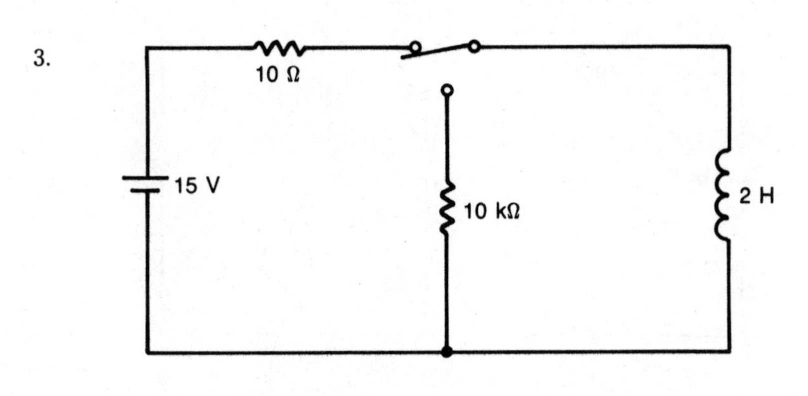
10 Ω
15 V
10 kΩ
2 H

4.

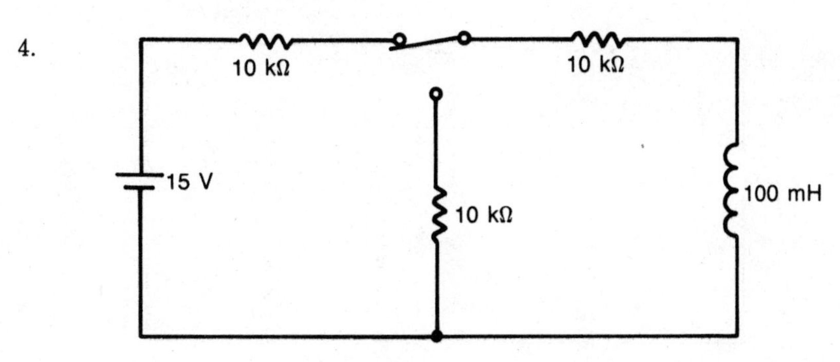
10 kΩ
10 kΩ
15 V
10 kΩ
100 mH

5.

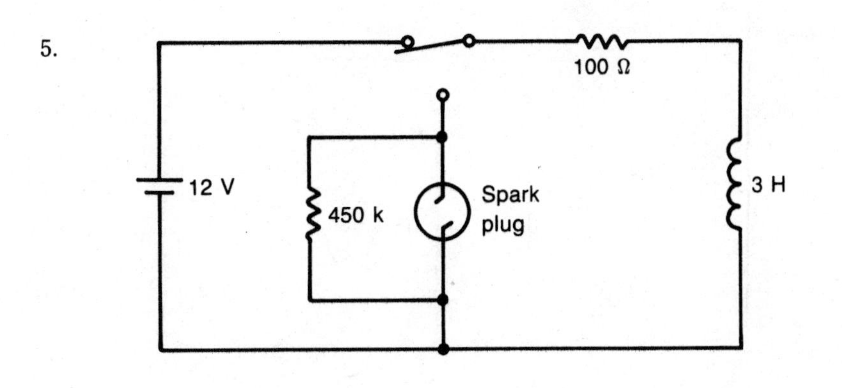
100 Ω
12 V
450 k
Spark plug
3 H

6.

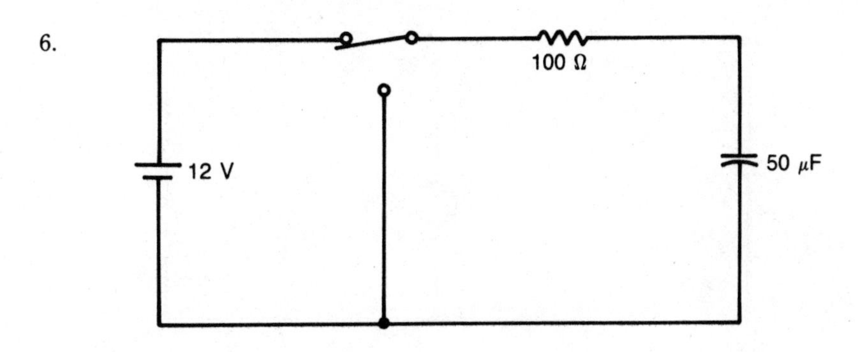
100 Ω
12 V
50 μF

7.

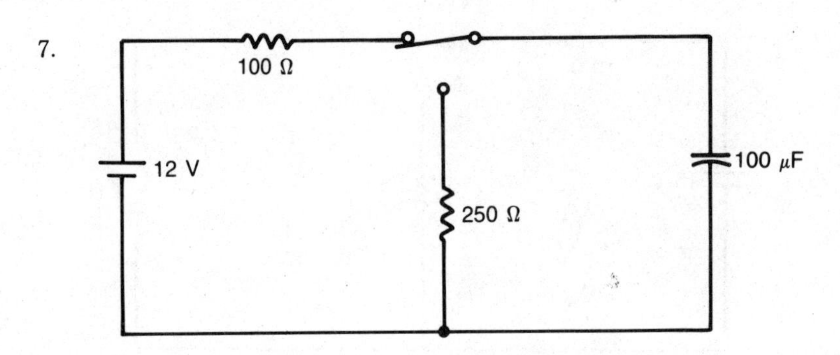
100 Ω
12 V
250 Ω
100 μF

8.

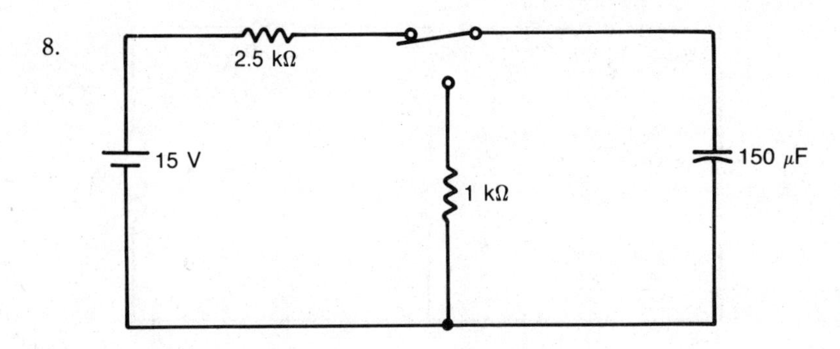
2.5 kΩ
15 V
1 kΩ
150 μF

9.

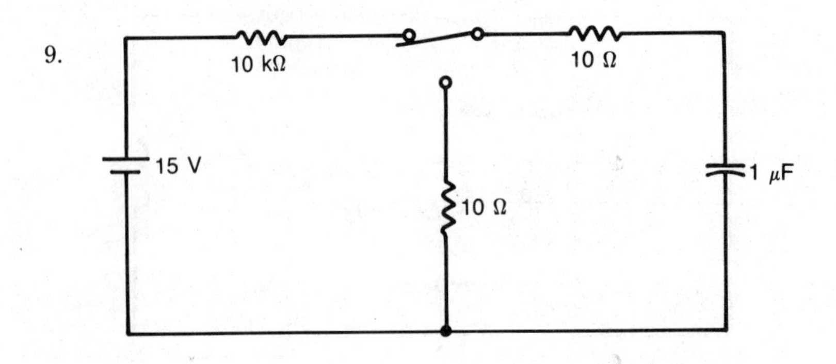
10 kΩ
10 Ω
15 V
10 Ω
1 μF

10.

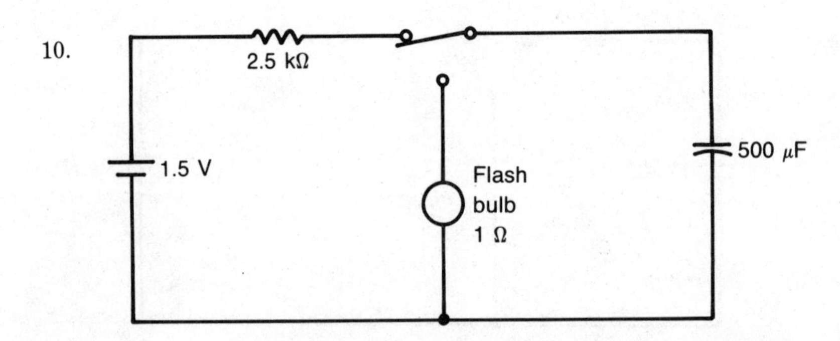
2.5 kΩ
1.5 V
Flash
bulb
1 Ω
500 μF

Chapter 7

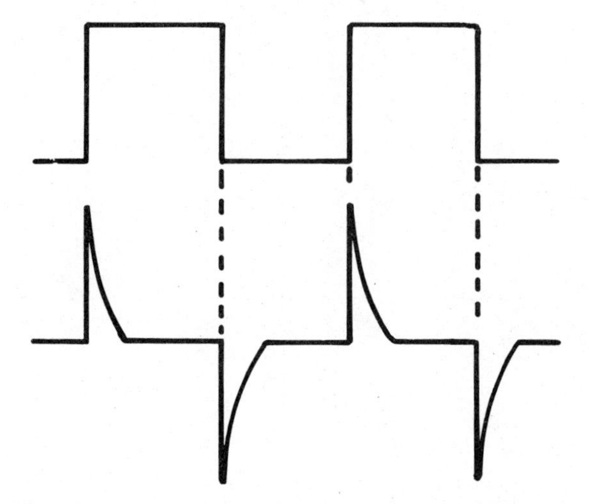

R-C Waveshaping

Waveshaping circuits are a useful application of the basic resistor/capacitor time constant. The greatest application is passing a square wave through the series circuit, with an output taken in parallel with the resistor, results in a negative spike. The negative spike has direct applications in digital circuits that require a negative trigger.

DIFFERENTIATION/INTEGRATION

Differentiation and integration are two words used to describe the R-C time constant circuit. This is the main topic of the chapter.

Compare the circuits shown in Figs. 7-1 and 7-2. Figure 7-1 is a differentiator circuit. It can be identified by where V_{out} is taken in the circuit. When the output is taken across, in parallel with the resistor, the circuit is called a differentiator. It is not necessary to use a square wave input, but it best demonstrates the function of the circuit.

The R-C combination is usually selected for a short time constant when it is to be used as a differentiator. The primary advantage is through the use of the spike seen across the resistor. Because the output is taken directly across the resistor, the output voltage will always be exactly the same as the voltage across the resistor. A resistor allows the voltage to change instantaneously.

Figure 7-2 shows an integrator circuit. The output is taken directly across the capacitor and will therefore, follow the charge and discharge of the capacitor. The word integrate means to accumulate or combine, and this is exactly what the capacitor is doing. An integrator circuit will usually use a medium to long time constant to allow a more pronounced curve of the charge and discharge of the capacitor.

Figures 7-3, 7-4, 7-5, and 7-6 show circuits with the output waveforms drawn. Notice that the circuit shows the output can be taken either across the capacitor or across

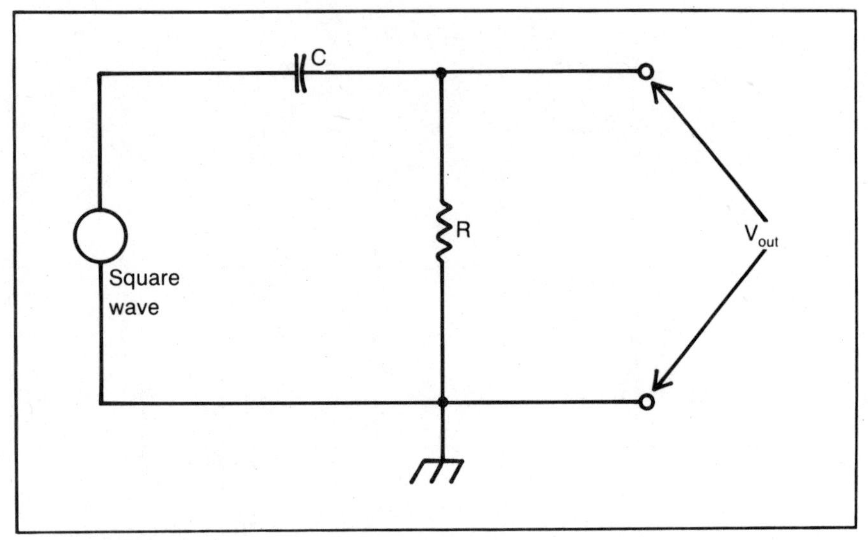

Fig. 7-1. Differentiator.

the resistor and the resultant waveform is shown. It has been drawn this way to simplify the drawings and explanation. Keep in mind, if a differentiator is desired, the output is across the resistor. If an integrator is desired, the output is across the capacitor.

A square wave is used in each of the four time constant circuits because it is a very effective way to use a battery that switches on and off at a precise frequency. The square wave used here is a 1 kHz wave. The square wave used is entirely positive. That means the waveform goes from zero to a peak positive voltage and returns to zero but does not go negative at all. The advantage of using this type of voltage is that it acts like a dc source that is switched on and off at a specified rate. During the time that the input is switched to 0 volts, is considered the off time. It will be treated as a short circuit to allow the capacitor a path to discharge.

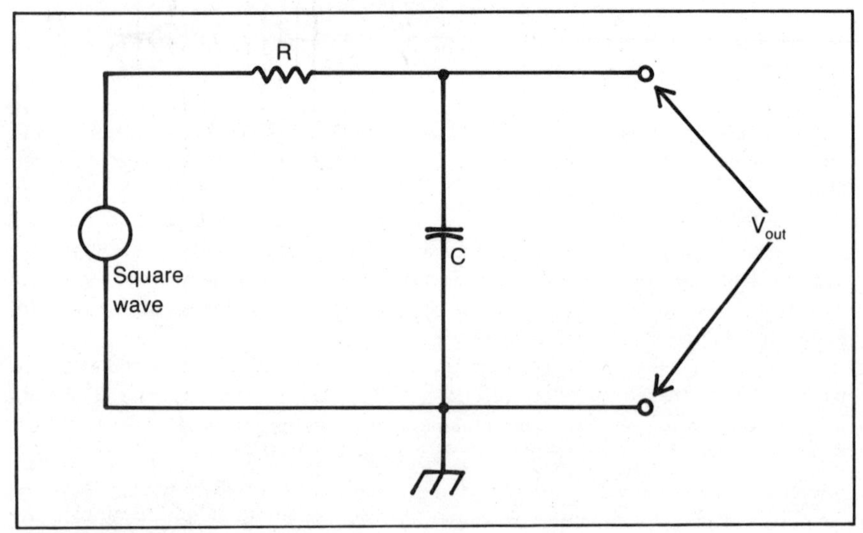

Fig. 7-2. Integrator.

SHORT TIME CONSTANT

Figure 7-3 shows a short time constant circuit. It is classified as a short time constant because the time for 5 time constants, full charge and discharge time, is shorter than the time allowed for charge and discharge. In other words, the on and off times of the square wave are each .5 ms long. One time constant for this circuit is .02 ms which makes 5 time constants .1 ms. The on and off times are 5 times longer than what is needed to reach full charge or discharge.

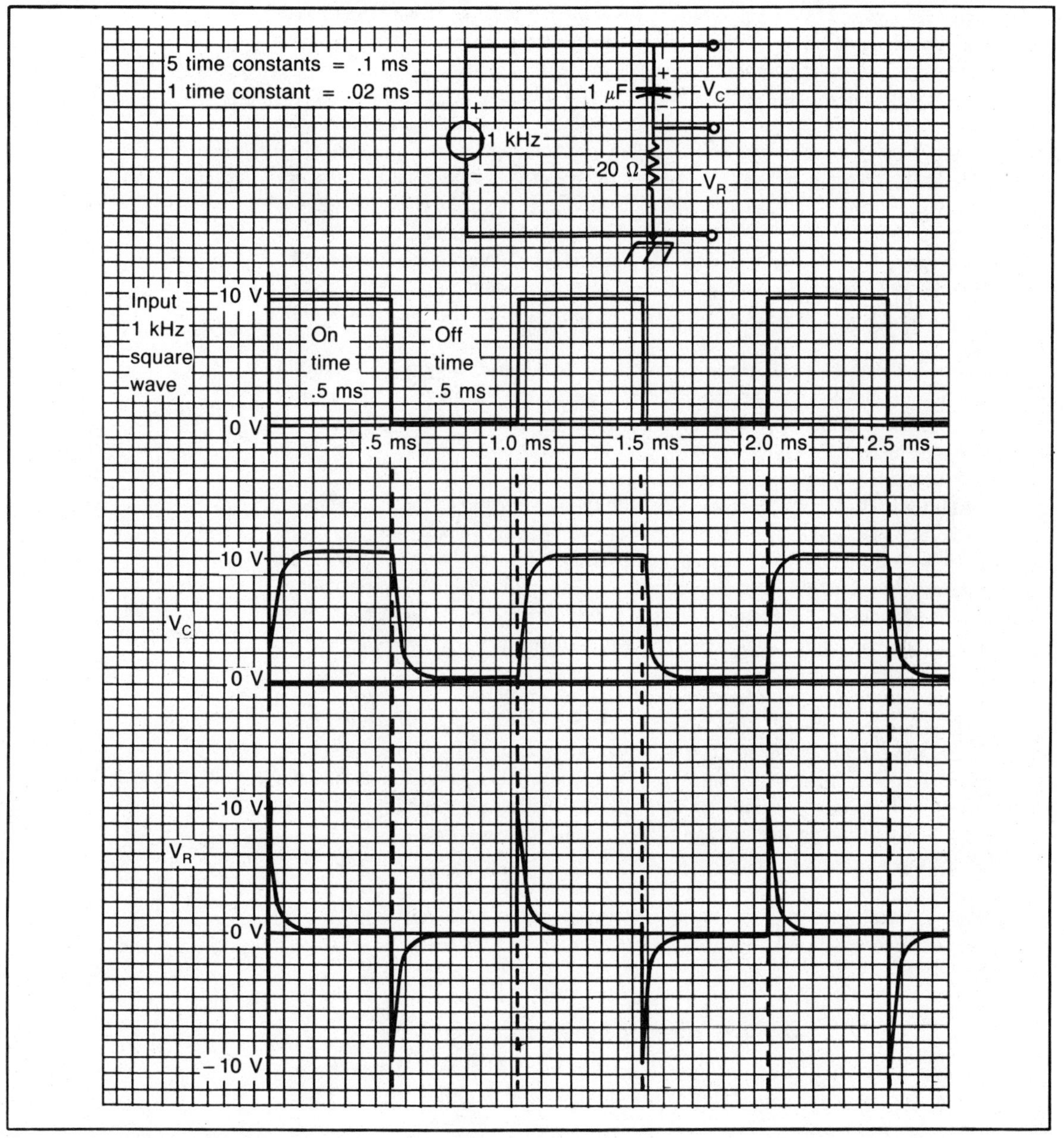

Fig. 7-3. Short time constant.

The curve of V_C follows the universal time constant curve used in Chapter 6. At the start of the on time, the capacitor starts its charge cycle through the 20 ohm resistor. The time constant is so short that the capacitor will reach full charge in short time. Once it reaches full charge, it will stay at full charge for the remainder of the on portion of the input square wave.

When the input square wave comes to the start of the off time, the capacitor will start its discharge, through the 20 ohm resistor and the power supply having what appears to be a short circuit. The time constant with this value of resistance is so short, the capacitor reaches full discharge in a very short period of time. It will remain discharged for the remainder of the off time. When the input changes again to the positive voltage, the capacitor will again charge and the cycle is repeated.

The output taken across the resistor, V_R, provides a unique output waveform, especially with the short time constant circuit. Because the input is a square wave, the voltage across the resistor is always whatever voltage is not across the capacitor, when the input is during on time. At the very start of the square wave, the capacitor is not charged and will therefore be zero volts. The resistor voltage is able to change instantly, therefore, the resistive voltage instantly goes to the applied voltage of 10 volts. As the capacitor charges, the capacitor will drop more of the applied voltage and less will be left for the resistor. The resistor voltage will decrease at the same rate that the capacitor voltage increases. The resistor voltage will go to zero when the capacitor is fully charged.

Keep in mind that the capacitor will charge with a positive voltage on the top of the capacitor in the drawing. When the input voltage goes to the off time, the polarity of the capacitor does not change. However, the relationship it had with the ground reference point does change.

During the charge time, the capacitor charged with a positive on the top and negative on the bottom. The resistor had a voltage with the positive on the top and the negative on the bottom. The ground reference point was then a negative polarity. Everything in the circuit, except the capacitor can change polarity instantly. Therefore, when the power supply is turned off, the capacitor acts like a battery. The polarity of the capacitor is with the positive on the ground reference point, during the discharge time. Having the ground reference point as a positive makes the polarity across the resistor instantly switch to the opposite of the charge cycle.

In reference to the ground reference point, the voltage across the resistor, during the off time, or discharge time of the capacitor, will have a polarity opposite to the polarity during the on time. The resistor voltage will instantly jump to the voltage across the capacitor. At the first moment of discharge, the capacitor is fully charged and the resistor voltage jumps to a negative 10 volts. As the capacitor discharges, the resistor voltage will exactly follow the capacitor voltage and return to the zero volt line.

When the input square wave again goes positive, the resistor voltage will again jump to positive 10 volts, then drop off as the capacitor charges. The cycle is repeated. The input wave switches to zero and the resistor voltage jumps to negative 10 volts because this is the voltage stored on the capacitor. The resistor voltage decreases as the capacitor voltage decreases.

MEDIUM TIME CONSTANT

Figure 7-4 shows a circuit with a medium time constant. It is classified as a medium time constant because 5 time constants of the circuit is equal to the on and off times of the input waveform.

The medium time constant circuit has a waveform that displays a universal time

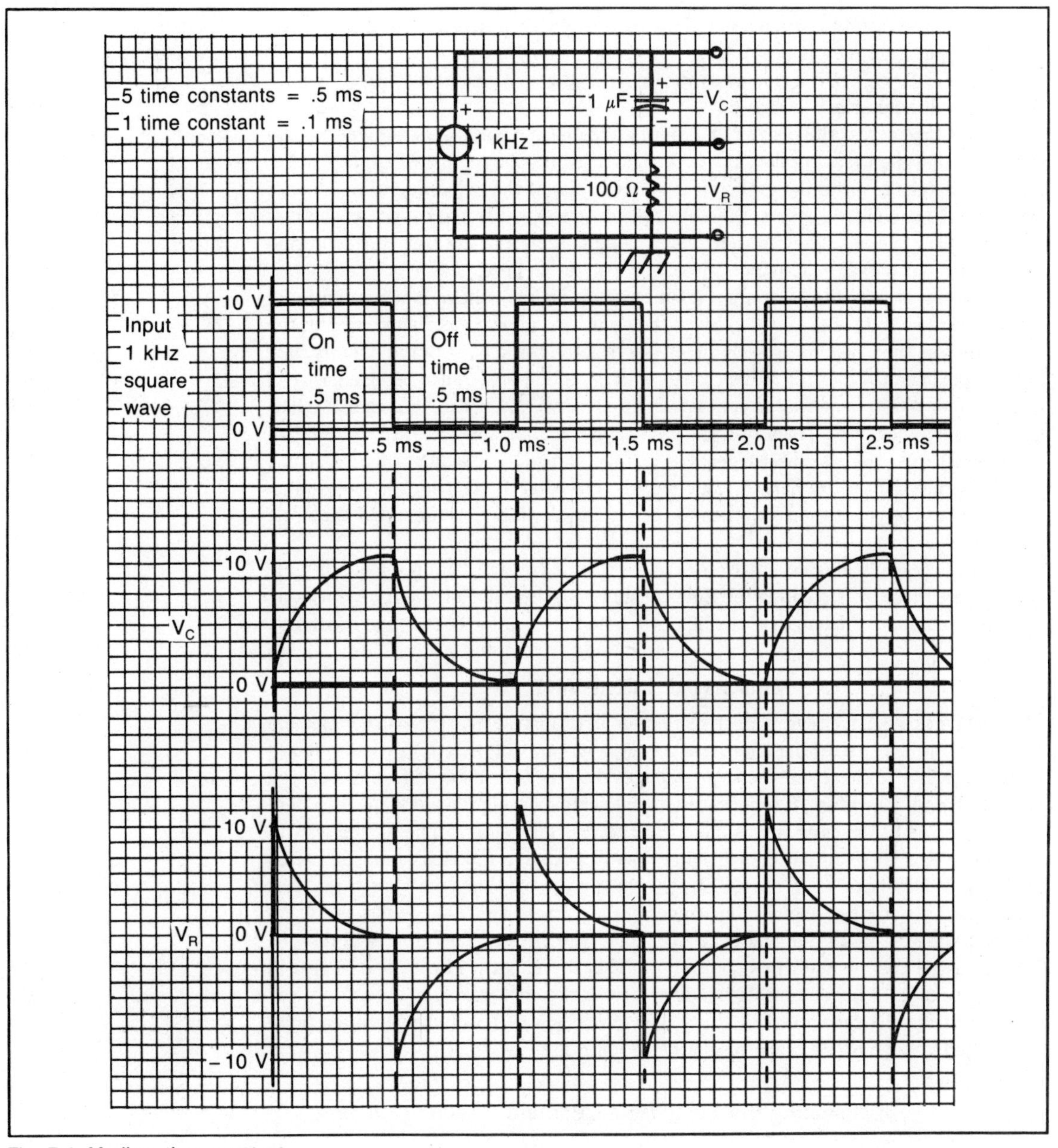

Fig. 7-4. Medium time constant.

constant curve. This waveform is useful when the positive or negative spikes need to be a longer duration than the short time constant would allow.

The voltage across the capacitor, V_C, starts at zero volts and when the input switches to the positive 10 volts, the capacitor will begin the charging process. The input is on for .5 ms, which is the same time required for the capacitor to reach a full charge. Therefore, at the end of the on time, the capacitor has charged to the full 10 volts. When the input switches to the off time, the capacitor is still with a full charge.

It will begin the discharge time and will be fully discharged in a period of .5 ms, the time the input is in the off position.

When the input again switches to the on time, the process repeats itself and the capacitor will again start to charge.

The voltage across the resistor, V_R, during the on time will take whatever voltage is applied, that the capacitor does not take. The resistor voltage jumps instantly to the full applied voltage since the capacitor is at zero volts at the beginning of the charge cycle. As the capacitor charges, the voltage is dropped across the capacitor and less is left for the resistor, therefore the resistor voltage will drop towards zero following the exponential curve, exactly opposite the capacitor's curve.

When the input switches to the off time, the capacitor voltage is momentarily at the full charged voltage of 10 volts. Since the input now appears as a short circuit, the voltage across the capacitor is seen directly across the resistor, with the polarity having a negative on the top side of the resistor and the ground reference being a positive. Because the voltage is measured compared to the ground reference, the resistor polarity switches to negative. Therefore, the resistor voltage instantly jumps to a negative 10 volts. As the capacitor discharges, the voltage across the resistor will also decay, until it reaches the full discharged value of zero volts after .5 ms, which is the off time of the input. The input then switches to the positive voltage, on time, and the cycle is repeated.

The only calculations necessary for the short time constant and the medium time constant are those of calculating the time constant and the value of five time constants in order to compare the length of the input cycle to the duration of the time constants.

LONG TIME CONSTANT

The long time constant circuit is much more difficult to analyze than the short or medium time constant circuits. In the long time constant circuit, the length of time required for the capacitor to charge or discharge is longer than the time allowed to either charge or to discharge. Therefore, it is necessary to calculate each significant step when plotting the curves of the capacitor voltage and the resistor voltage. The difficulty with these circuits is the fact that the capacitor does not reach a full charge during the on time, but it does have some charge stored. During the off time, the charge that has been stored is not given enough time to completely discharge. This results in the capacitor having a charge the next time the input switches on. The input will then be able to charge the capacitor to a somewhat higher level, but it also means there will be more voltage to try and discharge during the next off cycle.

Eventually, the capacitor will charge to a point where it will charge to not quite the applied voltage of 10 V and discharge to not quite 6 volts.

Calculating the Long Time Constant Circuit

charge voltage $V = 1 - e^{-T} \times V$
discharge voltage $V = e^{-T} \times V$

Note All calculations are made at a transition (change) in:
the input square wave. Each .5 ms is an input transition.
.5 ms = .5 time constants (1 time constant = 1 ms).
The T in the formula is replaced with .5 time constants.

At 0 ms the input switches to +10 volts. The capacitor starts to charge, starting from zero volts. The resistor at this instant in time will instantly go to the full applied volt-

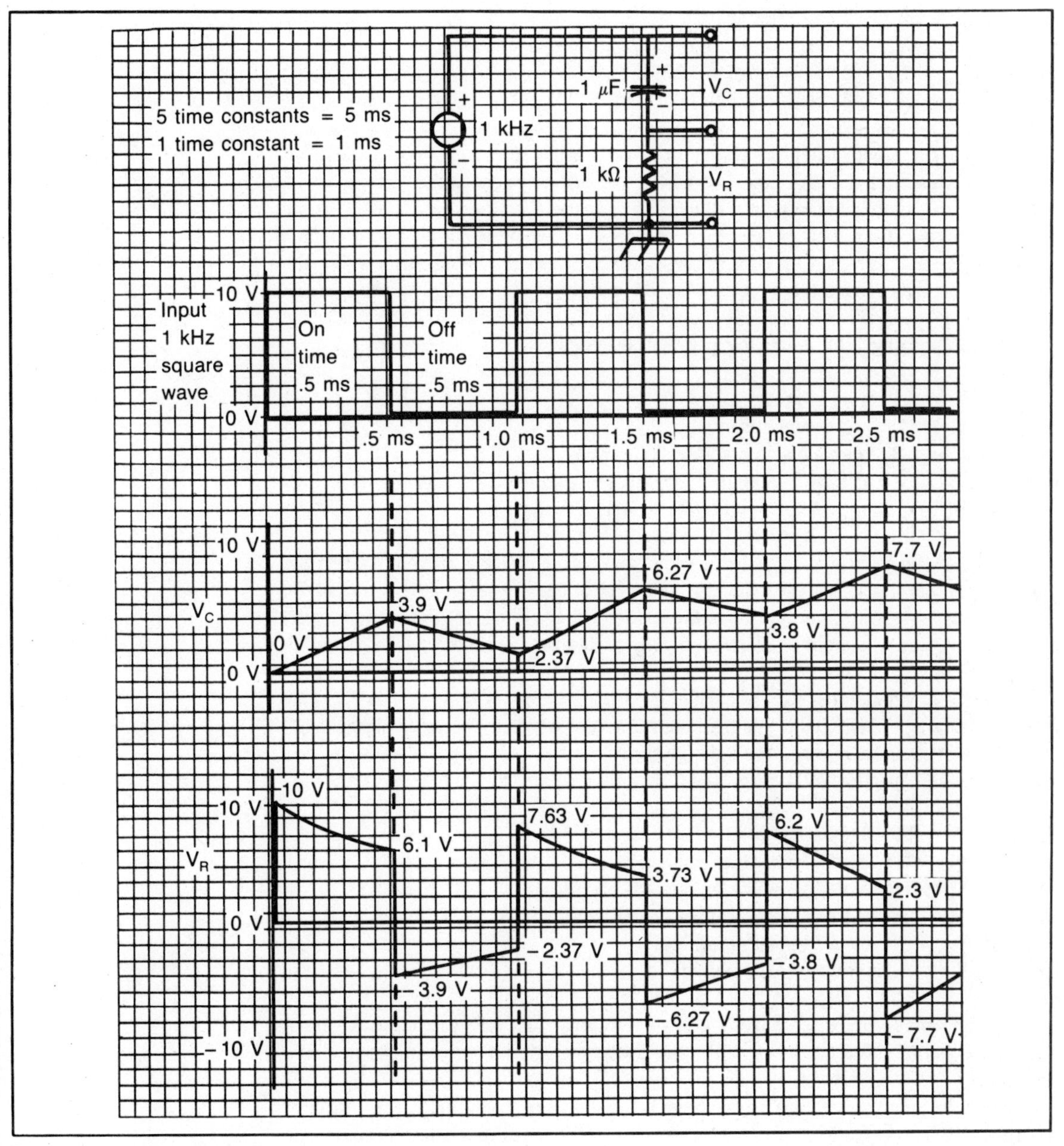

Fig. 7-5. Long time constant.

age since the capacitor has 0 volts across it.

At .5 ms the capacitor has had .5 time constants to charge and the input switches to zero volts. Because the capacitor has had .5 ms charge time constants, the voltage across the capacitor can be calculated with the charge formula.

Step 1 $V = 1 - e^{-T} \times V$ formula
Step 2 $V = 1 - e^{-.5} \times 10\ V$ substitute values

Step 3 $V = 3.9$ volts capacitor voltage at .5 ms
Step 4 $V_R = V - V_C$ resistor voltage = applied − capacitor
Step 5 $V_R = 10 - 3.9$ substitute values
Step 6 $V_R = 6.1$ volts resistor voltage at .5 ms

At .5 ms the input makes the transition from +10 to zero volts. When the input makes this transition, the capacitor will start to discharge. Because the voltage across the resistor can switch instantly, it will have the capacitor voltage across it—with the polarity reversed. Then, as the capacitor discharges to the next significant point, which is 1.0 ms, for a time period of .5 time constants, the resistor voltage will follow the capacitor voltage because the capacitor is now supplying the voltage to the circuit.

At 1.0 ms, the voltage across the capacitor can be calculated using the discharge formula, discharging from a starting point of 3.9 volts.

Step 1 $V = e^{-T} \times V$ formula
Step 2 $V = e^{-.5} \times 3.9$ substitute values
Step 3 $V = 2.37$ volts capacitor voltage at 1.0 ms
Step 4 $V_R = -V_C$ resistor voltage is negative V_C
Step 5 $V_R = -2.37$ resistor voltage at 1.0 ms

At 1.0 ms the input makes the transition from zero volts to +10 volts. When the input makes this transition, the capacitor will start to charge. Also, at this transition, the resistor voltage will instantly jump to the applied voltage minus the capacitor voltage.

The capacitor will again be allowed .5 time constants of charge time between 1.0 ms and the next transition at 1.5 ms.

The charge voltage that was calculated for all .5 ms steps will always be the same amount of charge voltage, with this time constant. The capacitor charge voltage is 3.9 volts. Notice, the capacitor started this charge cycle with a voltage of 2.37 already present. Therefore, the capacitor will charge the 3.9 volts starting from the 2.37 volts.

Step 1 $V_C = 3.9 +$ existing voltage formula
Step 2 $V_C = 3.9 + 2.37$ substitute values
Step 3 $V_C = 6.27$ volts capacitor voltage at 1.5 ms
Step 4 $V_R =$ applied voltage − capacitor voltage formula
Step 5 $V_R = 10 - 6.27$ substitute values
Step 6 $V_R = 3.73$ volts resistor voltage at 1.5 ms

At 1.5 ms the input switches from +10 volts to zero volts. The capacitor will again start to discharge and the resistor will jump to the capacitor voltage in the negative polarity.

Notice the trend of the capacitor is toward a full charge. Eventually, the capacitor will reach a full charge during the charge portion and then during the discharge portion it will discharge to approximately 6 volts. It will then continue to go between 10 volts and 6 volts as long as the input is allowed to switch at the present rate. It never does charge to the full applied voltage of 10 volts. However, it will get quite close.

The resistor voltage will become centered around the zero center line. The effect of this circuit is to pass the square wave, although its shape has been altered, and eliminate the center line being at a dc level.

Figure 7-6 shows a very long time constant circuit. Notice, the same results are developing as took place in Fig. 7-5, the long time constant circuit. The capacitor will eventually charge to very near the applied voltage and then discharge somewhat. The

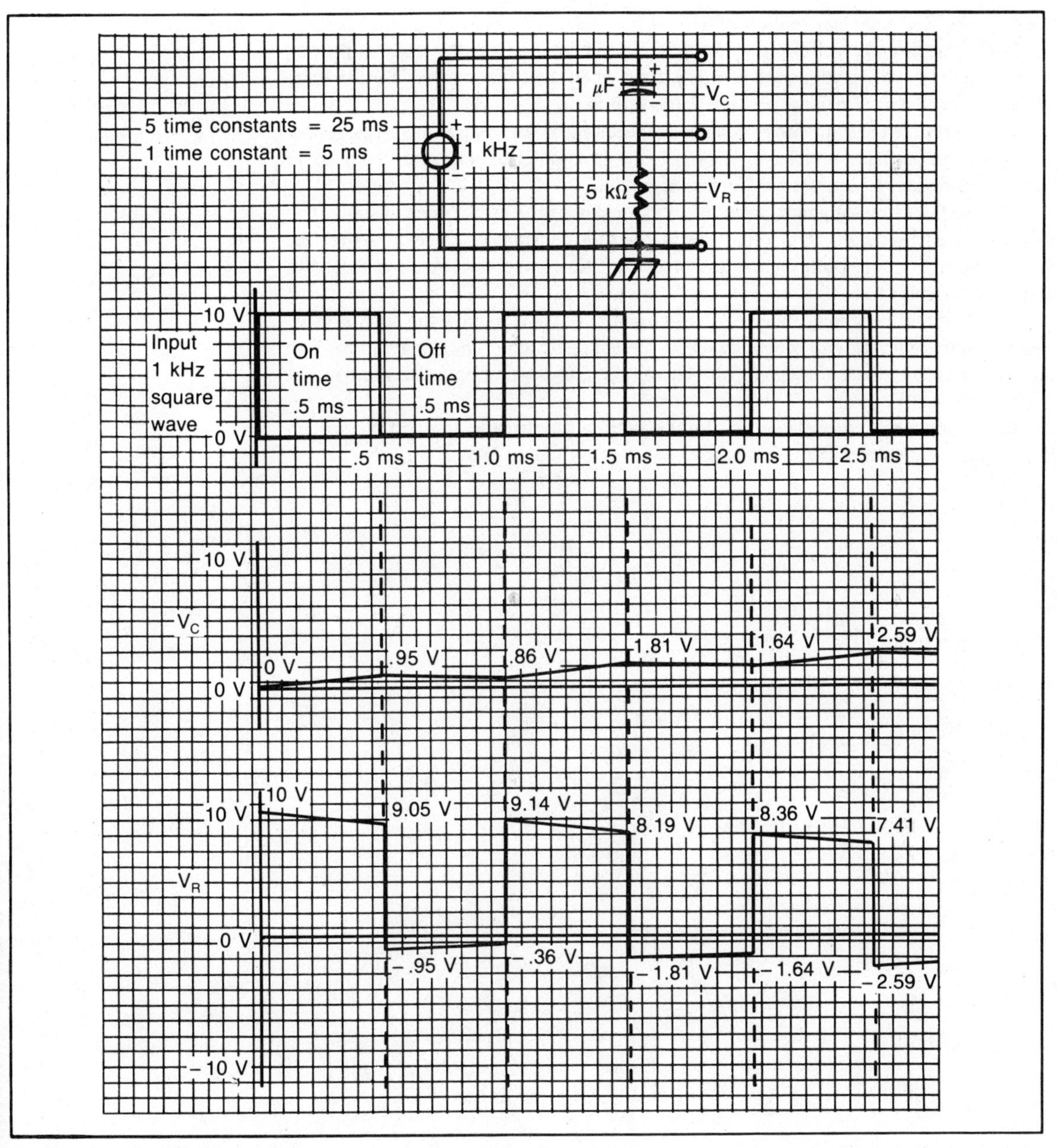

Fig. 7-6. Very long time constant.

resistor voltage will look more like the original square wave with a longer time constant. The resistor voltage will center around the zero volt axis, effectively passing the square wave, except removing the dc level from the original waveform.

CHAPTER SUMMARY

Following the universal time constant curve, it can be seen that there are some applications for the R-C time constant circuits.

In this chapter, the voltage developed across the resistor was the main subject by examining the negative voltage that is displayed when the capacitor holds the voltage and the input changes from a positive voltage to zero.

The short time constant circuit developed the sharpest negative (and positive spike). The spike can be made broader by making a longer time constant, up to the point of being classified as a medium time constant circuit.

The medium time constant circuit is an extension of the short time constant. The medium is classified right at, or very close to, the time constant being equal to the time of the applied voltage. Both the medium and short time constant circuits are used to generate the spikes.

The long and very long time constant circuits are used to pass the square wave through to the output, with some change in the shape of the original waveform. The capacitor has the effect of stopping the dc voltage and passing only the voltage that is changing.

Differentiation is when the output is taken across the resistor. Integration is when the output is taken across the capacitor.

Chapter 8

$$X_L = 2\pi fL$$

Inductive Reactance

When discussing reactance, whether it be inductive or capacitive, reactance is considered to be the ac resistance of the particular component. In other words, it will offer a different resistance to the flow of ac than it will to the flow of dc.

In addition to the fact that a reactive component will offer resistance to the ac signal, it will also cause a phase shift between the applied voltage and the voltage drop across the reactive component.

INDUCTIVE REACTANCE, X_L

$X_L = 2\pi fL$ **Formula 8-1**

The factor 2π is a constant in this formula.
f is frequency, measured in hertz.
L is inductance, measured in henries
X_L is inductive reactance, measured in ohms.

The formula states that there is a direct relationship between the inductive reactance, frequency, and inductance. Keep in mind, a direct relationship means that when one of the quantities (either frequency or inductance) is increased, the quantity it is compared to (inductive reactance) will also increase.

- Inductive reactance, L_L, is directly related to frequency and inductance.

The inductive reactance symbol, X_L, is read as X sub L. This means X, which stands for reactance, has the subscript L, which stands for inductance. When the section on capacitive reactance is discussed, it will be X_C, read as X sub C. These subscripts of X make it possible to use the same letter X to represent any reactance and through the use of the subscripts, distinguish between inductive and capacitive.

Linear Relationship of X_L Versus Frequency and Inductance

Refer to Fig. 8-1. The figure shows two curves. The first shows X_L plotted on the vertical axis with inductance plotted on the horizontal axis. The frequency for plotting this curve is 10 kHz. The second curve is X_L plotted on the vertical axis and frequency plotted on the horizontal axis, with the inductance for this curve being 16 mH.

The following is a sample calculation for the points on the curves:

Curve of X_L vs L. f = 10 kHz

For this curve, the X_L was selected for the points where the line would cross the major divisions and the equation was solved for the value of L at that point.

Step 1 $X_L = 2\pi fL$ formula

Step 2 $L = \dfrac{X_L}{2\pi f}$ rewrite to solve for L

Step 3 $L = \dfrac{600}{2\pi\ 10\ \text{kHz}}$ substitute values

Step 4 $L = 9.5$ mH inductance value when the frequency is 10 kHz and the X_L is 600 ohms

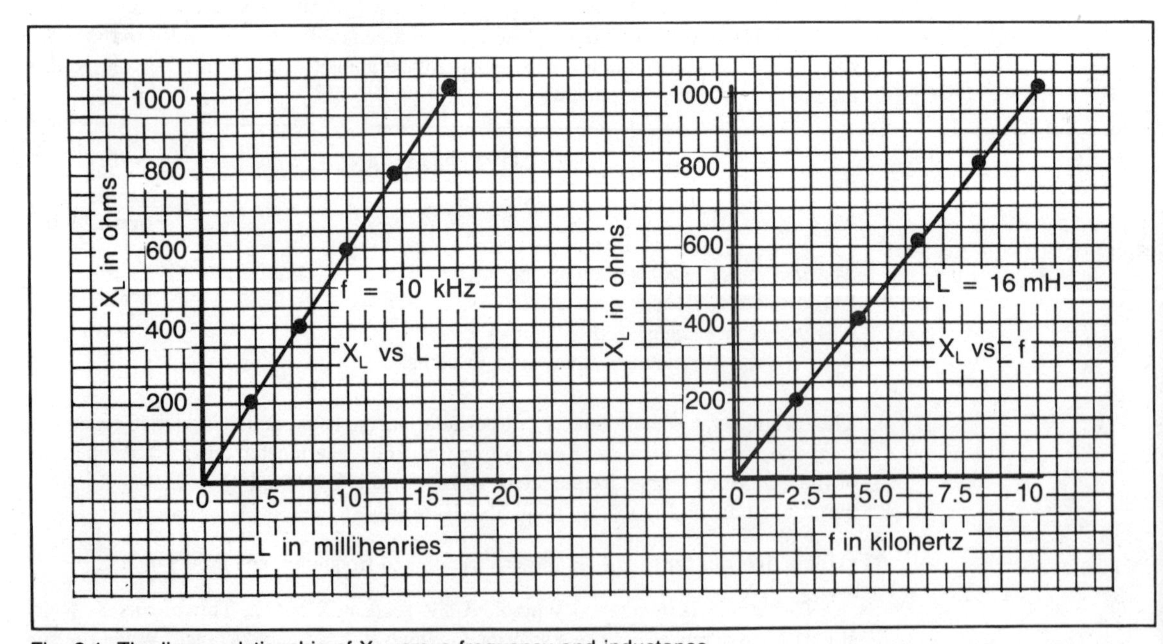

Fig. 8-1. The linear relationship of X_L versus frequency and inductance.

Note For calculations such as this one, or any that involve the constant π it is best to use the value of π in the calculator. In the event that the calculator being used does not have π it is acceptable to use 3.14 as the value of π.

Curve of X_L vs f. L = 16 mH

With this curve, the X_L was again selected so the points would be where the line crosses the major divisions. The equation is solved for f with L being equal to 16 mH.

Step 1 $X_L = 2\pi fL$ formula

Step 2 $f = \frac{X_L}{2\pi L}$ rewrite to solve for f

Step 3 $f = \frac{800}{2\pi\ 16\ \text{mH}}$ substitute values

Step 4 $f = 7960$ Hz frequency when the value of inductance is 16 mH and the inductive reactance is 800 ohms

The practice problems include some of the points for the two curves.

Practice Problems

Find the value of the unknown quantity using the information given.

1. f = 1000 Hz, L = 100 mH; find X_L
2. X_L = 1000 ohms, f = 10 kHz; find L
3. X_L = 200 ohms, f = 10 kHz; find L
4. X_L = 200 ohms, L = 16 mH; find f
5. X_L = 600 ohms, L = 16 mH; find f
6. f = 5000 Hz, L = 2 H; find X_L
7. X_L = 400 ohms, f = 10 kHz; find L
8. X_L = 1000 ohms, L = 16 mH; find f
9. L = 50 μH, f = 3 MHz; find X_L
10. L = 50 mH, f = 0 Hz (dc); find X_L

SERIES OR PARALLEL INDUCTIVE REACTANCES

Because reactance is measured in ohms, the same as resistance, it stands to reason that whenever a reactance, either inductive or capacitive, is connected with another one of the same type, it will follow the rules for resistors when they are in series or parallel.

Series X_L $$X_{Ls} = X_{L1} + X_{L2} + X_{L3} + \ldots$$ **Formula 8-2**

Parallel X_L $$\frac{1}{X_{Lp}} = \frac{1}{X_{L1}} + \frac{1}{X_{L2}} + \frac{1}{X_{L3}} + \ldots$$ **Formula 8-3**

When it is necessary to calculate a voltage drop or the current through a reactance, the same manner used for resistors will again be used, following Ohm's law. This only applies if it is only reactance in the circuit, not resistance and the reactances must all be the same type; either inductive or capacitive.

All the basic rules also apply. Current is the same throughout a series circuit. Current divides in a parallel circuit. Voltage drops in a series circuit are directly proportional to the size of the reactance. Voltage is the same throughout a parallel circuit.

Figure 8-2 shows a series circuit with two inductive reactances. To calculate the total inductive reactance, use the formula to add the series components, just like series resistors. Total current is calculated using Ohm's law with the applied voltage and total reactance. Voltage drops across each of the reactances is also found by using Ohm's law, using total current and the individual reactances.

Figure 8-3 shows two inductive reactances in parallel. Total reactance is found using the reciprocal formula. Total current is found using Ohm's law with the total reactance and the applied voltage. Voltage for each branch is the same since they are parallel branches. Branch current is calculated using Ohm's law with the branch voltages and the reactance of the individual branches. In this case branch voltage is the same as the applied voltage. Branch currents should add up to the total current.

SERIES X_L AND R

In all series circuits, the current is the same throughout the circuit. When an ac current is flowing in a circuit containing inductance, the inductor's magnetic field will expand and collapse following the applied sine wave. Because it is current that causes the magnetic field to develop and this requires a period of time, the voltage developed across the inductor will appear before the current can cause the magnetic field. The current flow, then, is delayed by the inductor voltage drop. This delay is measured in degrees of the sine wave. Inductive voltage is always 90 degrees ahead of the inductive current. This can also be said that the current lags the voltage by 90 degrees. Since the current in the series circuit is the same throughout the circuit, the current in the resistor must be the same as the current in the inductor. The current through a resistor causes a voltage drop that will have no phase shift, in comparison to the current. If the current through the resistor is the same as the voltage, then it stands to reason that the voltage drop across the resistor must be 90 degrees out of phase with the voltage developed across the inductor. The inductor voltage will be 90 degrees ahead or the voltage across the resistor lags by 90 degrees.

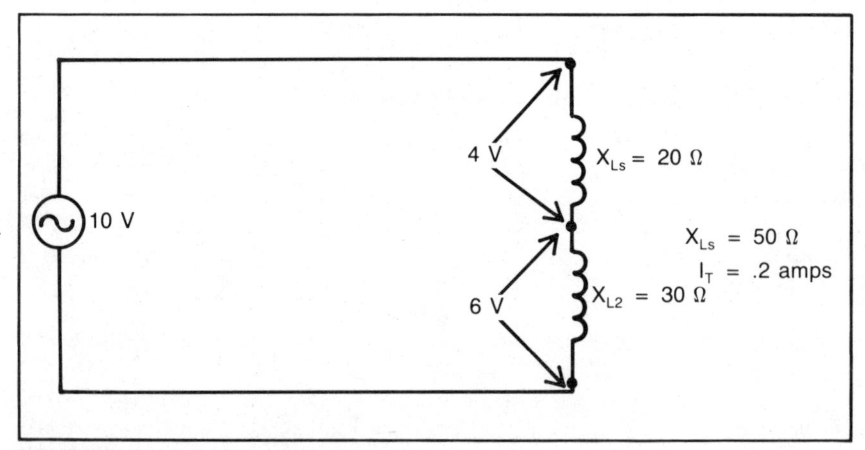

Fig. 8-2. Series inductive reactance.

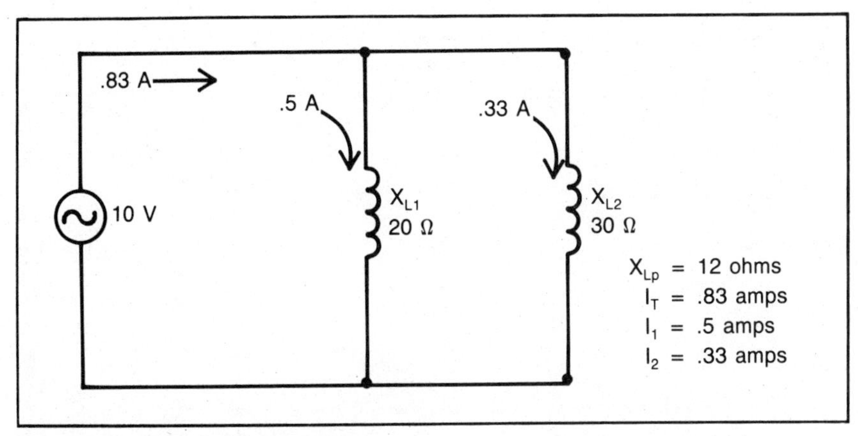

Fig. 8-3. Parallel inductive reactance.

- In a series circuit, I_L lags V_L, V_R lags V_L always by 90 degrees.

Figure 8-4 shows a sine wave analysis of the current through the series circuit, the voltage drops across the inductor, resistor, and the applied voltage. Notice how the current sine wave and the resistor sine wave, V_R, are exactly in phase, with only the amplitudes being different.

The phase relationship of two sine waves can be determined by the maximum and minimum values and when they occur in terms on the number of degrees of the sine wave.

Because the current and the resistive voltage are exactly in phase, they can be thought of as the reference. Usually the voltage across the resistor is chosen because it can be observed directly on an oscilloscope.

Notice in Fig. 8-4 how the sine wave representing the voltage across the inductor has its peak value occur at a point in time equal to 90 degrees earlier than the resistor voltage. Another way to say this is to say that the resistor voltage, V_R, has its peak value *lag* the inductor voltage, V_L, by 90 degrees.

The fourth sine wave in the Fig. 8-4 is the representation of the total voltage. This sine wave is actually the summation of each point along the curves of V_R and V_L. Notice the applied voltage has its peak value occur at a point between the inductor voltage and the resistor voltage. This will always be the case. The applied voltage in this particular circuit has a phase shift of 45 degrees.

- An inductive circuit will always have a phase shift value of between 0 degrees and 90 degrees, depending on circuit values.

Figure 8-5 is a little more difficult to visualize a relationship between it and an electrical circuit. The figure uses vectors or phasors to represent the relative magnitude (size) and phase relationship of the sine waves. The values of the sine waves are shown as peak values with the vectors. The relationship between the two or more vectors shows the phase.

In Fig. 8-5 there are two ways drawn to represent the methods of drawing the phasor diagrams. Either one is correct. Figure 8-5A uses the parallelogram method. Here the two basic vectors, V_R and V_L, are drawn at right angles to each other. V_L is plotted up to show that it leads the V_R by 90 degrees. V_R is always plotted on the horizontal to indicate a 0 degree phase shift, or the reference vector. Dotted lines are connected

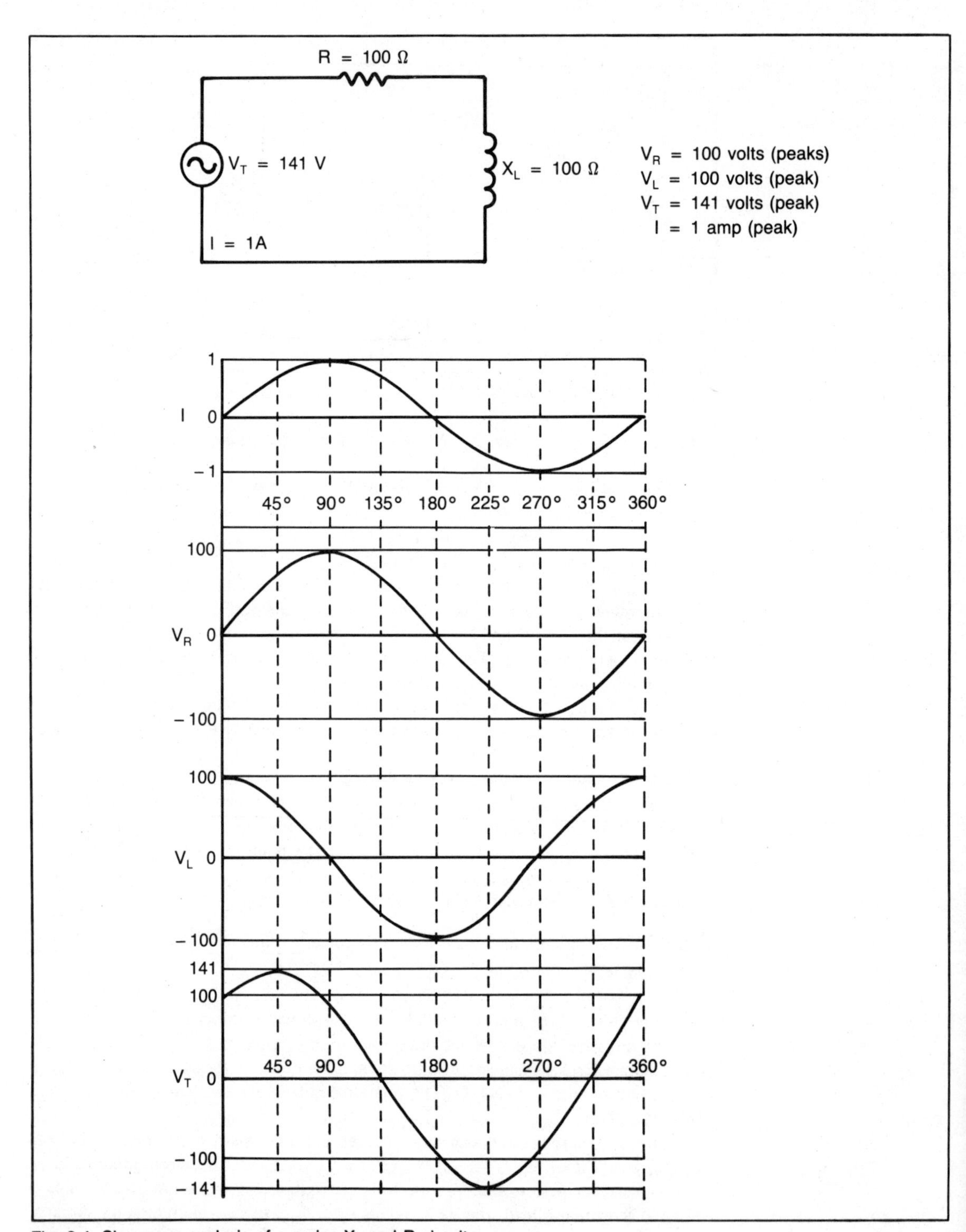

Fig. 8-4. Sine wave analysis of a series X_L and R circuit.

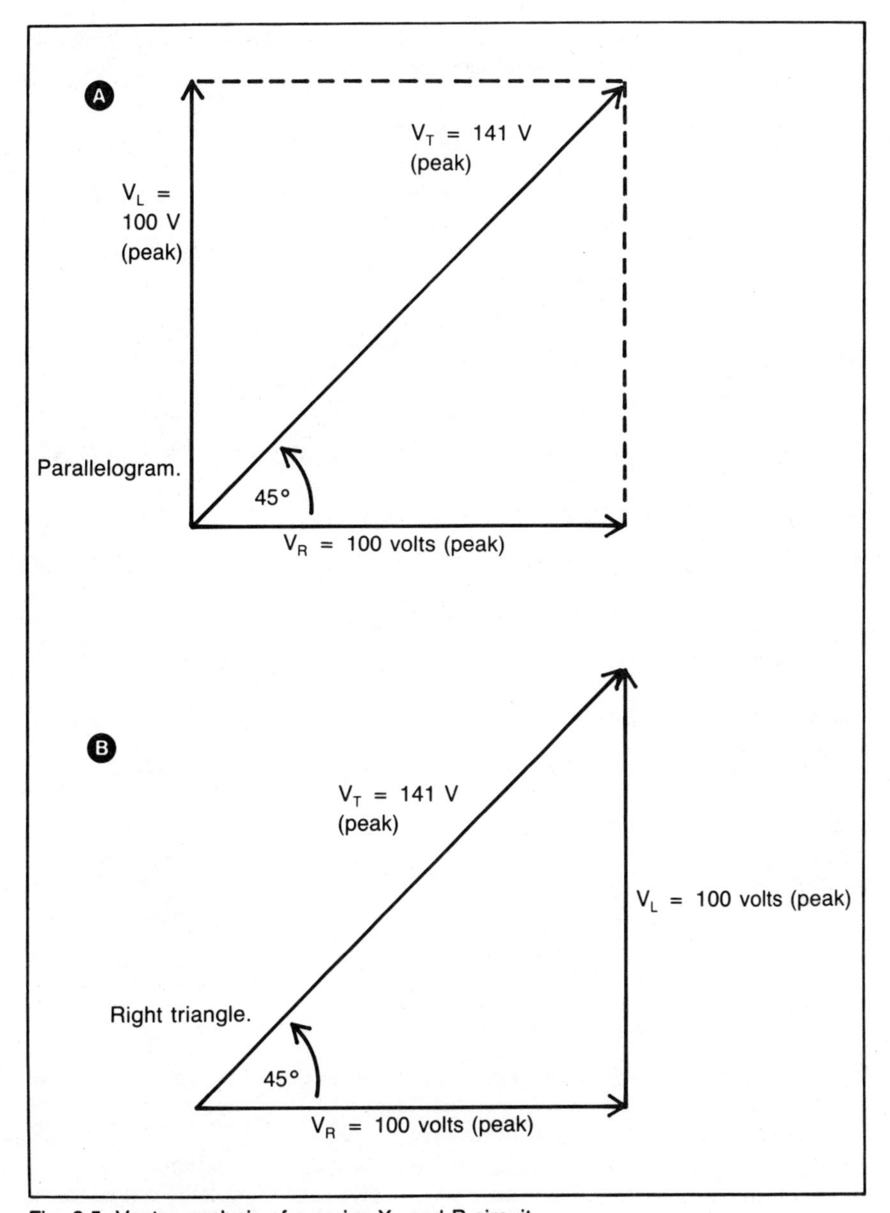

Fig. 8-5. Vector analysis of a series X_L and R circuit.

to the vectors to form a parallelogram (for an R-L circuit, it will be a rectangle). Then a new vector is drawn between the opposite angles of the parallelogram. The new vector represents the applied voltage. Notice how it will always be longer than either of the other two vectors, and the magnitude, or size, will not be simply the two sides added together.

Figure 8-5B shows the use of a right triangle to connect the vectors of the circuit voltages. The V_L vector is connected to the end of the V_R vector and the hypotenuse of the triangle is drawn by connecting the ends of the two voltage vectors. The hypotenuse forms the vector to represent the applied voltage. The angle formed by the hypotenuse is the operating voltage, called theta, Θ.

Calculating the Triangle

Figure 8-5 uses the same circuit shown in the schematic of Fig. 8-4.

V_R is plotted on the horizontal 0 degree axis.
V_L is plotted on the vertical 90 degree axis.
V_T is plotted to connect the two vectors.

To calculate the magnitude and length of V_T, use:

Pythagorean theorem $V_T^2 = V_R^2 + V_L^2$

total voltage in an inductive circuit $V_T = \sqrt{V_R^2 + V_L^2}$ **Formula 8-4**

Step 1 $V_T = \sqrt{100^2 + 100^2}$ substitute values

Step 2 $V_T = \sqrt{10{,}000 + 10{,}000}$ find the squares

Step 3 $V_T = 141$ V add and find the square root

To calculate the operating angle, the angle formed by the hypotenuse, use:

tangent of the angle equals the opposite over the adjacent $\tan \Theta = \frac{V_L}{V_R}$

operating angle formula arranged for use directly with a calculator. $\Theta = \tan^{-1} \frac{V_L}{V_R}$ **Formula 8-5**

Step 1 $\Theta = \tan^{-1} \frac{100}{100}$ substitute values

Step 2 $\Theta = 45°$ operating angle

Notice in the example shown above, the voltages of the inductor and resistor are equal. Whenever this occurs, the operating angle will be 45 degrees and the two voltages will each be 70.7 percent of the applied voltage.

When the value of inductor voltage and resistor voltage are different from each other, the operating angle will no longer be 45 degrees and the voltage will be calculated using the formula.

By observing the voltage triangle we can make some predictions. If the voltage of the inductor increases greater than the resistor voltage, the phasor representing the inductor will increase and become greater than the resistor phasor. The height of the triangle will become greater than its base. The operating angle will increase toward 90 degrees.

If the voltage of the inductor is made small compared to the resistor voltage, the base of triangle is greater than the height. The operating angle will decrease towards 0 degrees.

Figure 8-6A shows a triangle where the V_L is large compared to the V_R. Figure

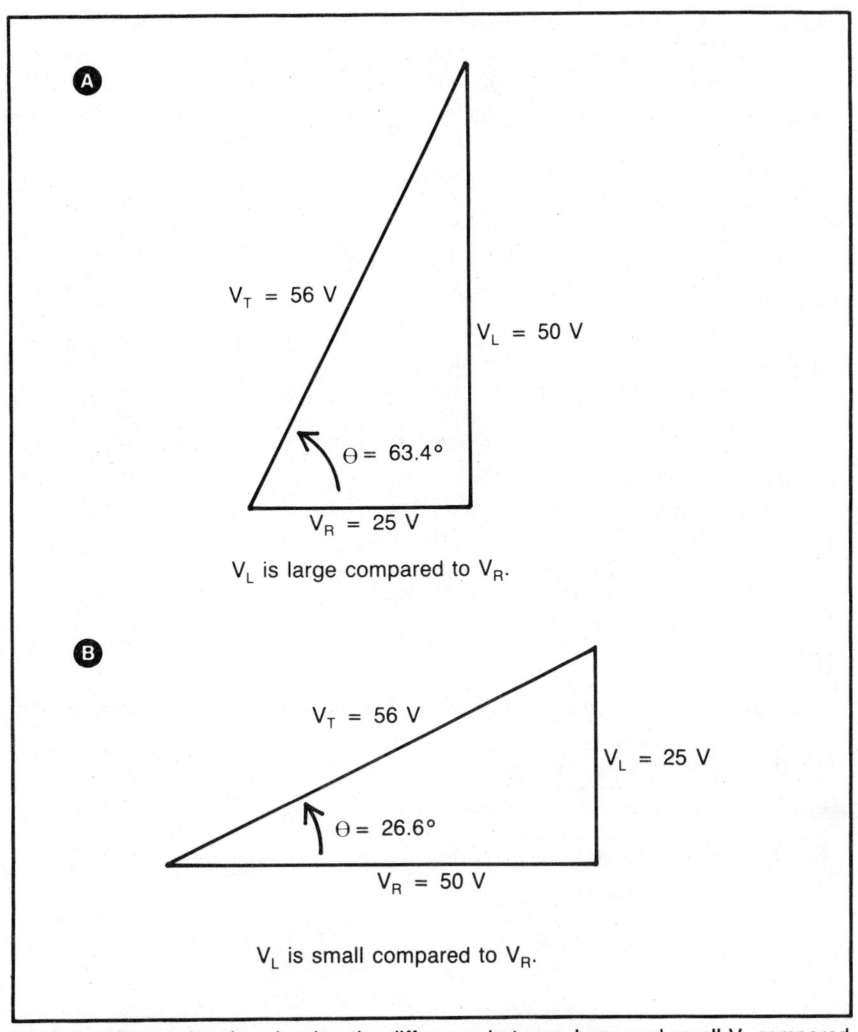

Fig. 8-6. Voltage triangles showing the difference between large and small V_L compared to V_R. A. V_L is large compared to V_R. B. V_L is small compared to V_R.

8-6B shows a triangle where the V_L is small compared to the V_R. Calculations are performed using the same methods as was previously demonstrated with the equal values of voltages.

Calculating the Impedance in a Series Circuit

In order to complete the calculations of the series circuit with inductance and resistance, it is necessary to determine the total impedance of the circuit.

The word impedance is used rather than resistance or reactance because neither of these terms would properly describe the total effect of the circuit.

- Impedance (Z) is the ac resistance of a circuit containing both resistance and reactance. It has an angle equal to the circuit phase angle.

It is necessary to determine the impedance of a series circuit through the use of vectors. The vector for impedance will look similar to the voltage triangle. Resistance is always plotted on the horizontal and reactance is always plotted on the vertical. Inductive reactance is plotted up. Impedance is the resultant hypotenuse, calculated by use of Pythagorean theorem:

impedance of a series circuit $Z = \sqrt{R^2 + X_L^2}$ **Formula 8-6**

The operating angle can be found using the impedance triangle. It will be exactly the same as the voltage triangle.

operating angle of a series circuit $\Theta = \tan^{-1} \frac{X_L}{R}$ **Formula 8-7**

Calculating a Complete Series Circuit

The circuit shown in Fig. 8-7 had only the inductance and frequency given, calculate the inductive reactance. The resistance and voltage of the power source were also given.

Calculate the X_L

Step 1 $X_L = 2\pi fL$ formula
Step 2 $X_L = 2\pi (60) (400 \text{ mH})$ substitute values
Step 3 $X_L = 150\ \Omega$

Calculate impedance (Z) using the impedance triangle. Plot X_L up and R horizontal.

Step 1 $Z = \sqrt{R^2 + X_L^2}$ formula

Step 2 $Z = \sqrt{100^2 + 150^2}$ substitute values

Step 3 $Z = \sqrt{10{,}000 + 22{,}500}$ find the squares

Step 4 $Z = \sqrt{32{,}500}$ add the squares

Step 5 $Z = 180\ \Omega$ find the square root

Calculate the operating angle, theta, Θ.

Step 1 $\Theta = \tan^{-1} \frac{X_L}{R}$ formula

Step 2 $\Theta = \tan^{-1} \frac{150}{100}$ substitute values

$\Theta = \tan^{-1} (1.5)$ reduce fraction

$\Theta = 56.3°$ use the calculator to find the angle

With the total impedance calculated and the supply voltage given, calculate the current.

Step 1 $I = \frac{V}{Z}$ Ohm's law modified

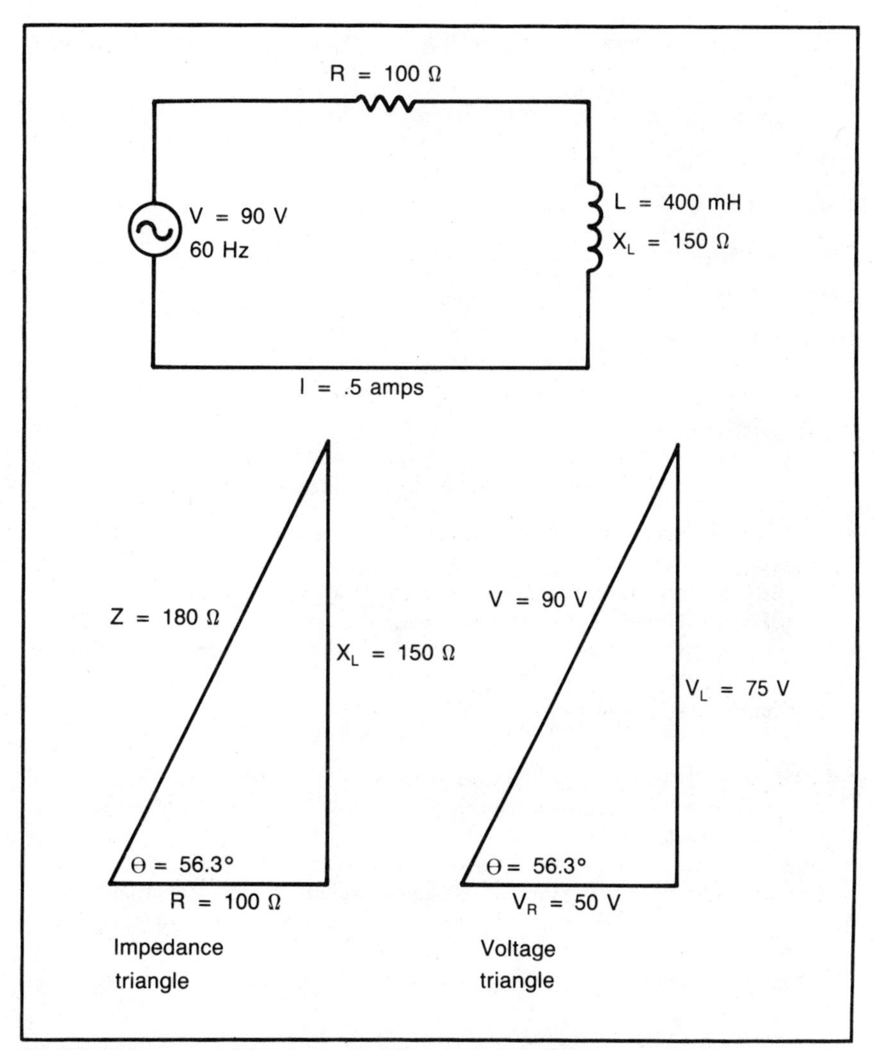

Fig. 8-7. Solving a series R-L circuit using phasor triangles.

Step 2 $I = \dfrac{90\ V}{180\ \Omega}$ substitute values

Step 3 $I = .5\ A$ current throughout the series circuit

Use the current to find the voltage drops.

Step 1 $V_R = I \times R$ formula
Step 2 $V_R = .5\ A \times 100$ ohms substitute values
Step 3 $V_R = 50\ V$ voltage across the resistor

Step 1 $V_L = I \times X_L$ formula
Step 2 $V_L = .5\ A \times 150$ ohms substitute values
Step 3 $V_L = 75\ V$ voltage across the inductor

- The voltage drops cannot be simply added. They must be added through vector addition.

Use vector addition to compare the calculated voltage drops with the given applied voltage.

Step 1 $V = \sqrt{V_R^2 + V_L^2}$ formula

Step 2 $V = \sqrt{50^2 + 75^2}$ substitute values

Step 3 $V = \sqrt{2500 + 5625}$ find the squares

Step 4 $V = 90$ add and find the square root

Numbers have been rounded for easy use. Without rounding, any calculations should be very close to what is expected.

Summary for a Series Circuit

- Calculate the inductive reactance, X_L, based on the frequency and inductance values. Usually X_L is given.
- Calculate the impedance, Z, using the impedance triangle. Vector addition requires the use of the Pythagorean theorem; $Z = \sqrt{R^2 + X_L^2}$.
- Calculate the operating angle, theta, Θ, based on the impedance triangle and the tangent function; $\Theta = \tan^{-1} \frac{X_L}{R}$
- Calculate the series current using the applied voltage and the impedance. If the applied voltage is not given, enough other information will be given to use Ohm's law to calculate voltage drops.
- Use the current to calculate the voltage drops across each component. If applied voltage is not given for calculating current, enough other information should be given.
- Use vector addition to calculate the applied voltage, based on the voltage drops. $V = \sqrt{V_R^2 + V_L^2}$

  Theta, Θ, can also be found using the voltage triangle rather than the impedance triangle.

Practice Problems

Use the two schematics shown to calculate the answers to the problems,

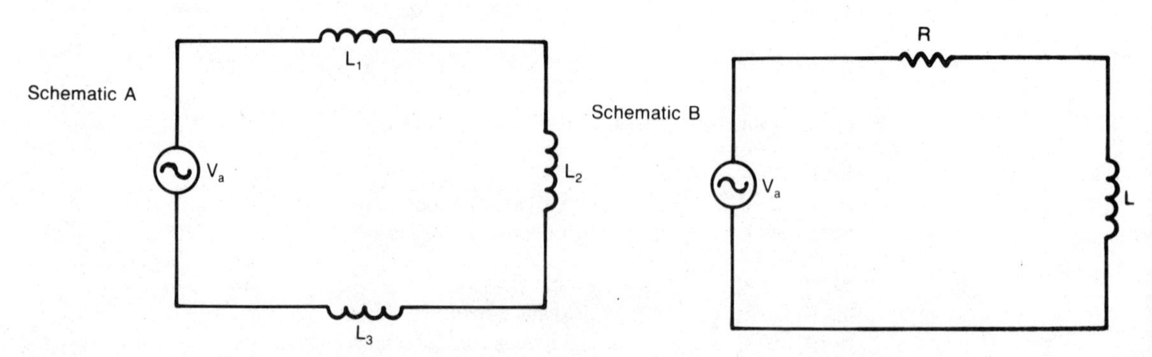

Use schematic A to answer problems 1 and 2. V_a = 100 volts/60 Hz

Find: L_T, X_{L1}, X_{L2}, X_{L3}, and X_{LT}.

1. L_1 = 1H, L_2 = 2H, L_3 = 3H
2. L_1 = 20 mH, L_2 = 40 mH, L_3 = 80 mH

Use schematic B to answer problems 3 through 8. V_a = 100 volts

Find: Z, I, V_R, V_L, Θ, draw impedance triangle

3. R = 100 ohms, X_L = 100 ohms
4. R = 25 ohms, X_L = 50 ohms
5. R = 10 ohms, X_L = 100 ohms
6. R = 75 ohms, X_L = 5 ohms
7. R = 75 ohms, X_L = 25 ohms
8. R = 0 ohms, X_L = 50 ohms

Use schematic B to answer problems 9 and 10.

Find: R, X_L, Z, V_a, Θ, draw voltage triangle.

9. V_R = 20 V, V_L = 40 V, I = .25 A
10. V_R = 25 V, V_L = 15 V, I = .333 A

PARALLEL X_L AND R

The basic rules of parallel circuits shows that the voltage is the same throughout the parallel branches and the current divides inversely to the resistance of the individual branches. Since the voltage is the same across all the branches, there cannot appear to be any phase shift in the voltage of the components compared to the applied voltage.

The current of the inductor will lag the current of the resistor by 90 degrees due to the fact that it is the current that causes the inductor to charge. Another way to think of this is to realize that the current of an inductor always lags the inductor's voltage by 90 degrees. The voltage cannot have a phase shift, the current of the inductor must lag the current of the resistor, which is always the reference, by 90 degrees.

- In a parallel circuit, I_L lags I_R by 90 degrees.

Figure 8-8 is the sine wave analysis of a circuit with equal values of R and X_L. Equal values were chosen to simplify the drawings. Since the values are equal, the current through each branch should be equal, as calculated by Ohm's law. I_T is the line current from the power source. I_T is the vector summation of I_R and I_L at every point along the sine wave. For ease of calculations, the peak value is chosen to perform any calculations.

Since I_L lags I_R by 90 degrees, the sine wave representing I_L starts 90 degrees later than the sine wave for I_R. The operating angle, theta, Θ, is determined by the sine wave for I_T, which will always be between 0 and negative 90 degrees. The negative angle indicates a lagging condition, or a delayed condition.

Figure 8-9 shows the vector diagram of the current in this same circuit. The resistive current is again plotted on the horizontal, as the reference line. The vector, or phasor, for I_L is plotted at a 90 degree angle to I_R. It is plotted down to show it is lagging

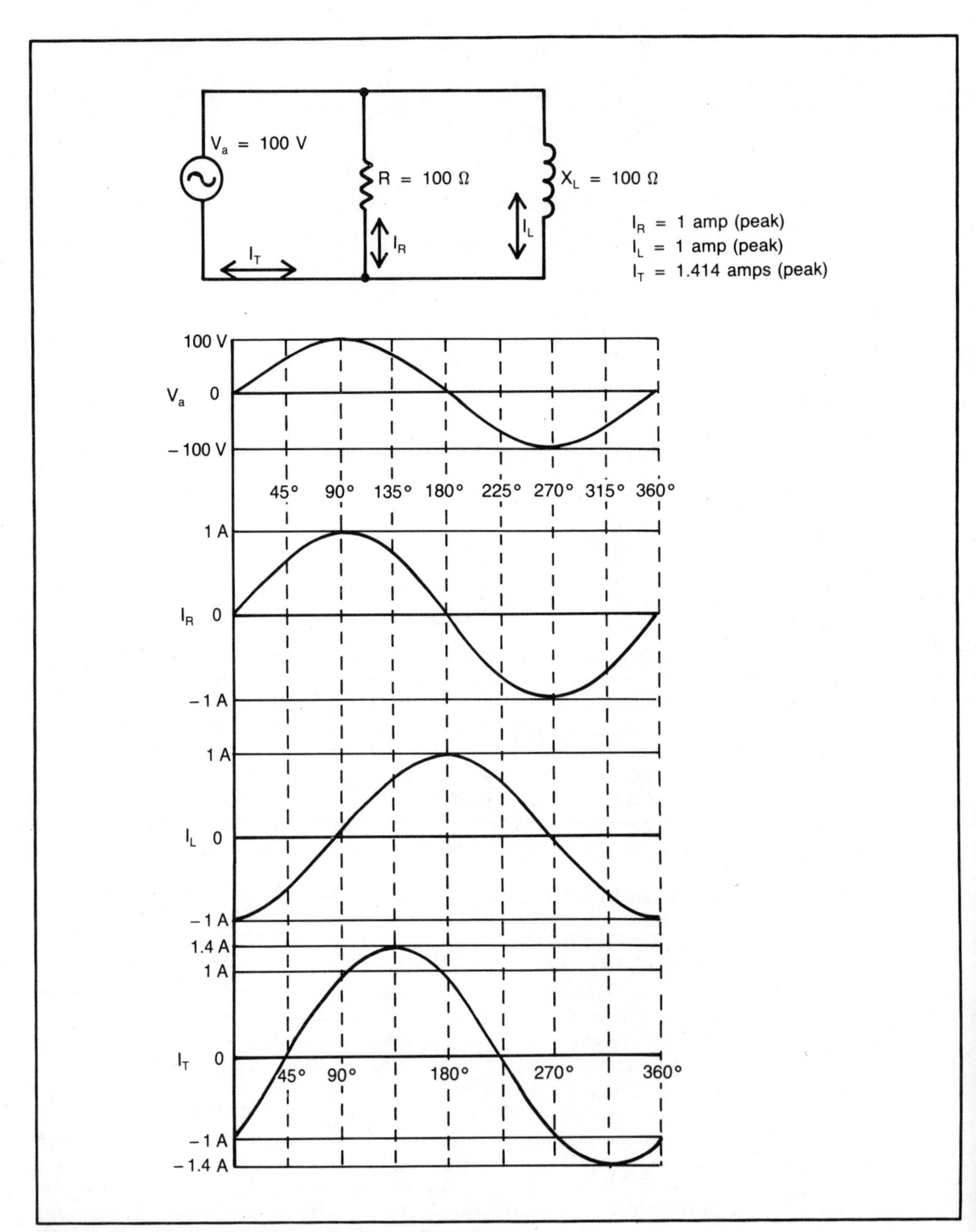

Fig. 8-8. Sine wave analysis of a parallel X_L and R circuit.

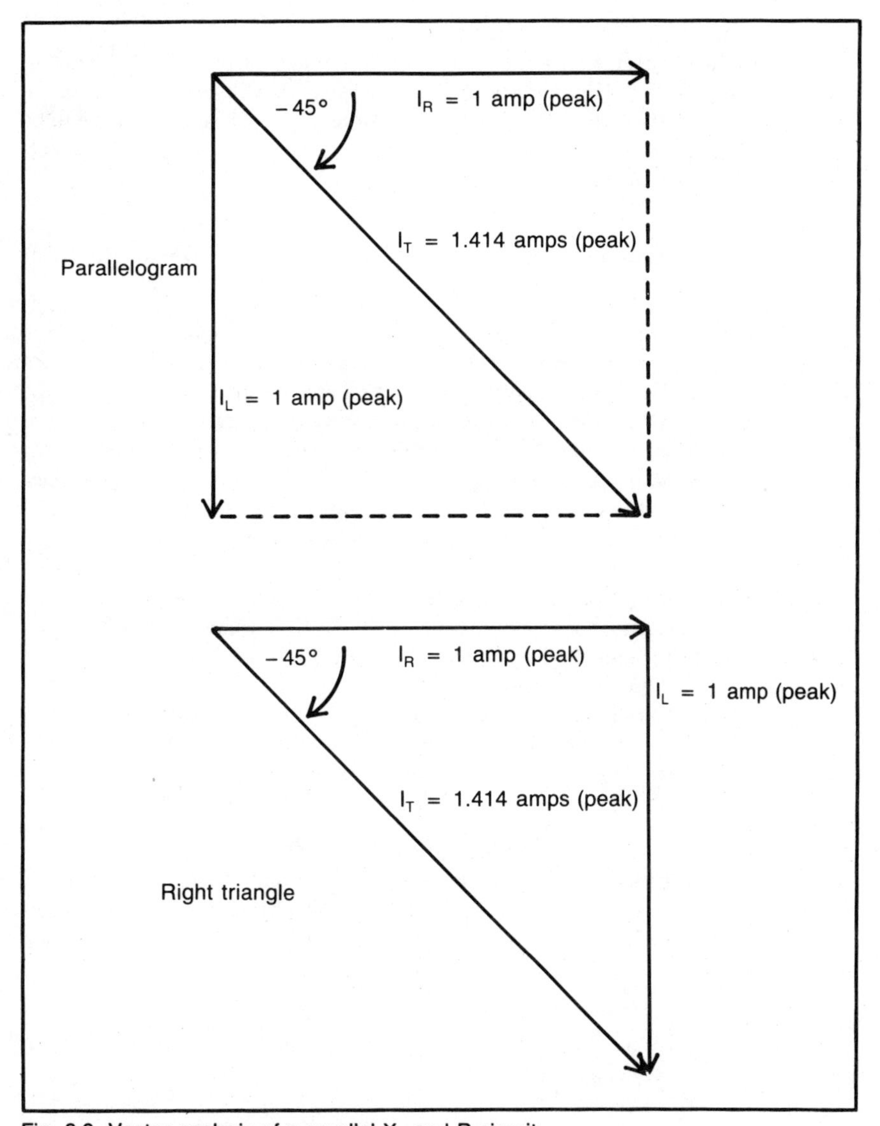

Fig. 8-9. Vector analysis of a parallel X_L and R circuit.

by 90 degrees. The hypotenuse of the triangle is completed and the result is I_T at the operating angle.

total current in a parallel circuit. $I_T = \sqrt{I_R^2 + I_L^2}$ **Formula 8-8**

operating angle $\Theta = \tan^{-1} \dfrac{-I_L}{I_R}$ **Formula 8-9**

When using the formula for the angle theta, Θ, if the current for the inductance is assigned a negative sign, as it should be, the calculator will give the angle as a negative angle. If the negative sign is neglected, the angle will be the same value, except it won't be negative from the calculator.

The only quantity remaining to be calculated in the parallel circuit is the value of total impedance. Since the resistor and reactance are in parallel, it becomes quite complicated to combine these through the reciprocal formula, considering they have to be combined with vectors. The easiest way to calculate the total Z is to use the total current and the applied voltage.

total impedance in a parallel circuit $Z = \frac{V_a}{I_T}$ **Formula 8-10**

In the section on series circuits, it was analyzed what would happen if the X_L were made greater or less than the value of R. Refer to Fig. 8-6, as a reminder. In a series circuit, the voltage drop is increased across the larger resistor. In a parallel circuit, the current is decreased with the larger value of resistance (or reactance). Therefore, if the X_L in a parallel circuit were made larger than the resistor, the current through the inductor would decrease. The operating angle would also decrease, toward 0 degrees. If the value of X_L were made smaller than the value of R, the current through the inductor would be larger than the current through the resistor and the operating angle would increase toward negative 90 degrees.

Figure 8-10 shows two current triangles. One triangle, Fig. 8-10A, shows a small inductive current in comparison to the resistive current. Because the current of the inductor is small, that means the X_L is large, in comparison to the resistor. Notice that the operating angle, theta, Θ, is smaller than 45 degrees.

Figure 8-10B shows a large inductive current compared to the resistive current. This means the inductive reactance, X_L, is smaller than the value of the resistor. Notice that the operating angle, Θ, is larger than 45 degrees.

Calculating a Parallel Circuit

Figure 8-11 shows a parallel circuit containing inductance and resistance. Included in the figure is the current triangle to determine the total amount of current using vector summation. When working with a parallel circuit the voltage across all branches is the same.

Calculate the resistive current, I_R.

Step 1 $I_R = \frac{V}{R}$ formula

Step 2 $I_R = \frac{10\ V}{10\ \Omega}$ substitute values

Step 3 $I_R = 1\ A$ plotted at 0°

Calculate the inductive current, I_L.

Step 1 $I_L = \frac{V}{X_L}$ formula

Step 2 $I_L = \frac{10\ V}{7\ \Omega}$ substitute values

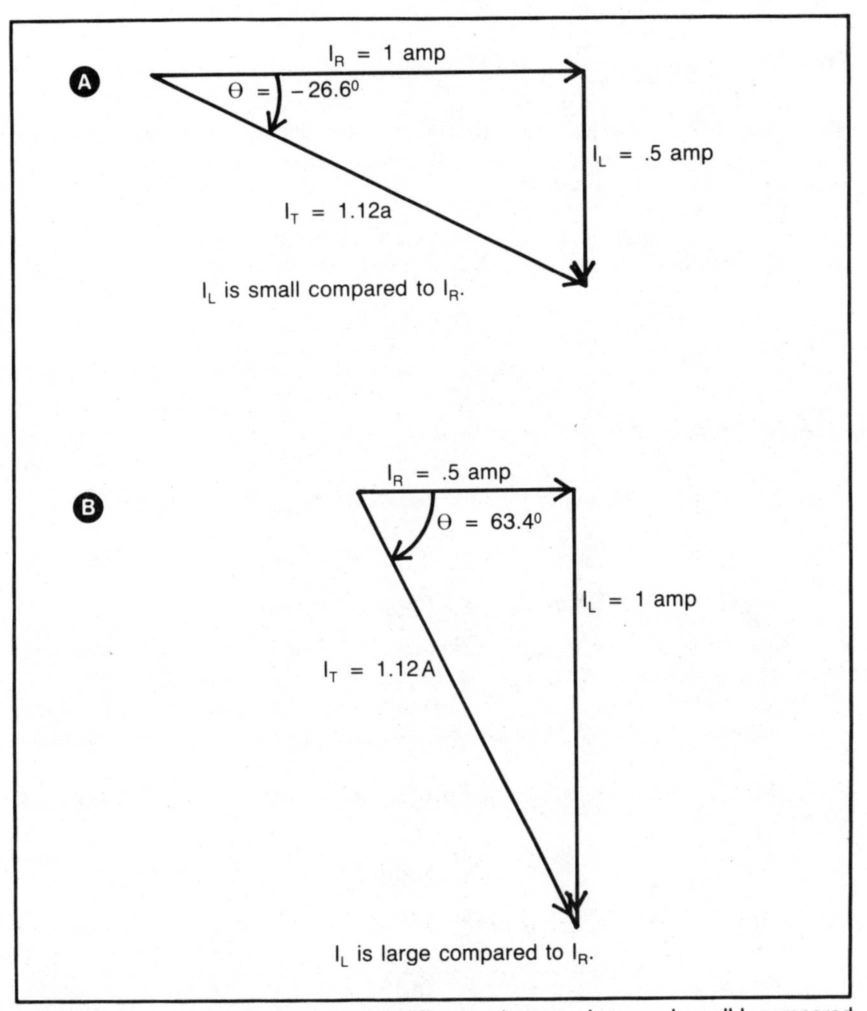

Fig. 8-10. Current triangles showing the difference between large and small I_L compared to I_R. A. I_L is small compared to I_R. B. I_L is large compared to I_R.

Step 3 $I_L = 1.43$ A plotted at $-90°$

Calculate the total current using Pythagorean theorem.

Step 1 $I_T = \sqrt{I_R^2 + I_L^2}$ formula

Step 2 $I_T = \sqrt{I^2 + 1.43^2}$ substitute values

Step 3 $I_T = 1.74$ A hypotenuse of the triangle

Calculate the operating angle, theta, Θ.

Step 1 $\Theta = \tan^{-1} \dfrac{-I_L}{I_R}$ formula

Step 2 $\Theta = \tan^{-1} \dfrac{-1.43}{1}$ substitute values

Step 3 $\Theta = -55°$ angle is negative because it is plotted down. Calculator can give a negative angle if I_L is used as a negative

Calculate total impedance, Z. Because the total current has been calculated and the applied voltage is given, this is the easiest method to find Z.

Step 1 $Z = \dfrac{V}{I_T}$ formula

Step 2 $Z = \dfrac{10\ V}{1.74\ A}$ substitute values

Step 3 $Z = 5.75$ ohms should be less than the smallest resistance of the parallel branches

Summary for a Parallel Circuit

- Calculate the values of the resistive current and the inductive current. These are also called the branch currents.
- Calculate the total current using the branch currents to form a current triangle.
- Calculate the operating angle based on the current triangle. The angle in an inductive circuit will be negative.
- Calculate the total impedance using the total current and the applied voltage with Ohm's law.

Practice Problems

Use the two schematics shown to calculate the answers to the problems.

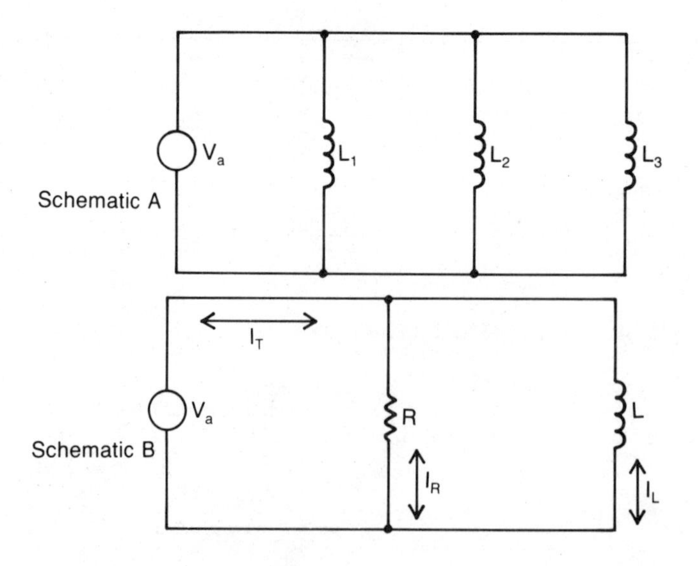

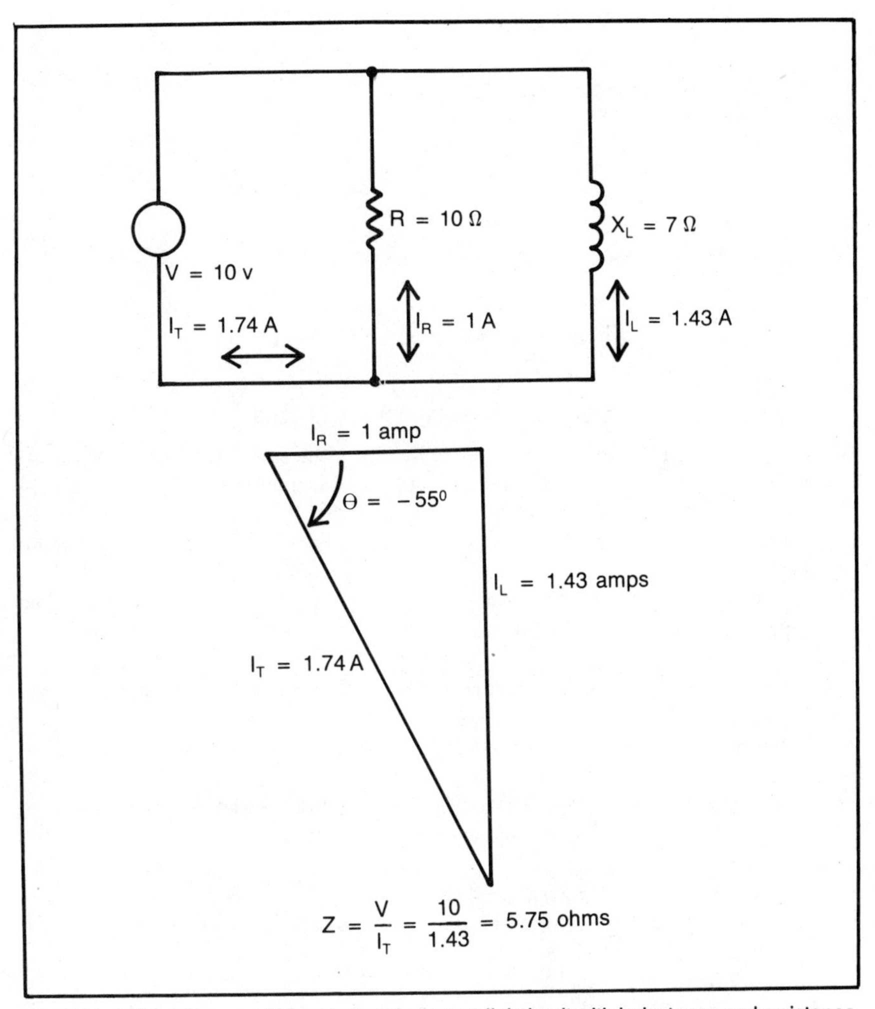

Fig. 8-11. Using the current triangle to solve a parallel circuit with inductance and resistance.

Use schematic A to answer problems 1 and 2.

1. L_1 = 1H, L_2 = 2H, L_3 = 3H, V_a = 100 V/50 Hz

   Find: L_T, X_{L1}, X_{L2}, X_{L3}, and X_{LT}

2. L_1 = 20 mH, L_2 = 40 mH, L_3 = 80 mH, V_a = 100 V/50 Hz

   Find: L_T, X_{L1}, X_{L2}, X_{L3} and X_{LT}

Use schematic B to answer problems 3 through 8. V_a = 100 volts

Find: Z, I_T, I_R, I_L, Θ, draw the current triangle

3. R = 100 ohms, X_L = 100 ohms
4. R = 25 ohms, X_L = 50 ohms

5. $R = 10$ ohms, $X_L = 100$ ohms
6. $R = 75$ ohms, $X_L = 5$ ohms
7. $R = 75$ ohms, $X_L = 25$ ohms
8. $R = 5$ ohms, $X_L = 5$ ohms

Use schematic B to answer questions 9 and 10. $V_a = 100$ volts

Find: R, X_L, Z, I_T, Θ

9. $I_R = 1$ A, $I_L = 3$ A
10. $I_R = 25$ mA, $I_L = 10$ mA

POWER IN A REACTIVE CIRCUIT

When working with circuits that contain a reactance, the calculations of voltage, current and total resistance, or impedance, become complicated in comparison to dc circuits, or any circuit containing only pure resistance. It stands to reason that the calculations for power in an ac circuit containing resistance and reactance are more complicated than when the circuit has only pure resistance.

In an ac circuit there are three ways of describing the power: real power (also called true power), reactive power, and apparent power.

- Real power (also called true power) is the power dissipated in pure resistance, unit is W (watts).
- Reactive power is the power dissipated in pure reactance, unit is VARS (Volt-Ampere-Reactive).
- Apparent power is the power of a circuit containing both resistance and reactance. It is calculated by: $I \times E$ (multiplying total current by the applied voltage), unit is VA (Volt-Ampere).
- Power factor is a ratio of the real power to the total power. It is a pure number, with no units and will always be between 0 and 1. It is calculated by taking the cosine of the operating angle; $\cos \Theta$.

When working with the power relationships in an ac circuit, the basic rules of power, in reference to series or parallel circuits, will be the same as a circuit with only resistors. The big difference is the fact that it is necessary to deal with the voltage or current triangles.

In a series circuit, the voltage triangle will give each of the three types of powers and the operating angle can be used to calculate the power factor.

In a parallel circuit, because voltage is the same throughout the parallel circuit, the current triangle is used to calculate the powers and the operating angle is again used to determine the power factor.

The following sample calculations are using the sample circuits done out in earlier sections of this chapter. Refer to Fig. 8-7, a series R-L circuit.

Calculate real power using the voltage triangle.

Step 1 $P = I \times E$ formula
Step 2 $P = .5\text{ A} \times 50\text{ V}$ substitute values
Step 3 $P = 25$ W resistive power

Calculate reactive power.

Step 1 P = I × E formula
Step 2 P = .5 A × 75 V substitute values
Step 3 P = 37.5 VARS inductive power

Calculate apparent power.

Step 1 P = I × E formula
Step 2 P = .5 A × 90 V substitute values
Step 3 P = 45 VA apparent power of total circuit

Calculate power factor.

Step 1 PF = cos Θ formula
Step 2 PF = cos 56.3° substitute values
Step 3 PF = .555 ratio of total power consumed by the resistor

Refer to Fig. 8-11, a parallel circuit, to calculate the power of the circuit.

Calculate resistive power.

Step 1 P = I × E formula
Step 2 P = 1 A × 10 V substitute values
Step 3 P = 10 W true power

Calculate reactive power.

Step 1 P = I × E formula
Step 2 P = 1.43 A × 10 V substitute values
Step 3 P = 14.3 VARS reactive power

Calculate apparent power.

Step 1 P = I × E formula
Step 2 P = 1.74 A × 10 V substitute values
Step 3 P = 17.4 VA total circuit power

Calculate power factor.

Step 1 PF = cos Θ formula
Step 2 PF = cos −55° substitute values
Step 3 PF = .574 ratio of resistive power to total power

The method shown above for both the series circuit and the parallel circuit is only one of the ways of performing the calculations. Following is a summary of the formulas and some alternative formulas.

resistive current times the resistive voltage — Real Power (watts) $= I_R \times E_R$ **Formula 8-11**

resistive current times the value of resistance — Real Power (watts) $= I^2 \times R$ **Formula 8-12**

Description	Quantity	Formula	
total circuit voltage times total circuit current times the cosine of the operating angle	Real Power (watts)	$= VI \cos \Theta$	**Formula 8-13**
inductive current times inductive voltage	Reactive Power (VARS)	$= I_L \times E_L$	**Formula 8-14**
total voltage times total current times the sine of the operating angle	Reactive Power (VARS)	$= VI \sin \Theta$	**Formula 8-15**
total voltage times total current	Apparent Power (VA)	$= V \times I$	**Formula 8-16**
cosine of the operating angle	Power Factor (No units)	$= \cos \Theta$	**Formula 8-17**
resistance divided by total impedance, applies to series circuits	Power Factor (No units)	$= \frac{R}{Z}$	**Formula 8-18**
resistive current divided by total current, applies to parallel circuits	Power Factor (No units)	$= \frac{I_R}{I_T}$	**Formula 8-19**

Practice Problems

Use the two schematics shown to calculate the answers to the problems.

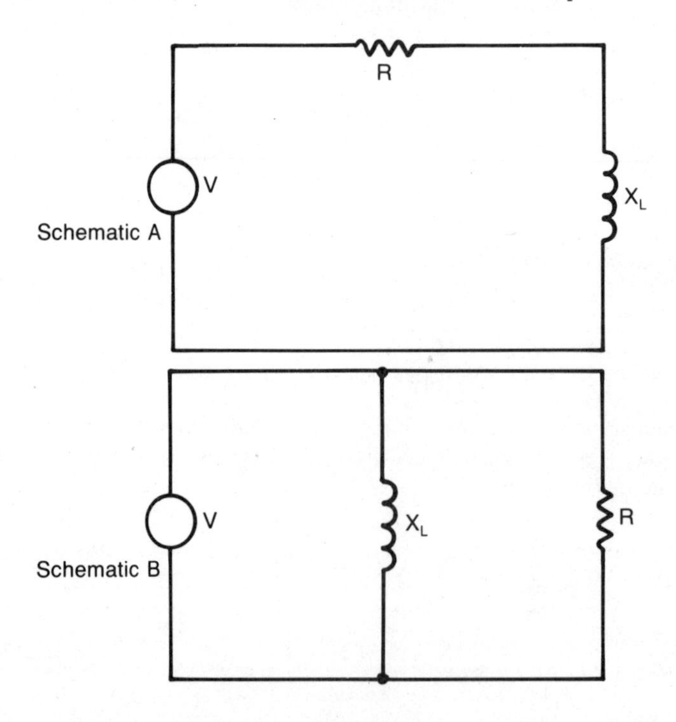

Use schematic A for problems 1 through 5. V = 100 V

Find: Z, I, V_R, V_L, Θ, Real Power, P_R, Reactive Power, P_X, Apparent Power, P_A, Power Factor, PF

1. R = 500 ohms, X_L = 500 ohms
2. R = 100 ohms, X_L = 250 ohms
3. R = 1000 ohms, X_L = 750 ohms
4. R = 1000 ohms, X_L = 500 ohms
5. R = 10 ohms, X_L = 20 ohms

Use schematic B for problems 6 through 10. V = 100 V

Find: Z, I_T, I_R, I_L, Θ, Real Power, P_R, Reactive Power, P_X, Apparent Power, P_A, Power Factor, PF

6. R = 500 ohms, X_L = 500 ohms
7. R = 100 ohms, X_L = 250 ohms
8. R = 1000 ohms, X_L = 750 ohms
9. R = 1000 ohms, X_L = 500 ohms
10. R = 10 ohms, X_L = 20 ohms

MEASURING PHASE ANGLE WITH AN OSCILLOSCOPE

The operating angle, or phase shift, of a circuit containing reactance is a measurement that is ideally suited for the dual trace oscilloscope. The drawings in the accompanying Figs. 8-12 through 8-17 are used as an explanation of calculating the phase shift with a dual trace oscilloscope. The circuit used is a simple series circuit with an inductor and resistor. Values of the individual components are not given because the calculations here are on the scope, not on the circuit. In the event that the student wishes to set up the circuit as an experiment and practice, the inductor would have to be obtained. The resistor value can easily be calculated. Calculate the resistor value by working with the frequency to be used and determining X_L. From there, use the impedance triangle for a series circuit to obtain the value of resistance needed to complete the triangle at the desired phase angle.

Note The next chapter deals with capacitive reactance and solving the capacitive circuits with the use of the triangles. There is also a section dealing with the use of the oscilloscope. It would probably be much easier and cheaper for the student to obtain a selection of capacitors, rather than an inductor, in order to experiment with the use of the oscilloscope.

Procedure for Calculating the Phase Angle on an Oscilloscope

This section is not intended to teach the use of the oscilloscope, therefore, it is assumed the student is familiar with the controls and basic operation.

Connect the dual trace scope as shown in the circuit diagrams; channel 1 across the entire circuit and channel 2 across the inductor. The scope ground is connected to the common ground point in the circuit.

The scope should display two sine waves. Channel 1 will display the applied voltage, V_a. Channel 2 will display the inductive voltage, V_L.

Because the two waveforms are being supplied a voltage from the same circuit, across different components, the frequency of both waveforms will be exactly the same. The amplitude of each waveform can be varied to make the measurements as easy as possible. The amplitude will not affect the phase angle to be measured. The waves should be as large as possible in order to have the best accuracy.

Adjust the scope to measure one full cycle, or a little more. Do not display too many cycles as this will make the readings difficult.

- Measurements on the scope should be taken where the sine wave cross the zero reference line.

At the zero reference line, the sine wave will be the straightest of any point along the waveform. Peaks can be used but the accuracy is much less.

Follow the steps and examples to determine the phase angle in degrees:

Step 1 Count the number of major divisions in one complete cycle (a major division is the darker or larger blocks).

Step 2 Because one complete cycle equals 360 degrees, the number of blocks counted also equals 360 degrees.

Example: Refer to Fig. 8-12. Using the V_a waveform, there are 4 major blocks in one cycle. (Notice the drawing is labeled 0 through 4 to show the major divisions.

Therefore: 360° = 4 blocks

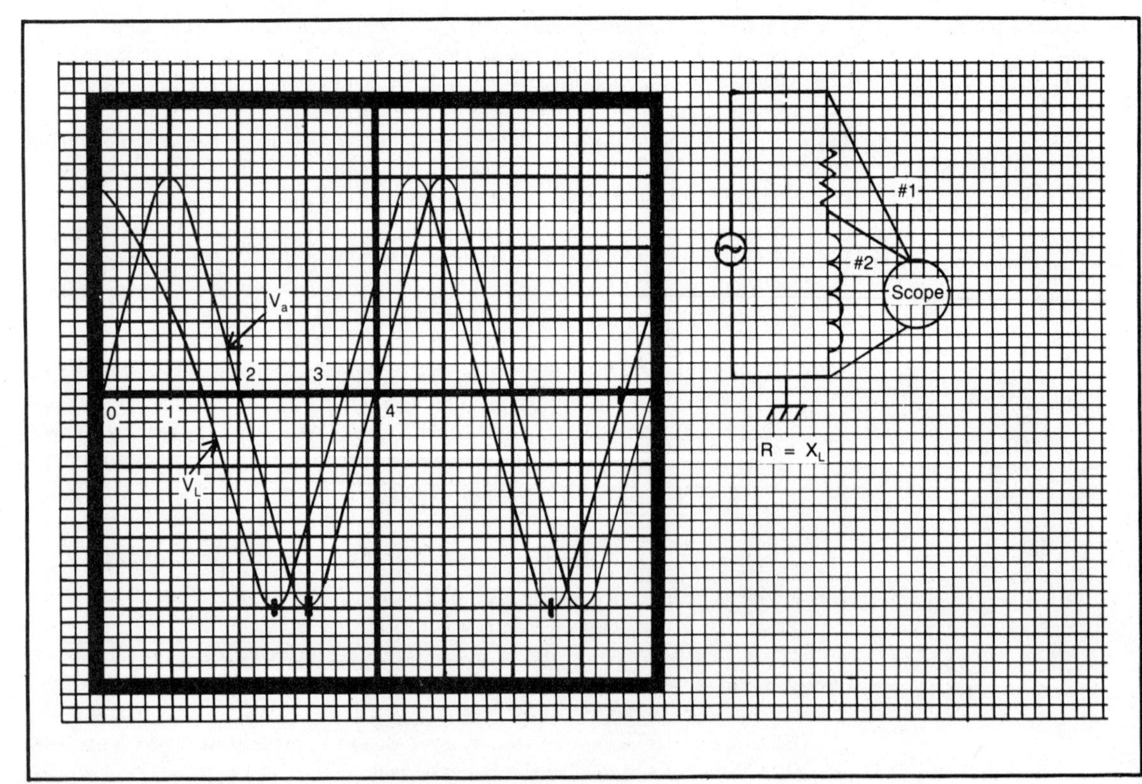

Fig. 8-12. 45° phase angle. Scope triggers on the input voltage. R = X_L

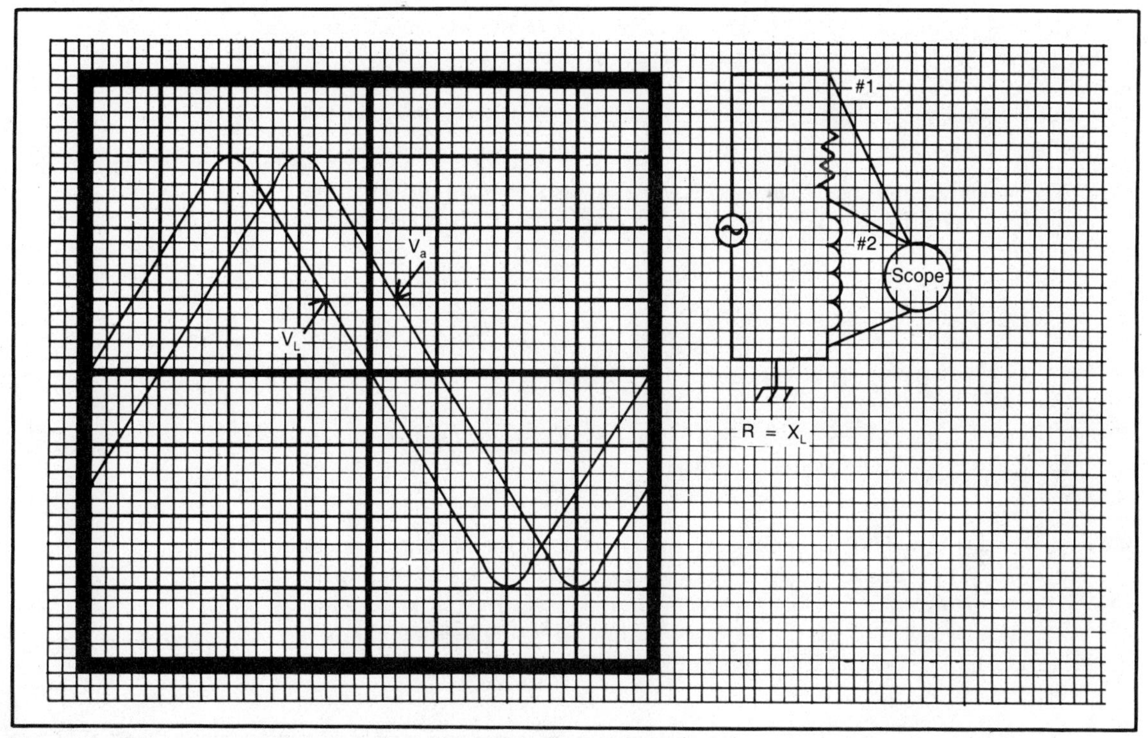

Fig. 8-13. 45° phase angle. Scope triggers on the inductor voltage. R = X_L

$$\frac{\text{degrees}}{\text{block}} = \frac{360\ \text{degrees}}{4\ \text{block}}$$

$$1\ \text{block} = 90\ \text{degrees}$$

Refer to Fig. 8-13. Count the blocks in one cycle.
Divide 360 by the number of blocks.

$$\frac{\text{degrees}}{\text{block}} = \frac{360\ \text{degrees}}{8\ \text{blocks}}$$

$$1\ \text{block} = 45\ \text{degrees}$$

All of the remaining figures use 8 blocks for one complete cycle. This is for simplicity.

Step 3 Count the number of major divisions (blocks) from where the first wave crosses the zero reference line to where the second wave crosses the zero reference line.

Each major division is divided into 5 other smaller blocks. The five smaller blocks represent 10 units or subdivisions of the major division. Therefore, each small block represents .2 major divisions.

When divisions are counted, they can be counted to the nearest .1 (one-tenth) division.

When counting divisions, or parts of a division, it is possible to use the sine wave anywhere it is the most convenient along the reference line.

Example: Refer to Fig. 8-12. V_L is the first wave and V_a is the second. That tells us that V_L leads V_a. The number of divisions between the two waves is .5 major divisions.

Figure 8-13 has V_L first and V_a second with 1.0 major divisions between the two waveforms.

Step 4 Combine Steps 2 and 3 together. In Step 2, it was found how many degrees per division. In Step 3, it was found how many divisions between the two waveforms.

Fig. 8-12:
90 degrees/division
.5 divisions difference (delay)
$90 \times .5 = 45$ degrees phase shift
Fig 8-13:
45 degrees/division
1 division difference
$45 \times 1 = 45$ degrees phase shift

Note Both Figs. 8-12 and 8-13 have a 45-degree phase shift, but the two drawings look different. The reason is the way the scope is triggered. Figure 8-12 is using the V_a as a reference and Fig. 8-13 uses V_L as the reference. The remainder of the sample drawings use V_a as the reference because this is the best way. Regardless of which method is used to trigger the scope, V_L will still be leading V_a by the phase shift.

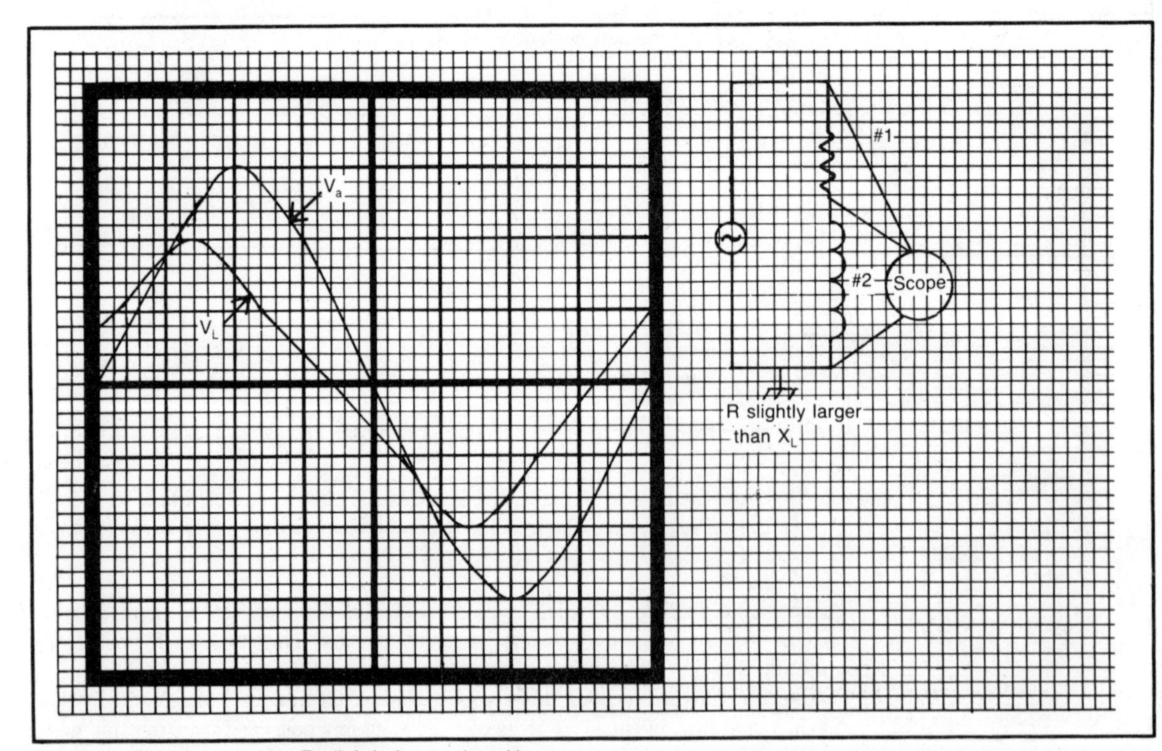

Fig. 8-14. 27° phase angle. R slightly larger than X_L.

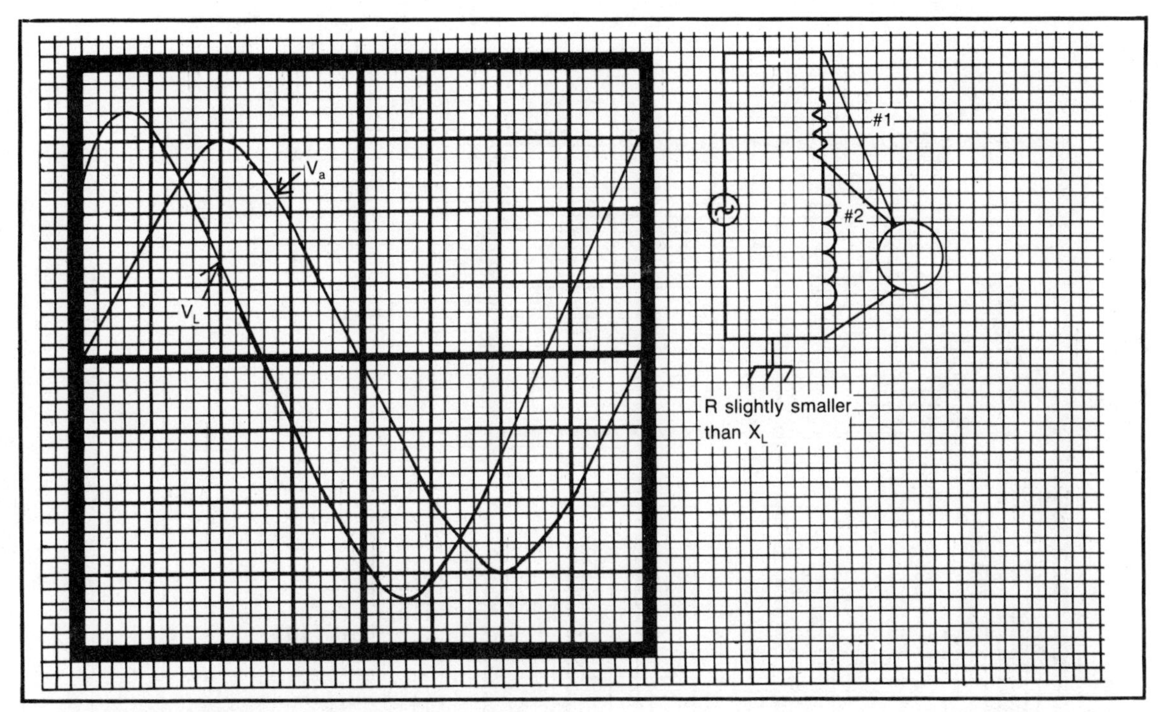

Fig. 8-15. 63° phase angle. R slightly smaller than X_L.

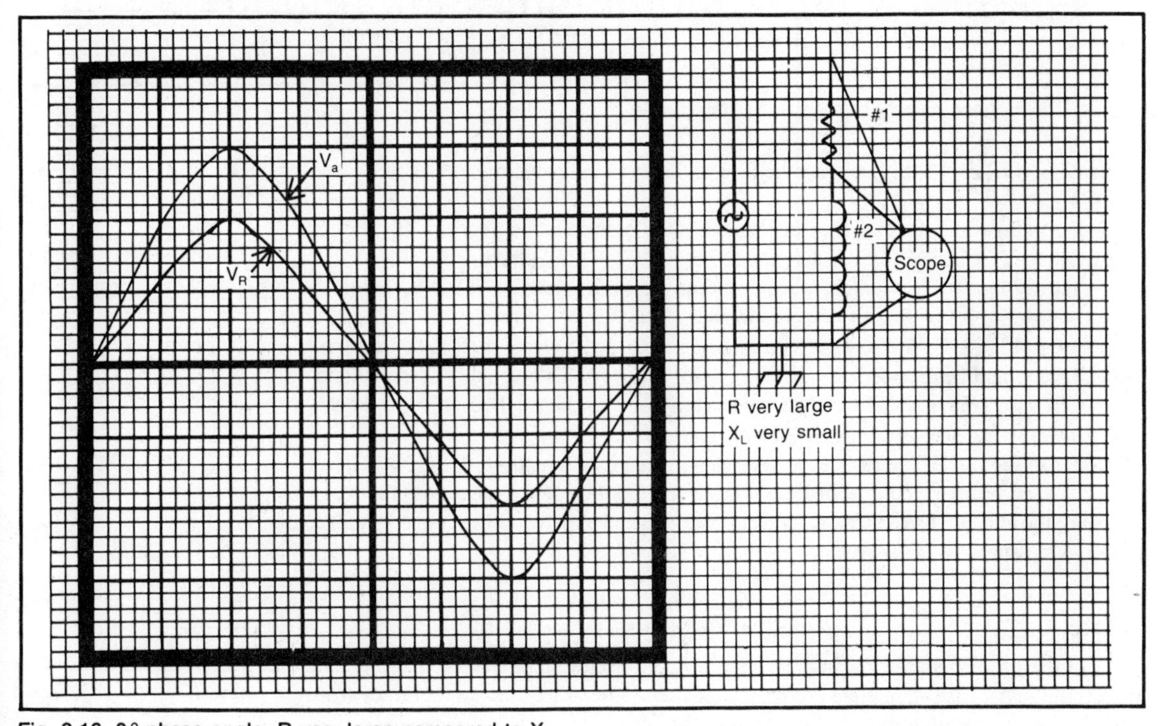

Fig. 8-16. 0° phase angle. R very large compared to X_L.

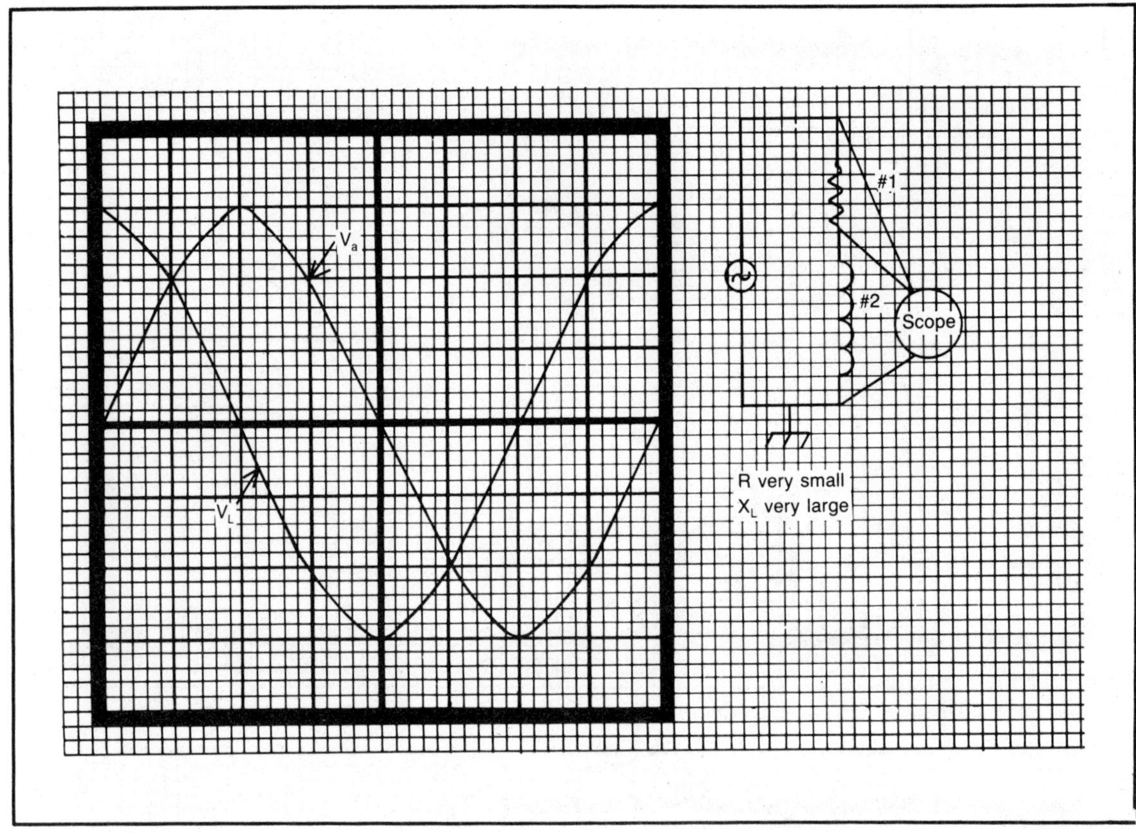

Fig. 8-17. 90° phase angle. R very small compared to X_L.

In Figs. 8-12 and 8-13, the phase shift of 45 degrees indicates the values of X_L and R are equal.

Refer to Fig. 8-14. R is slightly larger than X_L. 8 blocks in 360 degrees. 45 degrees per division. V_L leads V_a by 0.6 divisions. $0.6 \times 45° = 27°$ phase shift.

Refer to Fig. 8-15. R is slightly smaller than X_L. 8 divisions = 360°. V_L leads V_a by 1.4 divisions. $1.4 \times 360 = 63°$ phase shift.

Refer to Fig. 8-16. R is very large in comparison to X_L. Because R is so large, all the voltage would be dropped across the resistor and very little or none across the inductor. Resistors have no phase shift when compared to the applied voltage. Therefore, this drawing shows a 0 degree phase shift.

Refer to Fig. 8-17. The resistor is very small when it is compared to the X_L. Therefore, all of the voltage would be dropped across the inductor and none across the resistor. Since the inductor has a 90 degrees phase shift when compared to the applied voltage, the drawing shows a 90 degrees phase shift.

Practice Problems

Each of the problems are drawings representing an oscilloscope screen display. The waveforms shown are all intended to represent sine waves. Find the number of degrees per division (major divisions) and determine the phase angle.

1.

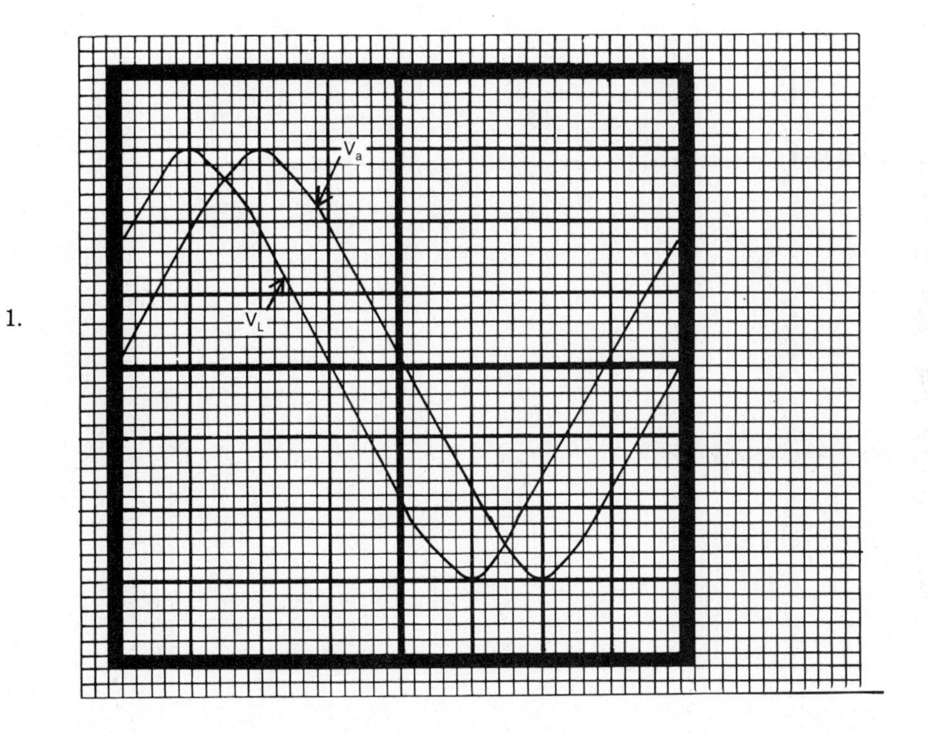

2.

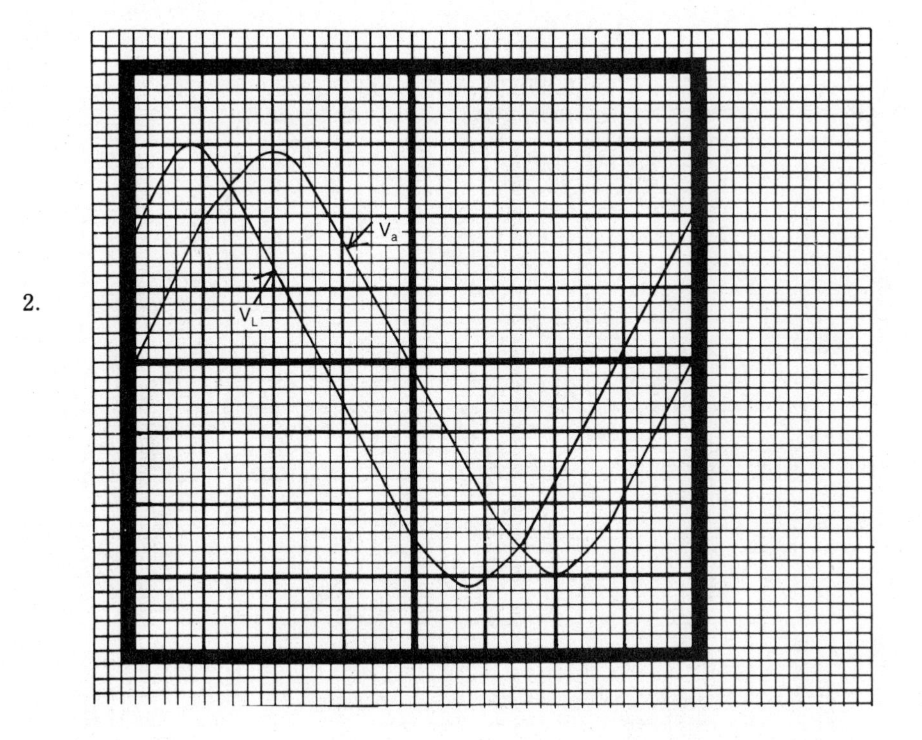

3.

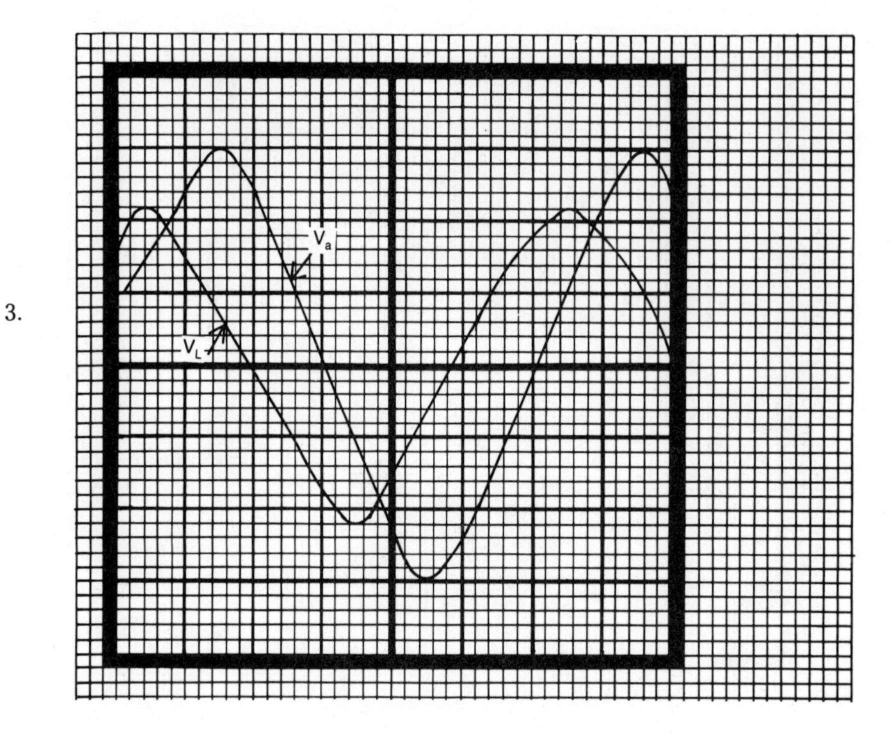

4.

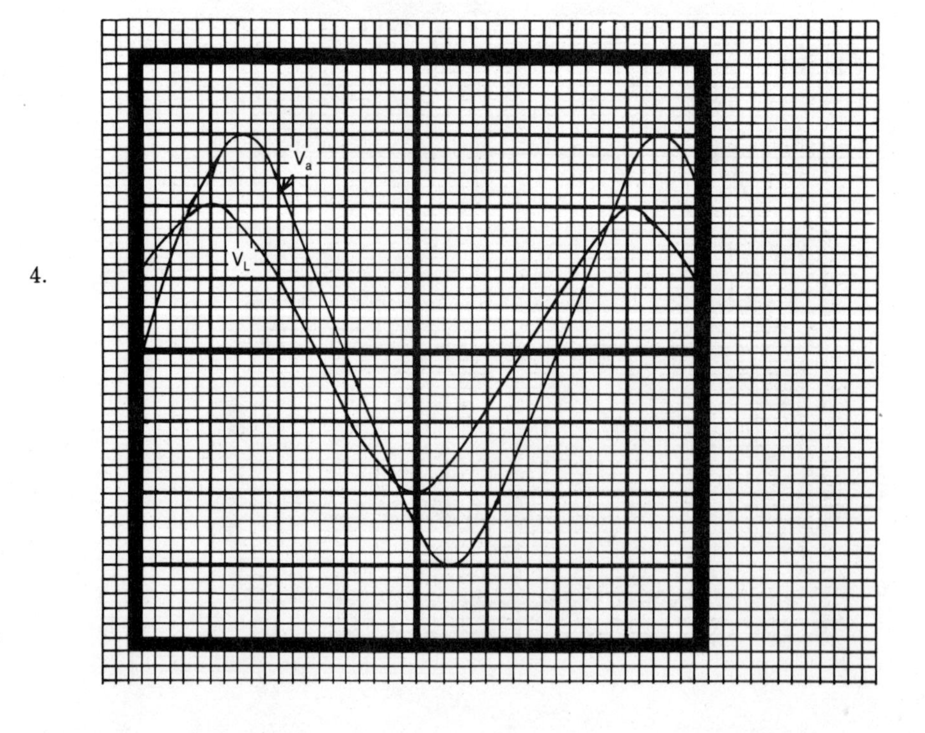

5.

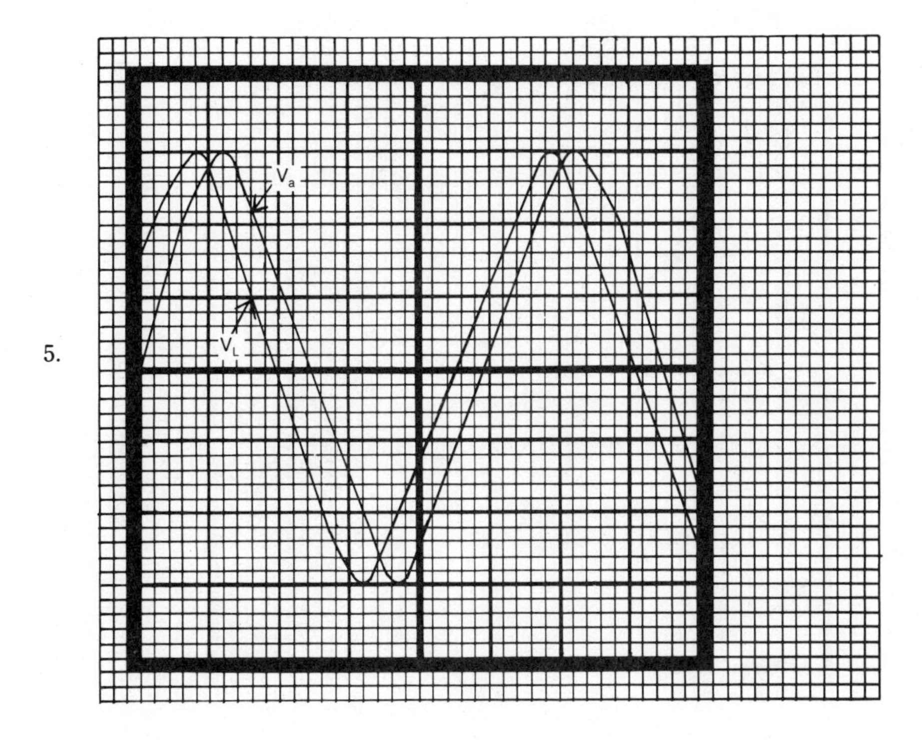

6.

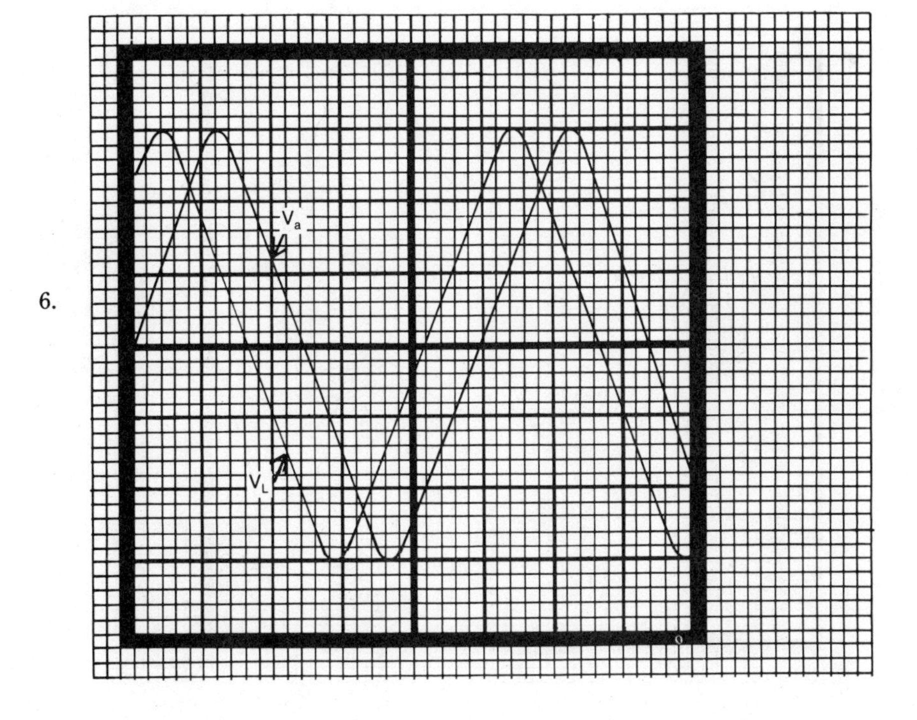

7.

V_a

V_L

8.

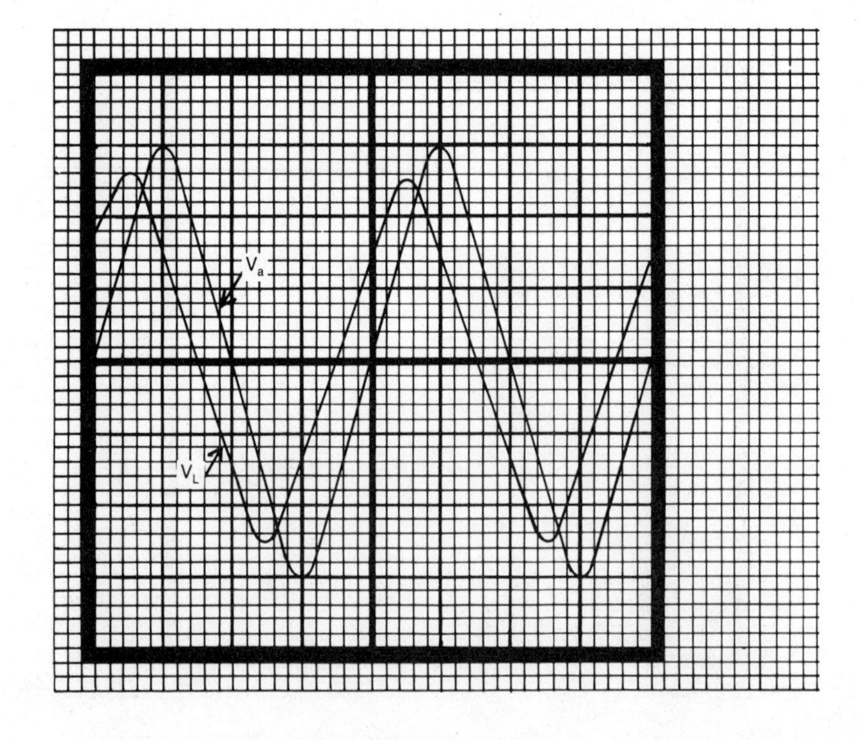

9.

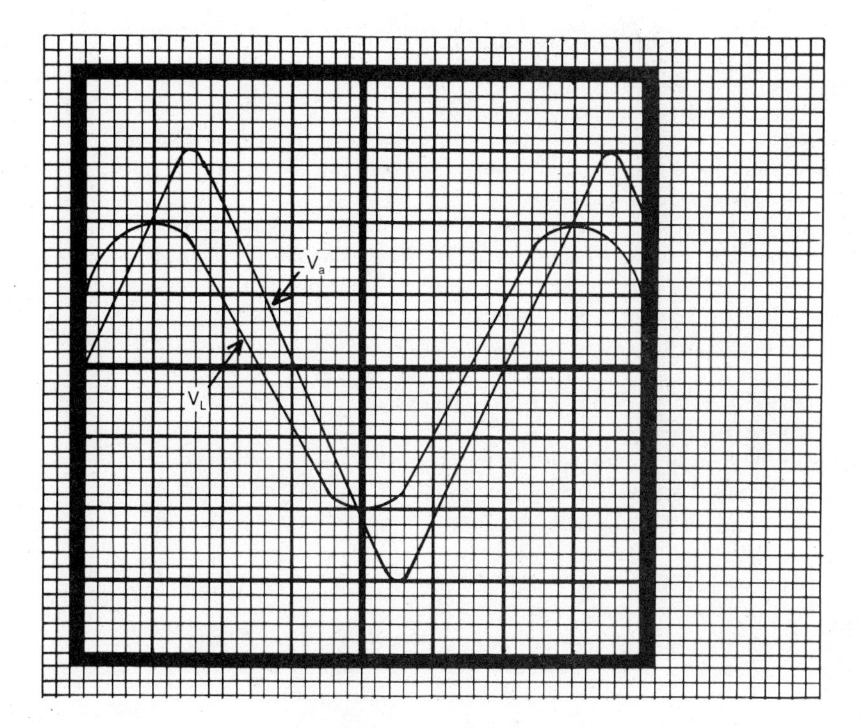

10.

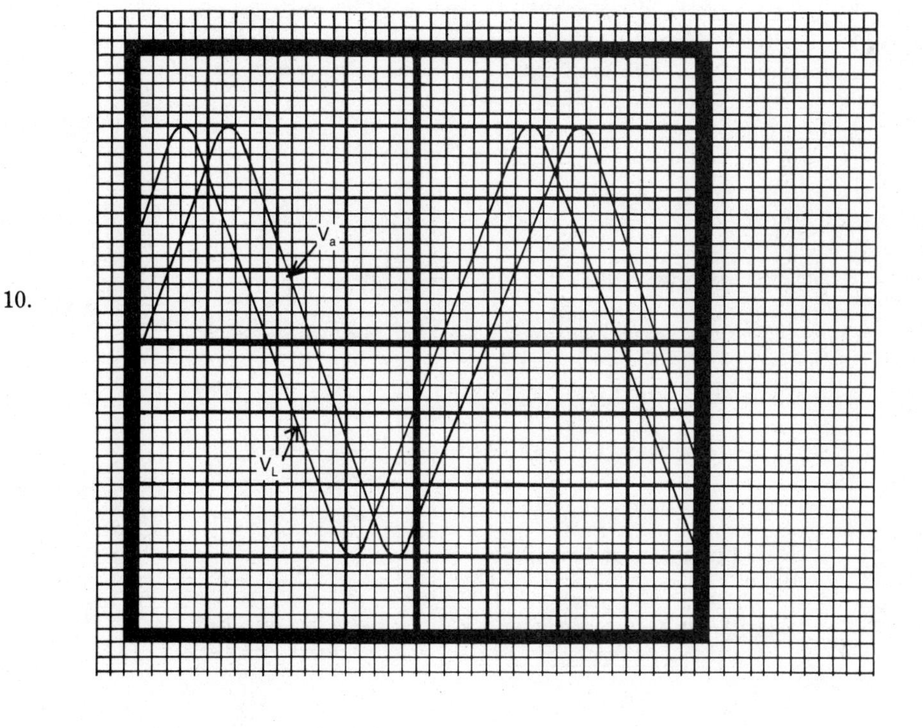

CHAPTER SUMMARY

The term reactance is used to describe the ac resistance of either a capacitor or an inductor. In this chapter, inductive reactance is investigated.

It is seen that inductive reactance varies with the frequency as well as the value of inductance. Inductors can be connected in series or parallel, just as resistors can, to arrive at different values of total reactance.

When an inductor is connected in series with a resistor there is a phase shift between the voltage drops across the two components. The voltage across the inductor always leads the voltage across the resistor by 90 degrees. Another way to say this is to say the current lags the voltage by 90 degrees. A triangle can be formed by the voltages to give the total voltage of the series circuit. Another triangle can be formed to give the total impedance of the series circuit. Voltage and resistance and reactance cannot be simply added as in a dc circuit with only resistance. The adding must be performed using phasor triangles.

In a parallel circuit consisting of an inductor and a resistor, the current splits to the different parallel branches and the current of the inductor will lag the current of the resistor by 90 degrees. This develops a current triangle to solve the circuit, in order to calculate the total current.

Both series and parallel circuits have an operating angle or a phase shift that is the angle theta, Θ, of the triangles used to solve the circuit.

When a circuit contains a reactive component, the power of the circuit is more difficult to calculate. The resistance of the circuit is the only true power dissipated, with the unit watts. The reactive component has reactive power, with a unit of VARS. The total power of the circuit is the apparent power, with a unit of VA.

- Inductive reactance, X_L, is directly related to frequency and inductance.
- In a series circuit, I_L lags V_L, V_R lags V_L always by 90 degrees.
- An inductive circuit will always have a phase shift of between 0 degrees and 90 degrees, depending on circuit values.
- Impedance (Z) is the ac resistance of a circuit containing both resistance and reactance. It has an angle equal to the circuit phase angle.
- The voltage drops of a circuit containing resistance and reactance cannot simply be added. They must be added through vector addition.
- In a parallel circuit, I_L lags I_R by 90 degrees.
- Real power (also called true power) is the power dissipated in a pure resistance, unit is W (watts).
- Reactive power is the power dissipated in pure reactance, unit is VARS (volt-ampere-reactive).
- Apparent power is the power of a circuit containing both resistance and reactance, unit is VA (volt-ampere).
- Power factor is a ratio of the real power to the total power. It is a pure number with no units.
- Measurements made on the scope should be taken where the sine wave crosses the zero reference line.

Summary of Formulas

inductive reactance	$X_L = 2\pi fL$	**Formula 8-1**
series reactances	$X_{LS} = X_{L1} + X_{L2} + X_{L3} + \ldots$	**Formula 8-2**

parallel reactances $\frac{1}{X_{LP}} = \frac{1}{X_{L1}} + \frac{1}{X_{L2}} + \frac{1}{X_{L3}} + \ldots$ **Formula 8-3**

total voltage in a series circuit $V_T = \sqrt{V_R^2 + V_L^2}$ **Formula 8-4**

phase angle of a series circuit, using voltages $\Theta = \tan^{-1} \frac{V_L}{V_R}$ **Formula 8-5**

total impedance of a series circuit $Z = \sqrt{R^2 + X_L^2}$ **Formula 8-6**

phase angle of a series circuit, using resistance $\Theta = \tan^{-1} \frac{X_L}{R}$ **Formula 8-7**

total current of a parallel circuit $I_T = \sqrt{I_R^2 + I_L^2}$ **Formula 8-8**

phase angle of a parallel circuit $\Theta = \tan^{-1} \frac{I_L}{I_R}$ **Formula 8-9**

total impedance of a parallel circuit $Z = \frac{V_a}{I_T}$ **Formula 8-10**

real (or true) power . . . (watts) $P_R = I_R \times E_R$ **Formula 8-11**

real power . . . (watts) $P_R = I^2 \times R$ **Formula 8-12**

real power . . . (watts) $P_R = VI \cos \Theta$ (total voltage) **Formula 8-13**

reactive power . . . (VARS) $P_X = I_X \times E_X$ **Formula 8-14**

reactive power . . . (VARS) $P_X = VI \sin \Theta$ (total voltage) **Formula 8-15**

apparent power . . . (VA) $P_A = VI$ (total voltage) **Formula 8-16**

power factor . . . (no units) $PF = \cos \Theta$ **Formula 8-17**

power factor . . . (no units) $PF = \frac{R}{Z}$ **Formula 8-18**

power factor . . . (no units) $PF = \frac{I_R}{I_T}$ (applies to parallel circuits) **Formula 8-19**

Chapter 9

$$X_C = \frac{1}{2\pi fC}$$

Capacitive Reactance

The previous chapter discusses inductive reactance as being an ac resistance. This same general statement also applies to a circuit containing a capacitor with an ac power source.

It will be demonstrated that many of the statements made about inductive circuits will also apply to capacitors. The big difference will be that capacitive reactance has an inverse relationship to frequency and capacitance, also the phasor triangles will be drawn opposite those of the inductor circuits.

CAPACITIVE REACTANCE

$$X_C = \frac{1}{2\pi fC} \qquad \textbf{Formula 9-1}$$

The factor 2π is a constant term in this formula.
f is the frequency measured in hertz.
C is capacitance measured in farads.
X_C is capacitive reactance measured in ohms.

Notice from the formula, the capacitive reactance varies inversely with the frequency and capacitance value. An inverse relationship means that the X_C is smaller with larger values of either f or C.

- Capacitive reactance, X_C, is indirectly related to frequency and capacitance.

Linear Relationship of X_C vs Frequency and Capacitance

Figure 9-1 shows two curves, one curve plotted with X_C vs capacitance and the other curve with X_C vs frequency.

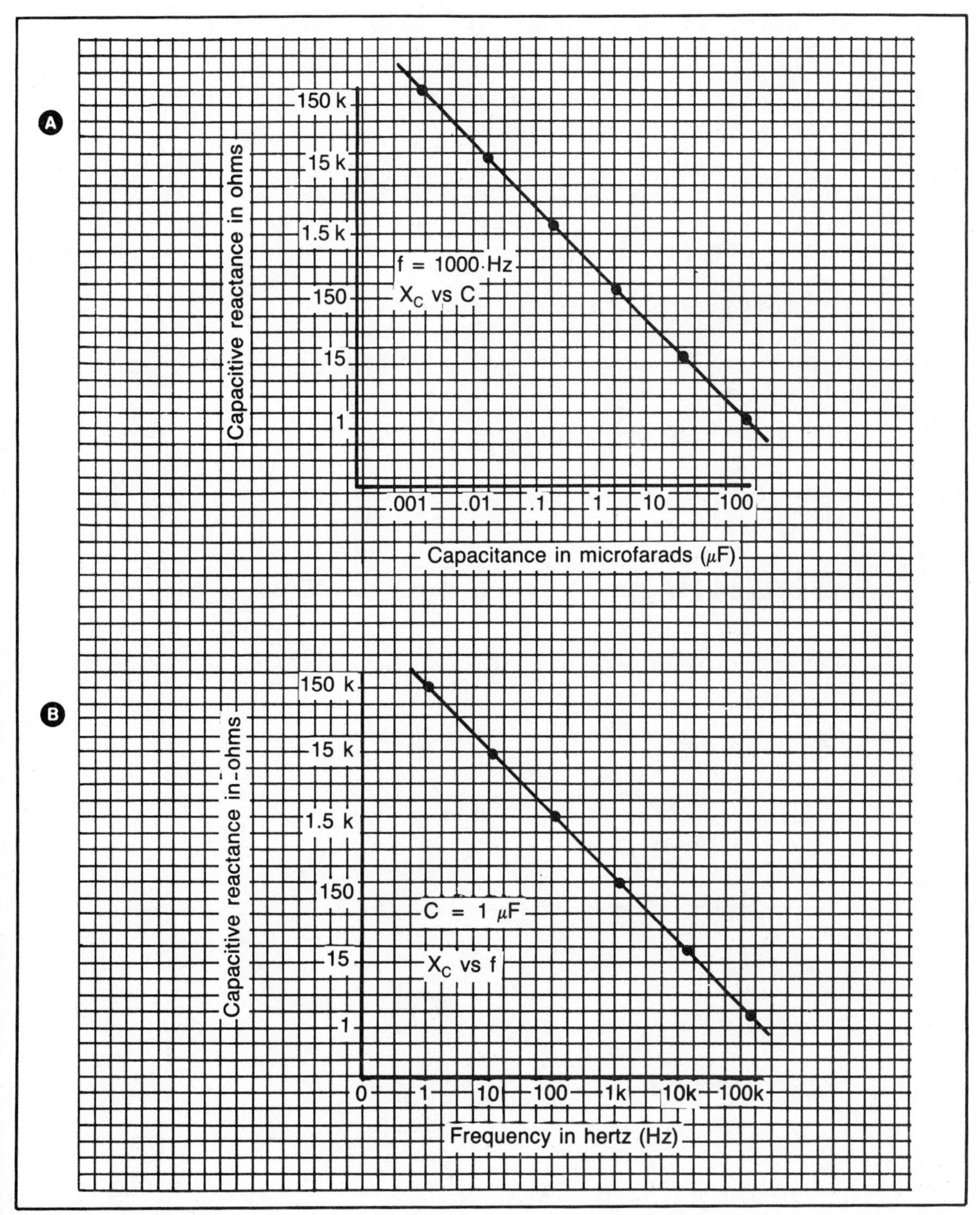

Fig. 9-1. Capacitive reactance versus A. capacitance and B. frequency. A. Capacitance in microfarads (μF). B. Frequency in hertz (Hz).

It has been stated that X_C is inversely (indirectly) related to frequency and capacitance. Looking at the curves, it can be seen that the reactance will decrease when there is an increase in either frequency or capacitance.

Curve 9-1A shows that when the frequency is held constant, the higher value of capacitance will result in the lower value of reactance. Figure 9-1B shows that when the capacitance is held constant, the higher value of frequency will result in the lower value of reactance. It can also be stated that for higher frequencies, a smaller value of capacitor would be satisfactory to produce a certain reactance. Smaller values of capacitance are less expensive and smaller in size than larger capacitances, but the smaller values have less storage capacity.

Sample Calculations of Capacitive Reactance

The following sample calculations are to show how to use the capacitive reactance formula. The practice problems for this section include calculations for some of the points for the curves of Fig. 9-1.

If $C = .22\ \mu F$ and $f = 2500$ Hz, find X_C.

Step 1 $X_C = \dfrac{1}{2\pi fC}$ formula

Step 2 $X_C = \dfrac{1}{2 \times \pi \times 2500\text{ Hz} \times .22\ \mu F}$ substitute values

Step 3 $X_C = 289$ ohms capacitive reactance

If $X_C = 100$ ohms and $f = 10$ kHz, find the capacitance.

Step 1 $X_C = \dfrac{1}{2\pi fC}$ formula

Step 2 $C = \dfrac{1}{2\pi f X_C}$ rearrange to solve for C

Step 3 $C = \dfrac{1}{2 \times \pi \times 10{,}000\text{ Hz} \times 100\ \Omega}$ substitute values

Step 4 $C = .159\ \mu F$ value of capacitance

If $X_C = 100$ ohms and $C = 15\ \mu F$, find the frequency.

Step 1 $X_C = \dfrac{1}{2\pi fC}$ formula

Step 2 $f = \dfrac{1}{2\pi C X_C}$ rearrange to solve for f

Step 3 $f = \dfrac{1}{2 \times \pi \times 15\ \mu F \times 100\text{ ohms}}$ substitute values

Step 4 f = 106 Hz value of frequency

Practice Problems

Find the value of the unknown quantity using the information given.

1. $f = 1000$ Hz, $C = .001\ \mu F$; find X_C
2. $X_C = 1.59\ k\Omega$, $f = 1000$ Hz; find C
3. $X_C = 15.9\ \Omega$, $C = 1\ \mu F$; find f
4. $X_C = 15.9\ k\Omega$, $f = 1000$ Hz; find C
5. $X_C = 159\ \Omega$, $C = 1\ \mu F$; find f
6. $f = 10$ kHz, $C = 150\ \mu F$; find X_C
7. $f = 0$ Hz (dc), $C = 10\ \mu F$; find X_C
8. $f = 100$ MHz, $C = 1\ \mu F$; find X_C
9. $f = 1$ GHz, $C = 1$ pF; find X_C
10. $f = 1$ Hz, $C = 1$ pF; find X_C

SERIES OR PARALLEL CAPACITIVE REACTANCE

Whenever two or more of the same type of reactance are connected together, the total value of the reactance is calculated the same as resistors in a dc circuit.

series X_C $$X_{Cs} = X_{C1} + X_{C2} + X_{C3} \ldots$$ **Formula 9-2**

parallel X_C $$\frac{1}{X_{Cp}} = \frac{1}{X_{C1}} + \frac{1}{X_{C2}} + \frac{1}{X_{C3}} \cdots$$ **Formula 9-3**

When it is necessary to calculate voltage drops, with only reactance in the circuit, follow Ohm's law. This only applies when there is only reactance in the circuit, not resistance and reactance.

Figure 9-2 shows a series circuit with two capacitors connected to a 10 volt ac power supply. The largest voltage drop is across the largest capacitive reactance. Keep in mind, however, the largest capacitive reactance results from the smallest value of capacitance.

Figure 9-3 shows a parallel circuit with two capacitors connected in parallel across and ac power supply. Notice, the larger current flows through the smallest value of reactance.

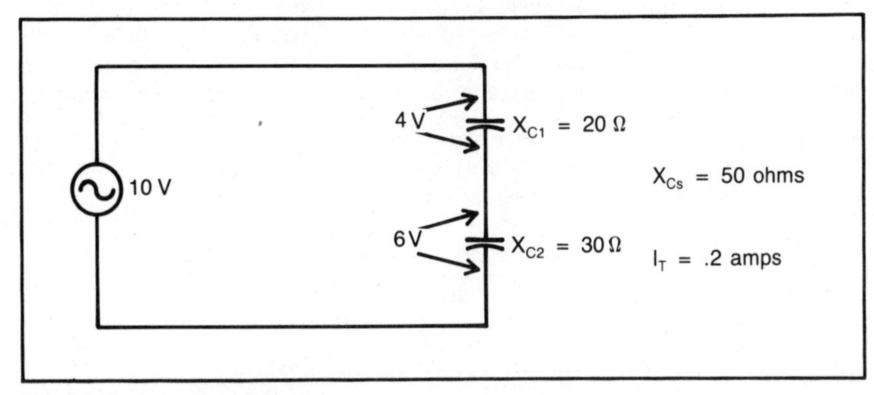

Fig. 9-2. Series capacitive reactance.

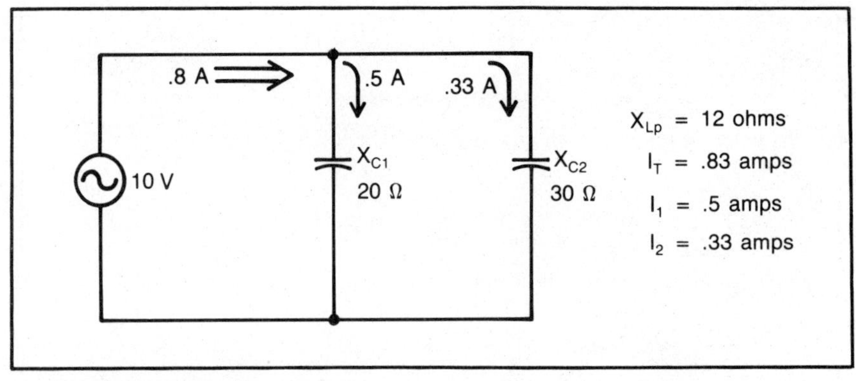

Fig. 9-3. Parallel capacitive reactance.

Because the two sample circuits, Figs. 9-2 and 9-3, contain only capacitance and no resistance or inductance, the values of voltage drops and current can be calculated based on Ohm's law. If the circuit contains any other component, it is necessary to use vector addition.

SERIES X_C AND R

In a series circuit, the current is the same throughout the series circuit, regardless of the types of components contained in the circuit.

In a circuit containing capacitance, the capacitor must charge and discharge. This charging, discharging process causes a delay or phase shift in the circuit. The current in the resistor is considered the reference point. The resistive voltage drop will be exactly the same phase as the circuit current. The current in the capacitor will be the same also since it is a series circuit. The capacitive voltage drop, however, will be 90 degrees out of phase with the current. This will be a lagging voltage or a leading current.

- In a series circuit, I_C leads V_C, V_R leads V_C always by 90 degrees.

Refer to Fig. 9-4 for the sine wave analysis of a series circuit containing capacitance and resistance. Figure 9-5 shows the vector analysis of the same circuit. Notice how the resistor voltage is plotted in phase with the current (Fig. 9-4) where the capacitive voltage is shown as lagging the resistor voltage by 90 degrees. The total voltage is the vector sum of these two. Figure 9-5 shows the vectors with the capacitive voltage drop plotted straight down. The line drawn down represents a phase angle of −90 degrees. The circuit phase angle (also called phase shift or operating angle) is a negative angle. Compare this drawing to the same drawing for an inductive circuit (Fig. 8-5).

- The voltage vectors for inductance and capacitance are opposite.
- The operating angle of a circuit with a capacitor in series with a resistor will be between 0 degrees and −90 degrees, depending on circuit values.

Calculating the Triangle

Figure 9-5 uses the same schematic diagram as Fig. 9-4.

V_R is plotted on the horizontal 0 degrees axis.
V_C is plotted down on the vertical −90 degrees axis.
V_T is plotted to connect the two vectors.

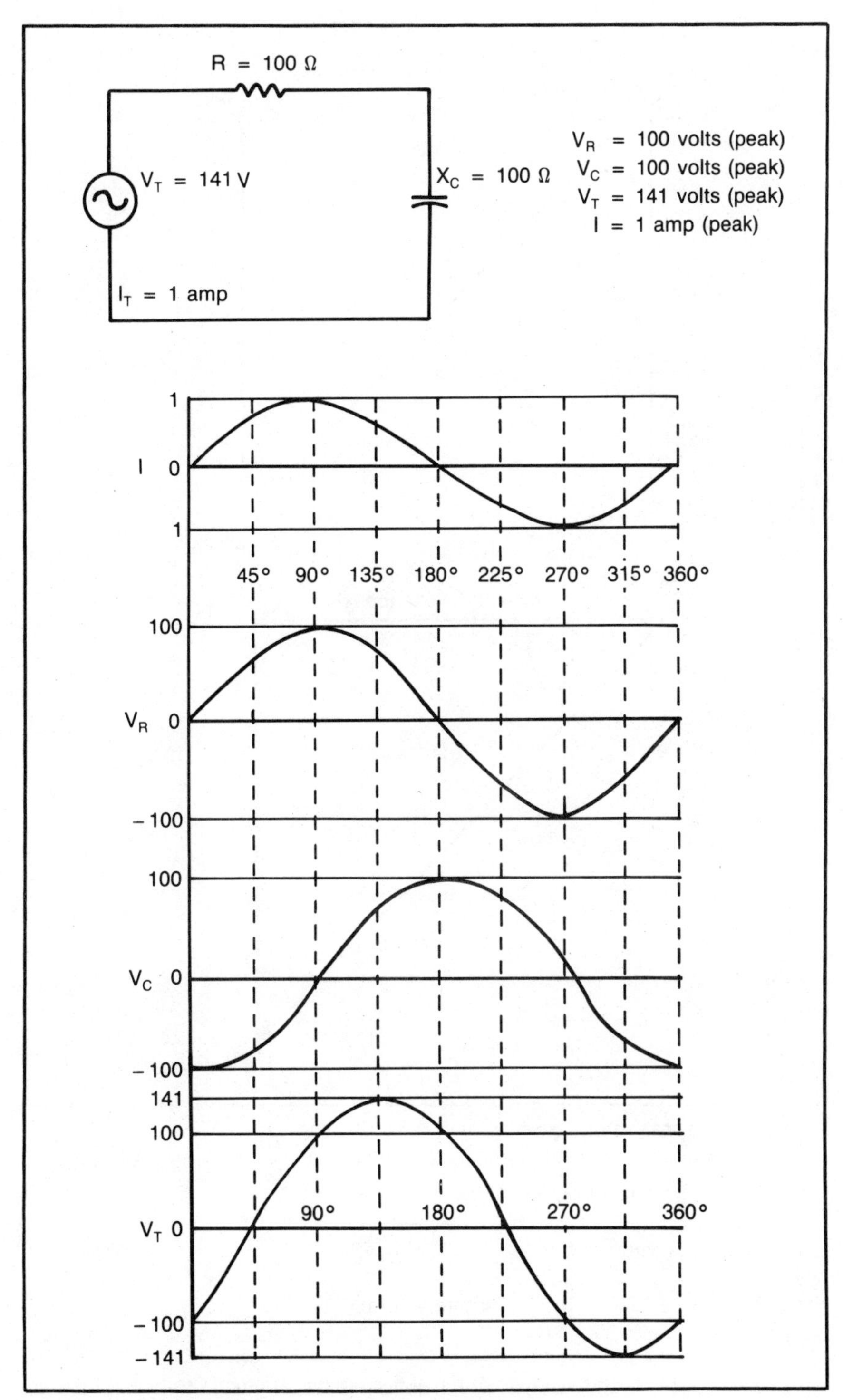

Fig. 9-4. Sine wave analysis of a series X_C and R circuit.

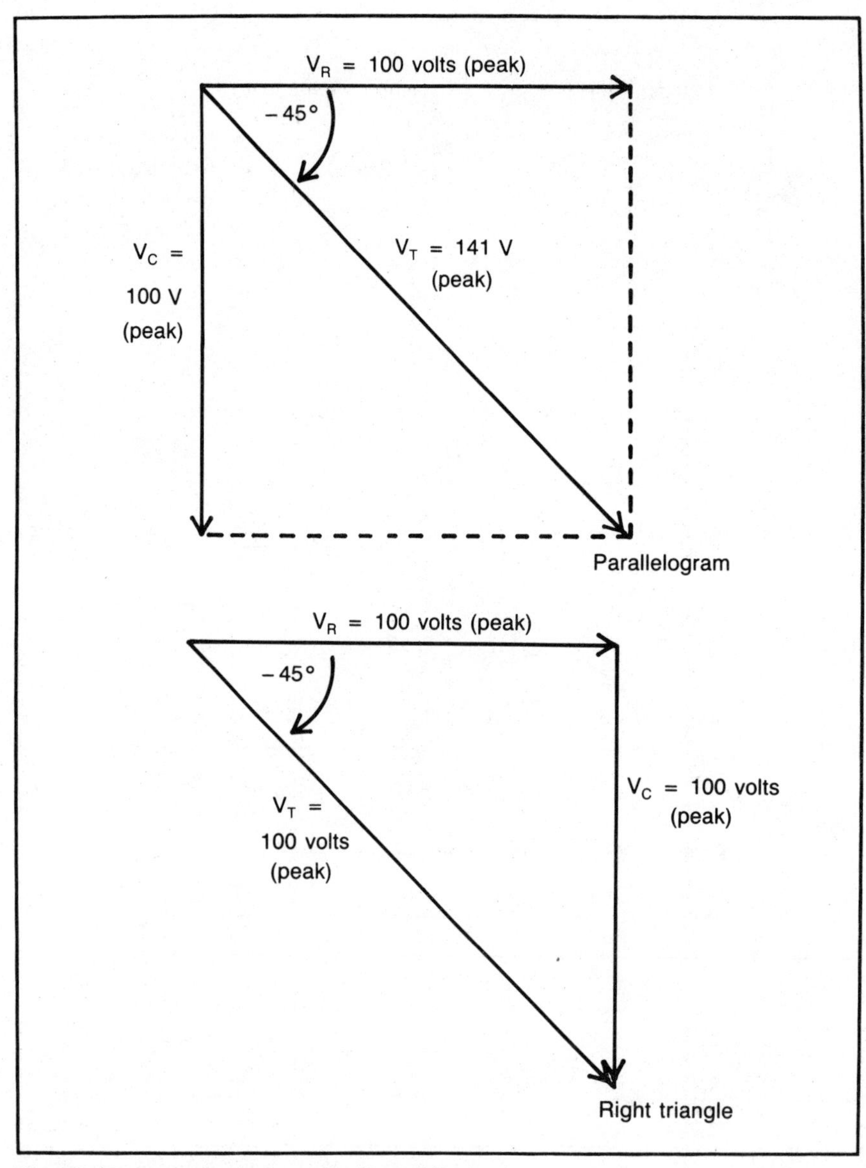

Fig. 9-5. Vector analysis of a series X_C and R circuit.

Calculate the magnitude, length, of V_T

total voltage in a series circuit $V_T = \sqrt{V_R^2 + V_C^2}$ **Formula 9-4**

Step 1 $V_T = \sqrt{100^2 + 100^2}$ substitute values
Step 2 V_T = 141 volts solve

Calculating the operating angle of the series circuit, in other words, the hypotenuse, V_T, of the voltage triangle forms the angle, theta, Θ.

Step 1 $\Theta = \tan^{-1} \dfrac{V_C}{V_R}$ operating angle theta, Θ **Formula 9-5**

Step 2 $\Theta = \tan^{-1} \dfrac{-100\ V}{100\ V}$ substitute values

Step 3 $\Theta = -45°$ solve

Note The angle will come out negative on the calculator if V_C is used as a negative voltage.

In the example shown above, the triangle has a 45 degree angle. This could have been predicted by the fact that the resistance and the capacitive reactance are equal. Therefore, the voltage drop across the resistor and capacitor would be the same value with 90 degrees phase difference between them. Having the two sides of the right triangle equal results in a 45 degree triangle.

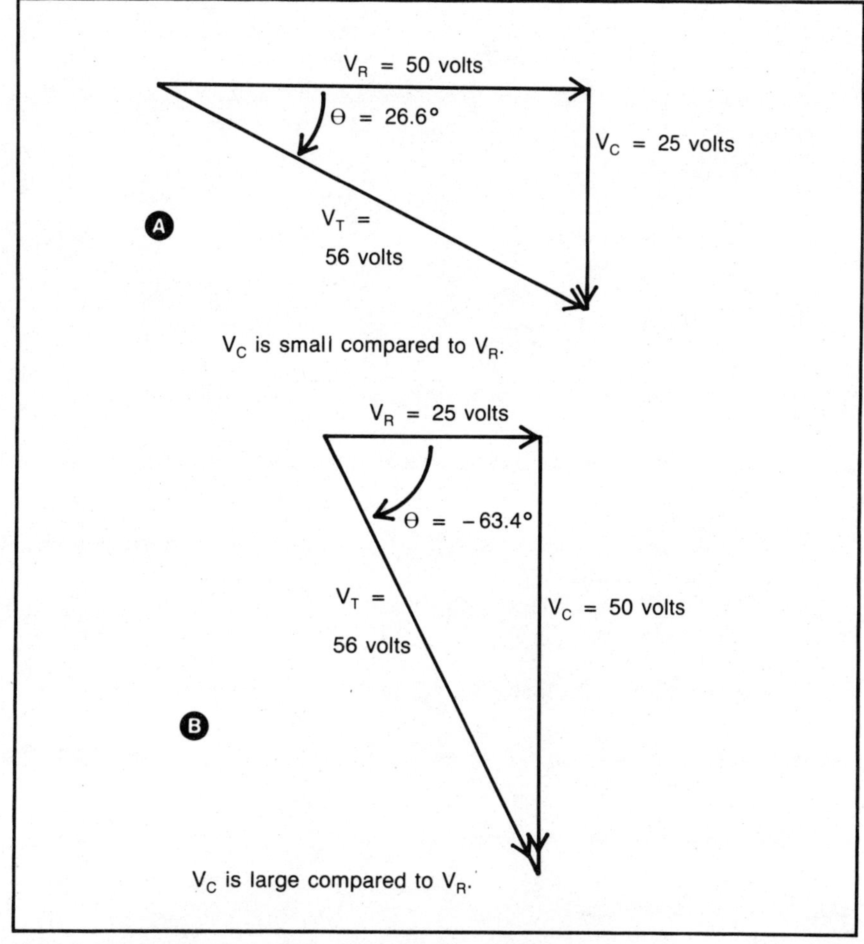

Fig. 9-6. Voltage triangles showing the difference between large and small V_C compared to V_R. A. V_C is small compared to V_R. B. V_C is large compared to V_R.

When the resistance and capacitive reactance are not equal, it is possible to predict if the angle will be less than or greater than 45 degrees. Circuits containing capacitive reactance will have a negative angle.

Figure 9-6 shows two triangles. The first has the resistance larger than the reactance, therefore, the phase shift is a negative angle less than −45 degrees. The second triangle has the resistance smaller than the capacitive reactance, therefore, the operating angle will have a negative angle, with a magnitude greater than −45 degrees.

Calculating the Impedance in a Series Circuit

The impedance of a series circuit with a reactive component is the same as the total resistance of a dc circuit. Impedance is the total ac resistance of a circuit. It will have an angle equal to the operating angle of the circuit. In other words, the operating angle can be found using either the voltage triangle or the impedance triangle.

Figure 9-7 shows a series circuit with its impedance and voltage triangles. Notice both triangles are plotted in the negative direction, which is opposite the triangles for an inductive circuit. See Fig. 8-7 to compare the inductive circuit.

Calculating a Complete Series Circuit

The circuit shown in Fig. 9-7 was originally given with the capacitance value and the frequency. It was necessary to calculate the capacitive reactance.

Calculate the X_C.

Step 1 $X_C = \dfrac{1}{2\pi fC}$ formula

Step 2 $X_C = \dfrac{1}{2 \times \pi \times 100\text{ Hz} \times 10.6\ \mu\text{F}}$ substitute values

Step 3 $X_C = 150$ ohms

Calculate impedance (Z) using the impedance triangle. Plot R on the horizontal and X_C down.

impedance of a series circuit $Z = \sqrt{R^2 + X_C^{\,2}}$ **Formula 9-6**

Step 1 $Z = \sqrt{100^2 + 150^2}$ substitute values
Step 2 $Z = 180$ ohms find the square

Calculate the phase angle, theta, Θ.

operating angle of a series circuit $\Theta = \tan^{-1} \dfrac{X_C}{R}$ **Formula 9-7**

Step 1 $\Theta = \tan^{-1} \dfrac{-150}{100}$ substitute values

Step 2 $\Theta = -56.3°$ operating angle

Calculate the total current using the calculated impedance and the applied voltage.

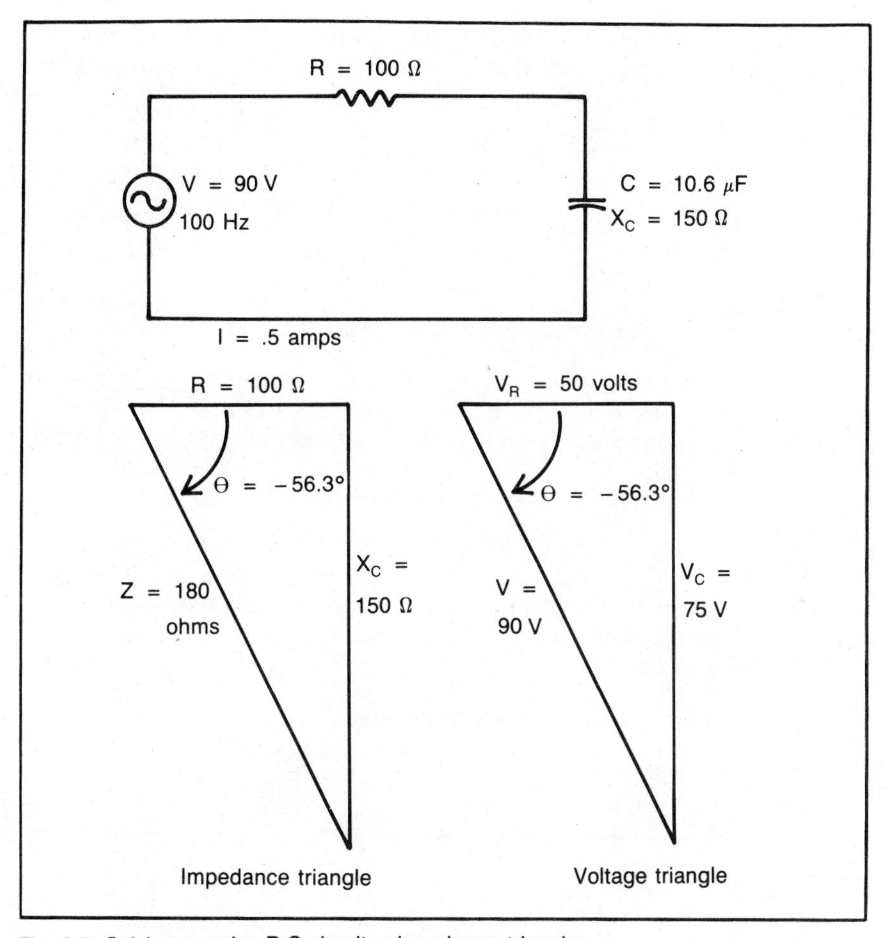

Fig. 9-7. Solving a series R-C circuit using phasor triangles.

Step 1 $I = \frac{V}{Z}$ Ohm's law modified

Step 2 $I = \frac{90\ V}{180\ \Omega}$ substitute values

Step 3 $I = .5\ A$ total current

Use the current to find the voltage drops across the resistor and capacitor.

Step 1 $V_R = I \times R$ using Ohm's law
Step 2 $V_R = .5\ A \times 100$ ohms substitute values
Step 3 $V_R = 50\ V$ resistive voltage drop

Step 1 $V_C = .5\ A \times 150$ ohms substitute values
Step 2 $V_C = 75\ V$ capacitive voltage drop

Refer to Fig. 9-7 to see the voltage drops plotted in the voltage triangle. The hypotenuse of the voltage triangle should calculate to be equal to the voltage given as the applied voltage.

Step 1 $V = \sqrt{V_R^2 + V_C^2}$ formula

Step 2 $V = \sqrt{50^2 + 75^2}$ substitute values

Step 3 $V = 90\ V$ checks with given voltage

Summary for a Series Circuit

- Calculate the capacitive reactance if it is not already given.
- Calculate the impedance, Z, using the impedance triangle. $Z = \sqrt{R^2 + X_C^2}$
- Calculate the operating angle, theta, Θ, based on the impedance triangle. The angle can also be found using the voltage triangle. $\Theta = \tan^{-1} \dfrac{X_C}{R}$
- Calculate the total circuit current using the applied voltage and the impedance.
- Use the current to calculate the voltage drops across each component. Use Ohm's law.
- Compare the voltage drops to the given applied voltage by using vector addition with the voltage triangle. Theta can also be found using the voltage triangle, rather than the impedance triangle.

Practice Problems

Use the two schematics to calculate the answers to the problems.

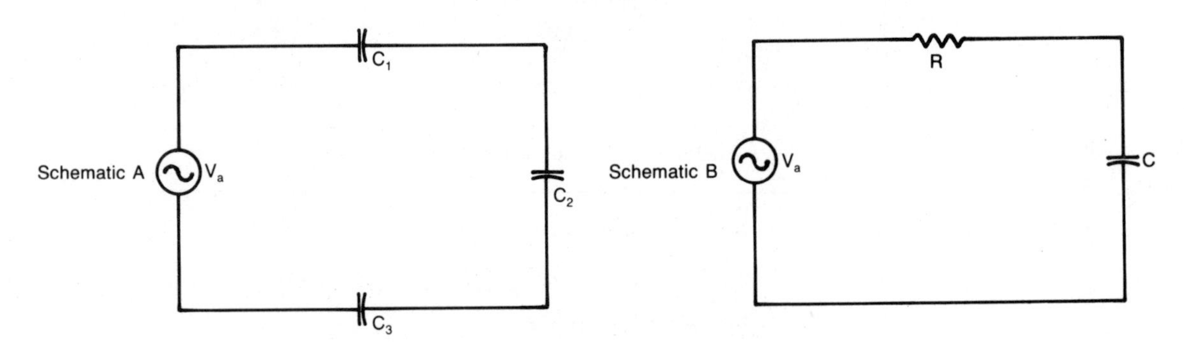

Use schematic A for problems 1 and 2. $V_a = 100\ V/60\ Hz$

Find: C_T, X_{C1}, X_{C2}, X_{C3}, and X_{CT}

1. $C_1 = 1\ \mu F$, $C_2 = .01\ \mu F$, $C_3 = .1\ \mu F$
2. $C_1 = 47\ \mu F$, $C_2 = 150\ \mu F$, $C_3 = 100\ \mu F$

Use schematic B to answer problems 3 through 8. $V_a = 100\ V$

Find: Z, I, V_R, V_C, Θ, draw the impedance triangle.

3. $R = 100$ ohms, $X_C = 100$ ohms

4. $R = 25$ ohms, $X_C = 50$ ohms
5. $R = 10$ ohms, $X_C = 100$ ohms
6. $R = 75$ ohms, $X_C = 5$ ohms
7. $R = 75$ ohms, $X_C = 25$ ohms
8. $R = 0$ ohms, $X_C = 50$ ohms

Use schematic B to answer problems 9 and 10.

Find: R, X_C, Z, V_a, Θ, draw voltage triangle.

9. $V_R = 20$ V, $V_C = 40$ V, $I = .25$ A
10. $V_R = 25$ V, $V_C = 15$ V, $I = .333$ A

PARALLEL X_C AND R

The rules of a parallel circuit state that the voltage throughout a parallel circuit is the same and the current will divide to the individual branches. Because the voltage is the same throughout the parallel circuit, there cannot be any phase shift. There can however, be a phase shift in the current to the individual components.

Because a capacitor causes the voltage to have a lagging relationship, it makes sense then that the current will be a leading current when compared to the current of a resistor in parallel.

- In a parallel circuit, I_C leads I_R by 90 degrees.

Figure 9-8 shows the sine wave analysis of a circuit with equal values of R and X_C. The equal values have been chosen to simplify the drawings, using a 45 degree phase angle. Notice the resistive current is in phase with the applied voltage, as would be expected since resistors do not cause any phase shift. The capacitive current is then shown leading the resistive current by 90 degrees.

Because the two currents are out of phase, the total current must be between the two currents, with its phase shift. Each point along the total current line is found by simply adding the individual points along the sine waves of the two other currents. The resultant total current will have a 45 degree phase shift, as shown, and this phase shift will be leading the applied voltage. The phase angle is therefore considered a positive angle.

Figure 9-9 is the vector analysis of a parallel R and X_C circuit. It is of the same circuit used in Fig. 9-8. Figure 9-9A is the parallelogram drawing of the vectors and Fig. 9-9B is the right triangle method of drawing the vectors. The only difference between the two drawings is the point at which the capacitor current vector has its starting point. In the parallelogram, it starts at the same point as the resistor vector and in the triangle, it starts at the tail end of the resistor vector. The triangle method is the preferred method in this book, although it is a matter of personal preference.

Figure 9-9 shows the capacitor current is plotted up to show the leading condition. Resistive current is always plotted on the horizontal, to the right. The phase angle is a positive angle.

Calculating the hypotenuse of the right triangle is the square root of the sum of the squares formula and the phase angle is the inverse tangent function as follows:

total current for a capacitor in parallel with a resistor

$$I_T = \sqrt{I_R^2 + I_C^2}$$

Formula 9-8

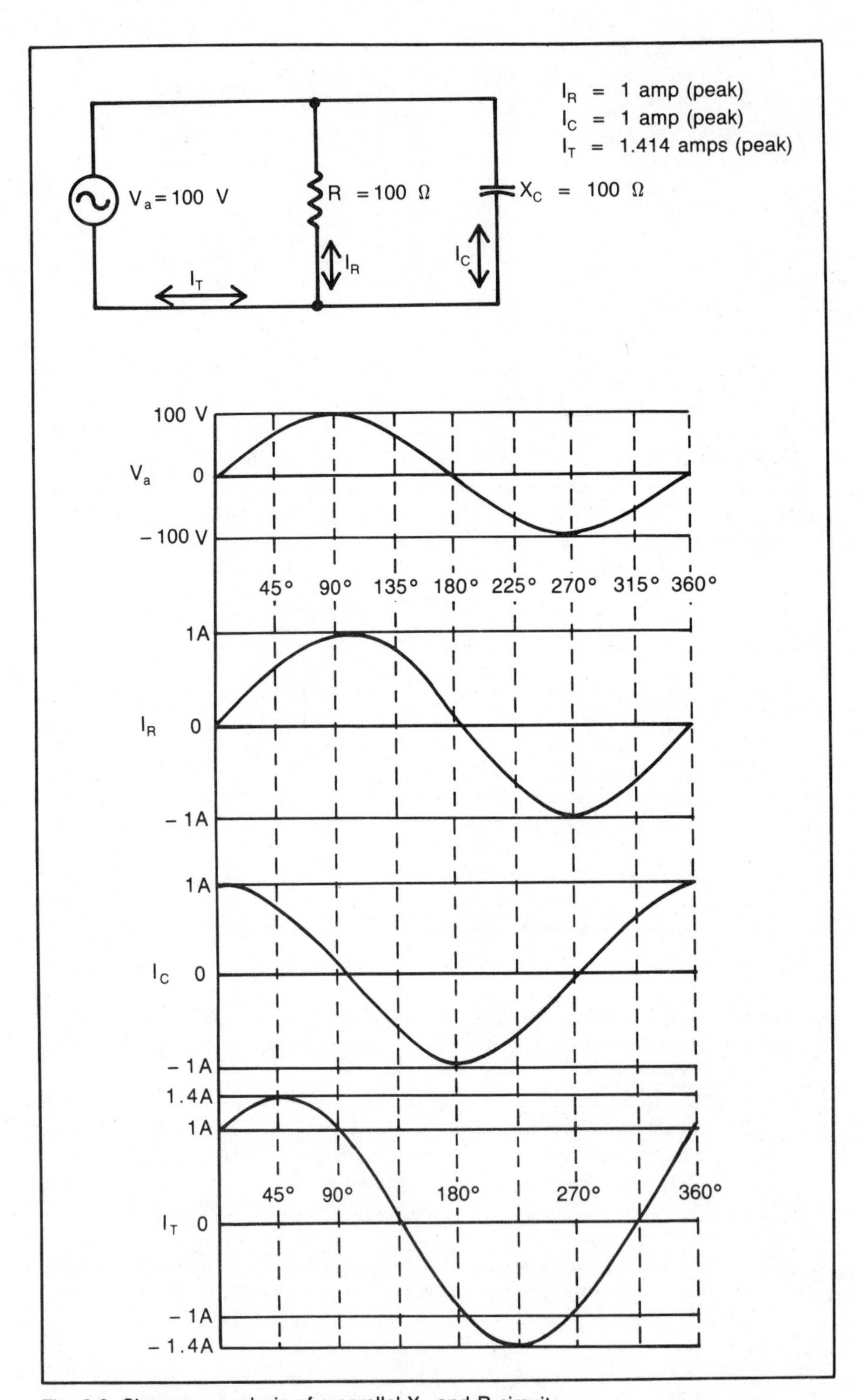

Fig. 9-8. Sine wave analysis of a parallel X_C and R circuit.

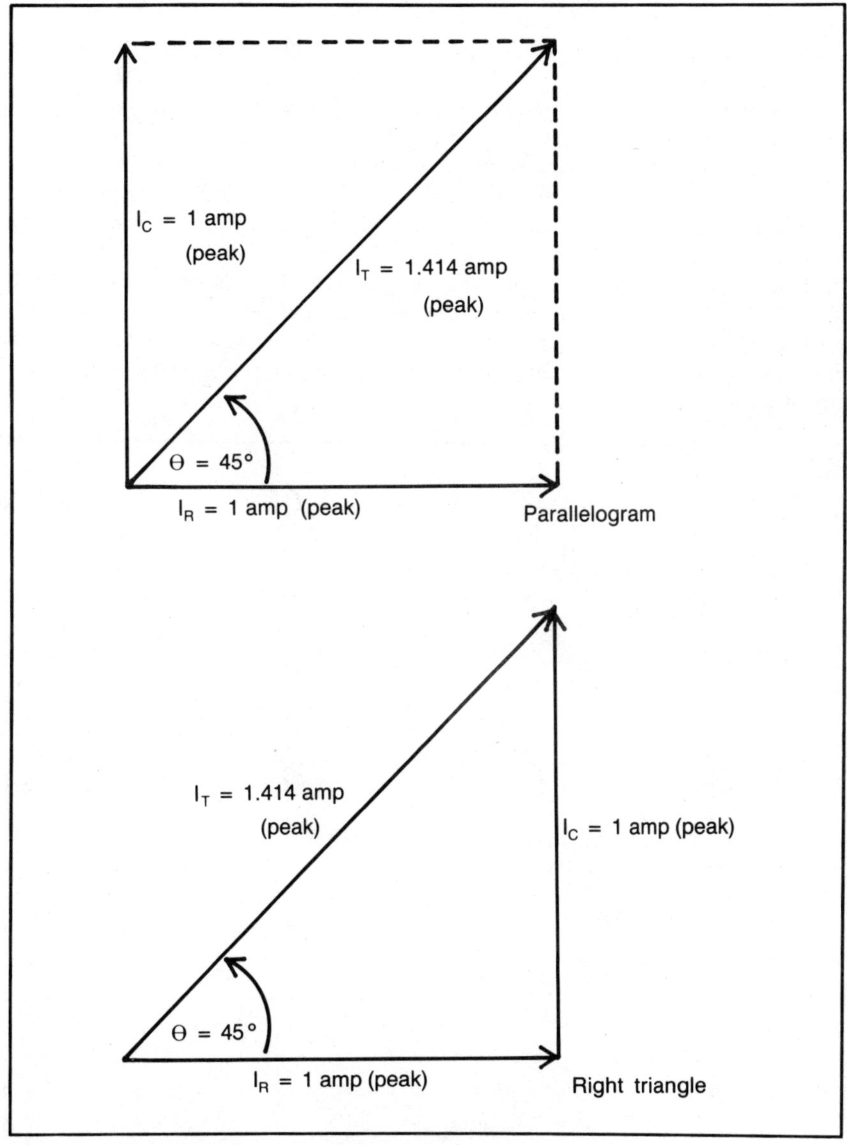

Fig. 9-9. Vector analysis of a parallel X_C and R circuit.

phase angle of an R-C parallel circuit $\quad \Theta = \tan^{-1} \dfrac{I_C}{I_R}$ **Formula 9-9**

Because the total current determines the phase angle of a parallel circuit, the impedance is determined by using Ohm's law with the total current and the applied voltage.

impedance of a parallel R-C circuit $\quad Z = \dfrac{V}{I_T}$ **Formula 9-10**

Figure 9-10 shows two different triangles to demonstrate the difference between an I_C larger than I_R and an I_C smaller than I_R. The larger phase angle, Fig. 9-10A, is caused by a larger amount of I_C. The smaller phase angle results when the resistive current is larger.

To continue the analysis of the larger and smaller currents shown in Fig. 9-10, consider which one would have the larger or smaller value of capacitance. Let's take the larger capacitive current. A larger current means the value of the capacitive reactance must be smaller. A smaller value of reactance means the value of capacitance is larger, assuming there has been no change in frequency. Another way to say it, would be to say the capacitance value stays the same and the frequency is caused to vary. The larger the frequency, the smaller the capacitive reactance. Therefore, the two triangles shown in Fig. 9-10 could represent a circuit where the values of resistance and capacitance are not changed, only the frequency changes. Figure 9-10B would be the current triangle for a parallel R-C circuit with a low frequency and Fig. 9-10A would be the same circuit for a higher frequency.

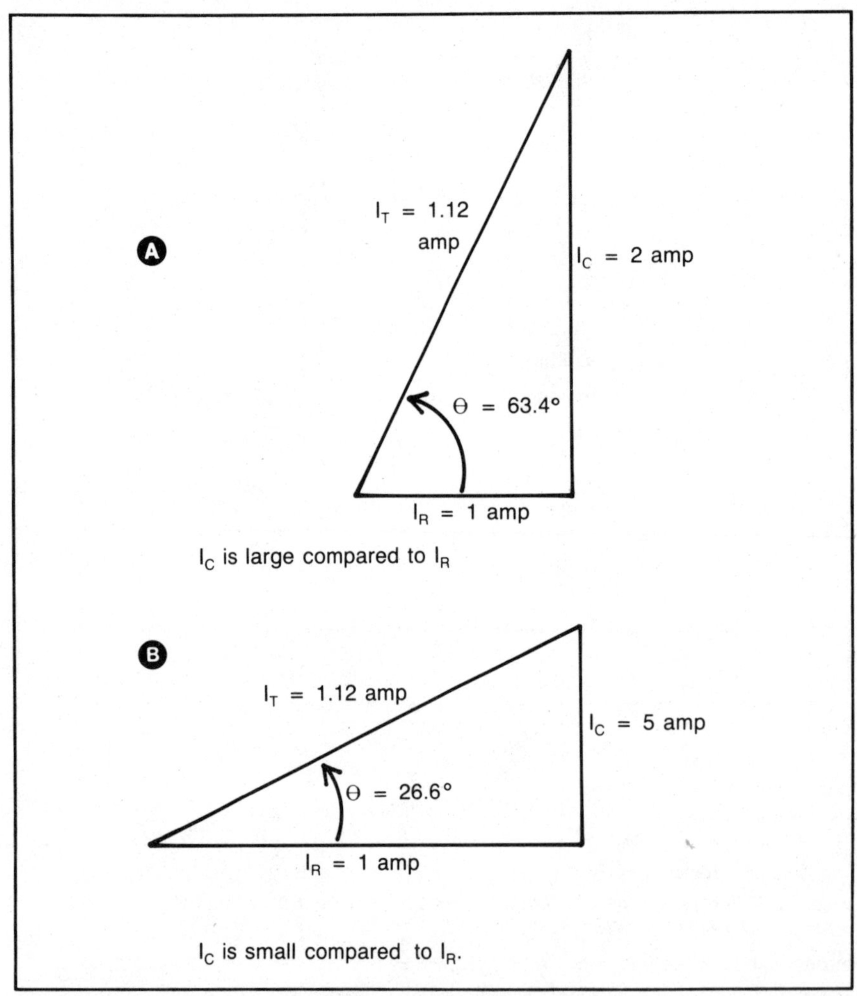

Fig. 9-10. Current triangles showing the difference between large and small I_C compared to I_R. A. I_C is large compared to I_R. B. I_C is small compared to I_R.

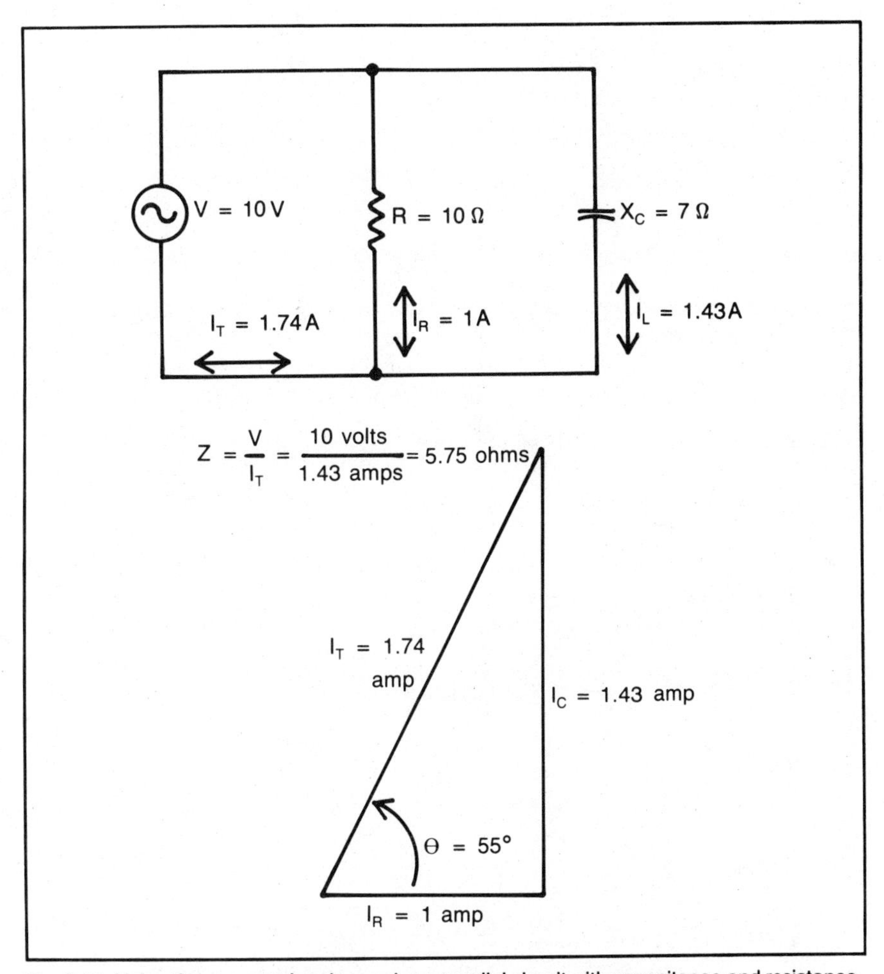

Fig. 9-11. Using the current triangle to solve a parallel circuit with capacitance and resistance.

Calculating a Parallel Circuit

Figure 9-11 shows a parallel R-C circuit and the current triangle used in performing the calculations. It is very important to keep in mind the fact that the total current is the vector sum of the branch current, not simply an arithmetic sum.

In a parallel circuit, the applied voltage is the same across all branches of the parallel circuit.

Calculate the resistive current, I_R.

Step 1 $I_R = \frac{V}{R}$ use Ohm's law

Step 2 $I_R = \frac{10\ V}{10\ \Omega}$ substitute values

Step 3 I_R = 1 amp I_R is plotted in the triangle at 0°

Calculate the capacitive current, I_C.

Step 1 $I_C = \frac{V}{X_C}$ Ohm's law

Step 2 $I_C = \frac{10\ V}{7\ \Omega}$ substitute values

Step 3 $I_C = 1.43$ amps plot at $+90°$ in the triangle

Calculate the hypotenuse of the current triangle, the total current, using Pythagorean's theorem.

Step 1 $I_T = \sqrt{I_R^2 + I_C^2}$ Pythagorean theorem

Step 2 $I_T = \sqrt{1^2 + 1.43^2}$ substitute values

Step 3 $I_T = 1.74$ A total circuit current

Calculate the operating angle, theta, Θ.

Step 1 $\Theta = \tan^{-1} \frac{I_C}{I_R}$ formula

Step 2 $\Theta = \tan^{-1} \frac{1.43\ A}{1\ A}$ substitute values

Step 3 $\Theta = 55°$ operating angle

Calculate the total impedance, Z, using the calculated total current and the given applied voltage.

Step 1 $Z = \frac{V}{I_T}$ Ohm's law

Step 2 $Z = \frac{10\ V}{1.74\ A}$ substitute values

Step 3 $Z =$ 5.75 ohms total impedance of a parallel circuit should be less than the smallest resistance, or reactance, in the parallel branches

Summary for a Parallel Circuit

- Calculate the current for each of the parallel branches, using Ohm's law.
- Calculate the total current using the branch currents to form the current triangle.
- Calculate the operating angle based on the current triangle. The angle will be positive for a capacitive circuit.
- Calculate the total impedance using the total current and the applied voltage.

Practice Problems

Use the schematics to calculate the answers to the problems.

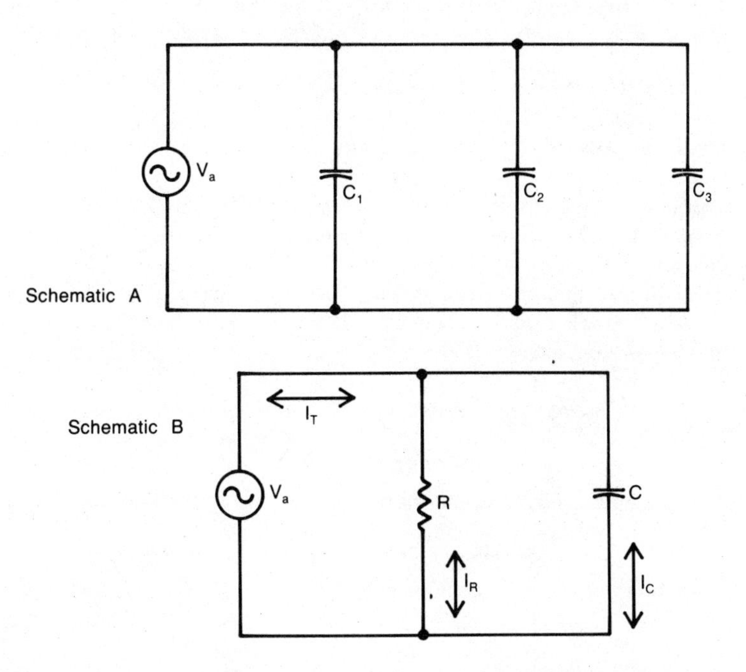

Use schematic A to answer problems 1 and 2. V_a = 100 V/100 Hz

Find: C_T, X_{C1}, X_{C2}, X_{C3}, X_{CT}

1. C_1 = 1 μF, C_2 = 2 μF, C_3 = 3 μF
2. C_1 = 50 μF, C_2 = 100 μF, C_3 = 75 μF

Use schematic B to answer questions 3 through 8. V_a = 100 V.

Find: Z, I_T, I_R, I_C, Θ, draw the current triangle.

3. R = 100 ohms, X_C = 100 ohms
4. R = 25 ohms, X_C = 50 ohms
5. R = 10 ohms, X_C = 100 ohms
6. R = 75 ohms, X_C = 5 ohms
7. R = 75 ohms, X_C = 25 ohms
8. R = 5 ohms, X_C = 5 ohms

Use schematic B to answer questions 9 and 10. V_a = 100 V

Find: R, X_C, Z, I_T, Θ, draw current triangle.

9. I_R = 1 A, I_C = 3 A
10. I_R = 25 mA, I_C = 10 mA

POWER IN A REACTIVE CIRCUIT

Power in a reactive circuit is the same whether it is an inductive circuit, as in Chapter 8, or it is a capacitive circuit. In fact, all the calculations are the same. Because it is a fairly important concept, the key points are shown here as they were in Chapter 8.

- Real power (also called true power) is the power dissipated in pure resistance, unit is watts.
- Reactive power is the power dissipated in pure reactance, unit is VARS (volt-ampere-reactive).
- Apparent power is the power of a circuit containing both resistance and reactance. It is calculated by: I × E (multiplying total current by the applied voltage), unit is VA (volt-ampere).
- Power Factor is a ratio of the real power to the total power. It is a pure number, with no units and will always be between 0 and 1. It is calculated by taking the cosine of the operating angle; cos Θ.

Note Sample calculations and a summary of formulas can be found in Chapter 8.

Practice Problems

Use the two schematic diagrams shown to calculate the answers to the problems.

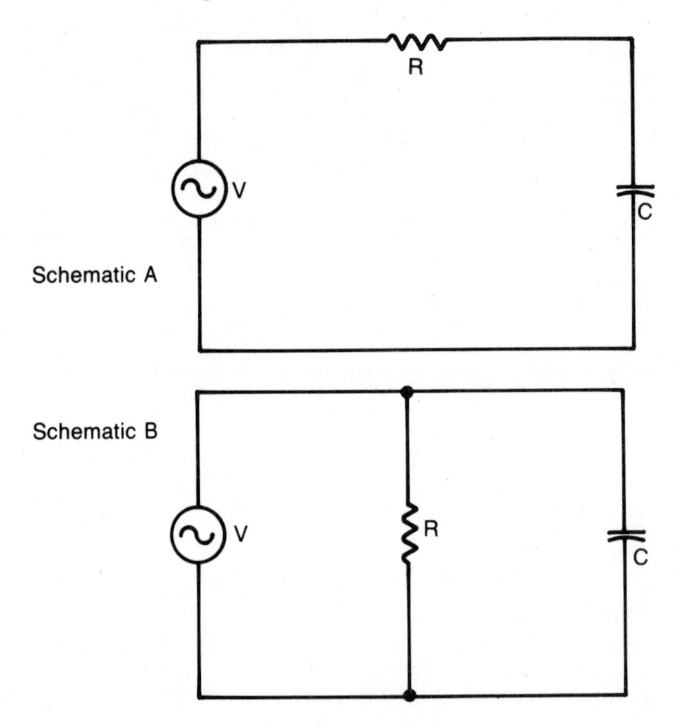

Use schematic A for problems 1 through 5. V = 100 volts.

Find: Z, I, V_R, V_C, Θ, Real Power, P_R, Reactive Power, P_X, Apparent Power, P_A, Power Factor, PF.

1. R = 500 ohms, X_C = 500 ohms

2. R = 100 ohms, X_C = 250 ohms
3. R = 1000 ohms, X_C = 750 ohms
4. R = 1000 ohms, X_C = 500 ohms
5. R = 10 ohms, X_C = 20 ohms

Use schematic B for problems 6 through 10. V = 100 volts.

Find: Z, I_T, I_R, I_C, Θ, Real Power, P_R, Reactive Power, P_X, Apparent Power, P_A, Power Factor, PF.

6. R = 500 ohms, X_C = 500 ohms
7. R = 100 ohms, X_C = 250 ohms
8. R = 1000 ohms, X_C = 750 ohms
9. R = 1000 ohms, X_C = 500 ohms
10. R = 10 ohms, X_C = 20 ohms

MEASURING PHASE ANGLE WITH AN OSCILLOSCOPE

The end of Chapter 8 discusses how to make phase angle measurements using an oscilloscope. To perform phase angle measurements with inductors or capacitors is exactly the same with the exception being that the capacitor has a lagging phase angle (for voltage) and the inductor has a leading phase angle.

In order for this section not to be a repeat of the section in Chapter 8, this section explains the calculations necessary for the student to set up experiments in measuring the phase angle using an oscilloscope.

Keep in mind, phase angle is found by first counting the number of divisions in one cycle and dividing that into 360 degrees to determine the degrees per division. Then multiply the degrees per division by the number of divisions to determine the phase angle. Frequency is found by counting the number fo divisions in one cycle and multiplying that by the time per division and then taking the reciprocal.

$$\frac{\text{degrees}}{\text{division}} = \frac{360°}{\text{\# of divisions in one cycle}}$$ **Formula 9-11**

$$\text{phase angle} = \text{\# of divisions} \times \frac{\text{degrees}}{\text{division}}$$ **Formula 9-12**

$$\text{period} = \frac{\text{\# divisions in}}{\text{one cycle}} \times \frac{\text{time per division}}{\text{(scope setting)}}$$ **Formula 9-13**

$$\text{frequency} = \frac{1}{\text{period}}$$ **Formula 9-14**

Materials Required for Phase Angle Measurements

Sine wave signal generator with variable frequency output
100 ohm resistor, 1/4 watt
1 μF capacitor, disc type
Dual trace oscilloscope

Figures 9-12 through 9-17 show the series circuit connected to the oscilloscope and the resultant oscilloscope displays. The same circuit will be used with only the frequency of the variable sine wave generator being varied. The following calculations are to predict the required frequency needed to arrive at the approximate phase angle.

Each time the frequency is adjusted, the voltage across the capacitor should change its relative amplitude in comparison to the applied voltage. The applied voltage amplitude should change only slightly. As the capacitor voltage changes, adjust the oscilloscope to try and maintain the two waveforms with approximately the same amplitude.

Figures 9-12 and 9-13 are the same frequency setting with the oscilloscope being adjusted differently.

Find X_C, R = 100. A 45 degree phase angle is desired.

Step 1 $\tan \Theta = \dfrac{X_C}{R}$ formula

Step 2 $\tan 45° = \dfrac{X_C}{100\ \Omega}$ substitute values

Step 3 $1 \times 100 = X_C$ finding tangent and rearranging equation
Step 4 $X_C = 100$ ohms value needed for 45° phase angle

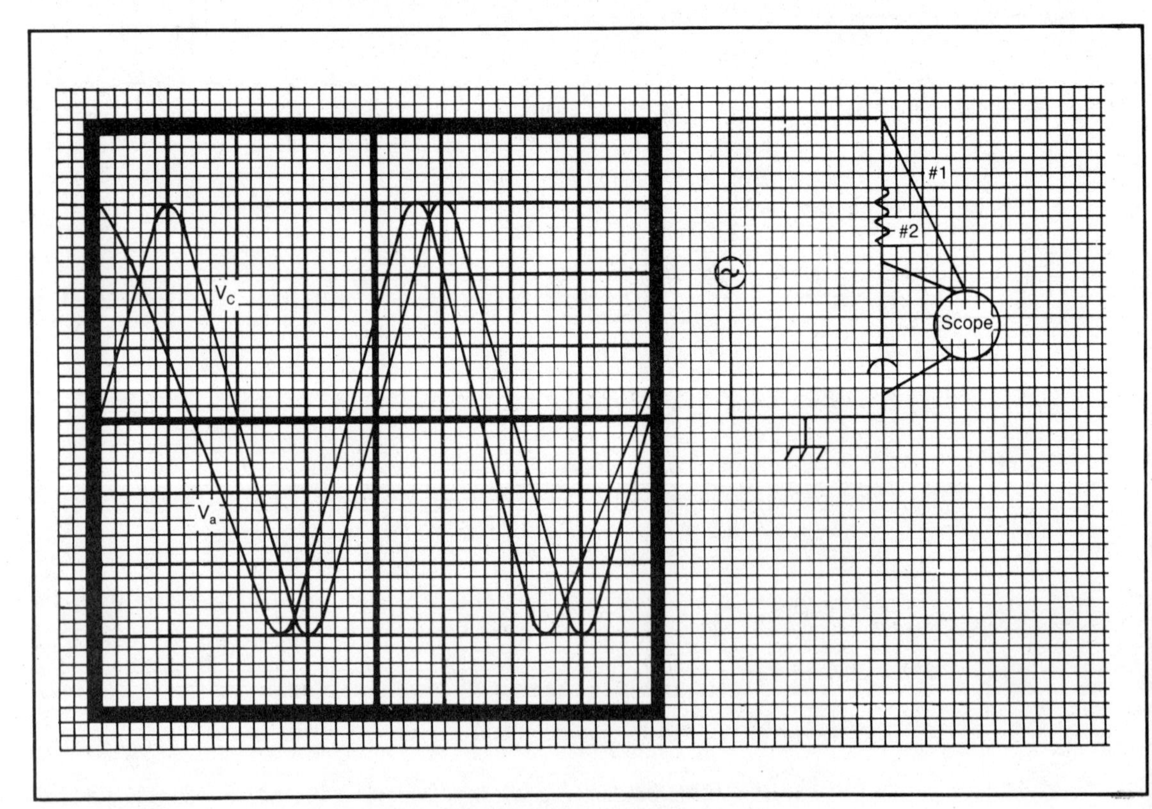

Fig. 9-12. 45° phase angle. Scope triggered on the capacitor voltage. $R = X_C$

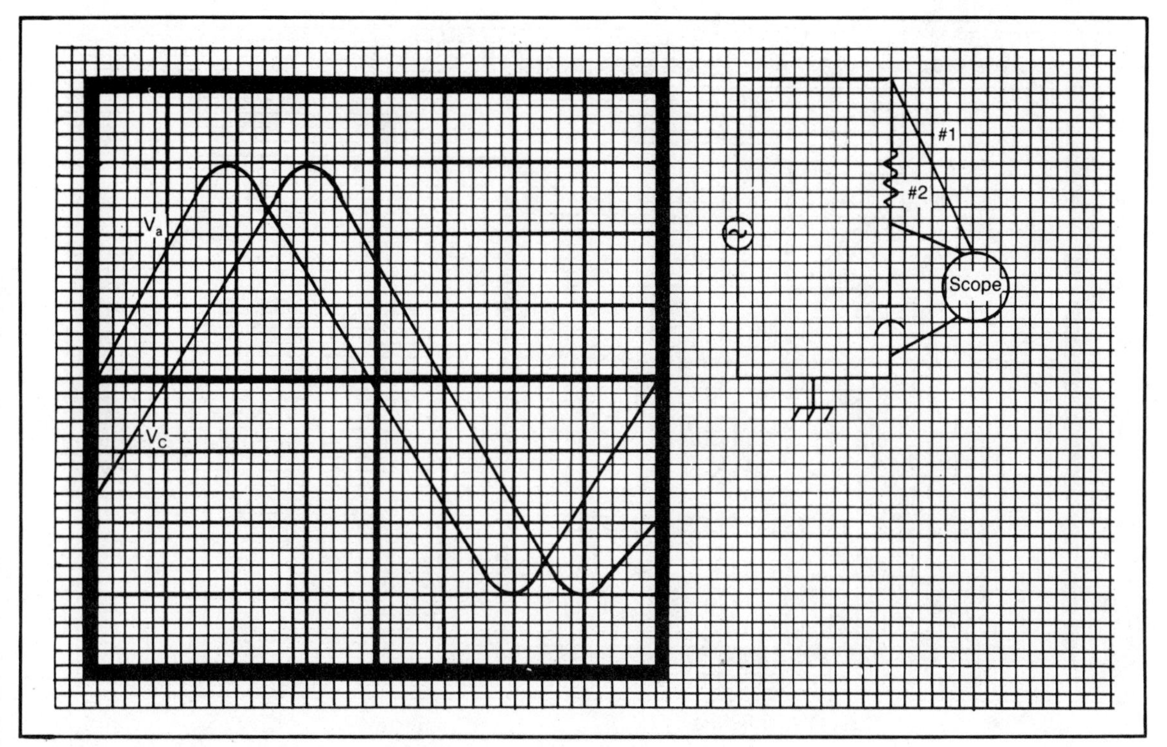

Fig. 9-13. 45° phase angle. Scope triggered on the input voltage. $R = X_C$

If $C = 1\ \mu F$, $X_C = 100$ ohms; find the frequency.

Step 1 $X_C = \dfrac{1}{2\pi fC}$ therefore; $F = \dfrac{1}{2\ \pi\ C\ X_C}$

Step 2 $f = \dfrac{1}{2 \times \pi \times (1 \times 10^{-6}) \times 100}$ substitute values

Step 3 f = 1600 Hz approximate frequency needed for 45 degree phase shift

Figure 9-14 shows approximately a 27 degree phase shift between the two sine waves. Use the outline for the calculations shown above to determine:

X_C = 51 ohms f = 3120 Hz

Figure 9-15 shows a phase angle of approximately 63 degrees. Use the above outline to find:

X_C = 193 ohms f = 825 Hz

Figure 9-16 shows a phase angle of 0 degrees. It is possible to arrive at a phase angle so close to 0 degrees that it could not be seen on the scope but it would require

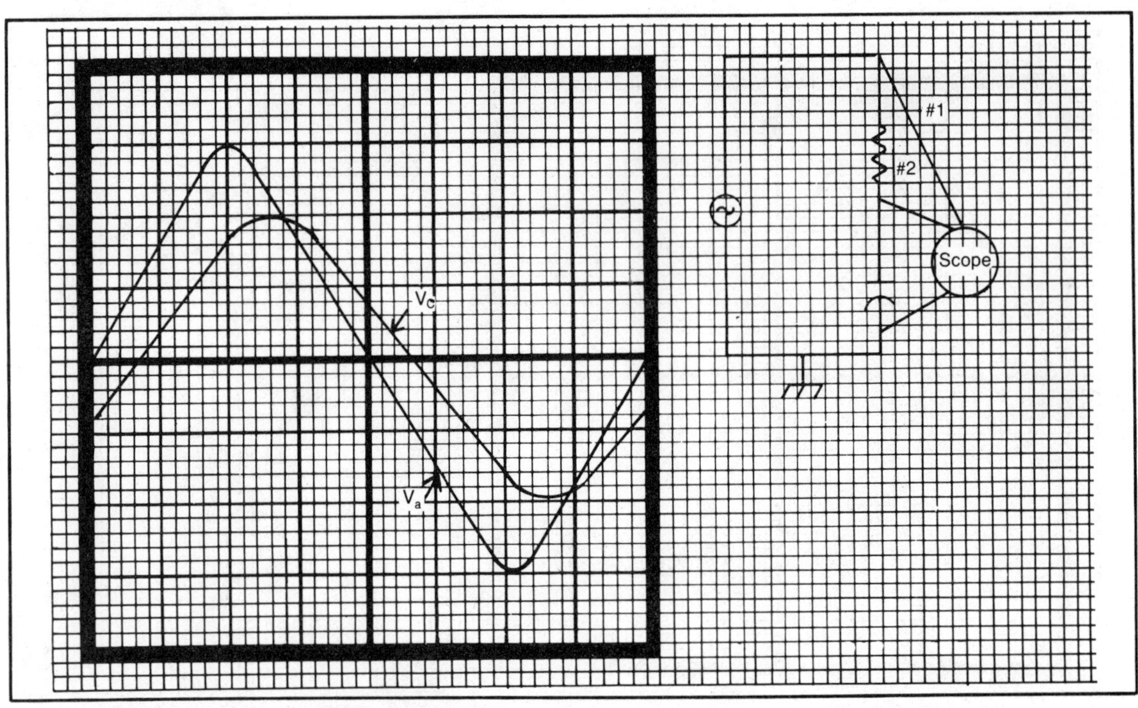

Fig. 9-14. 27° phase angle. R slightly larger than X_C.

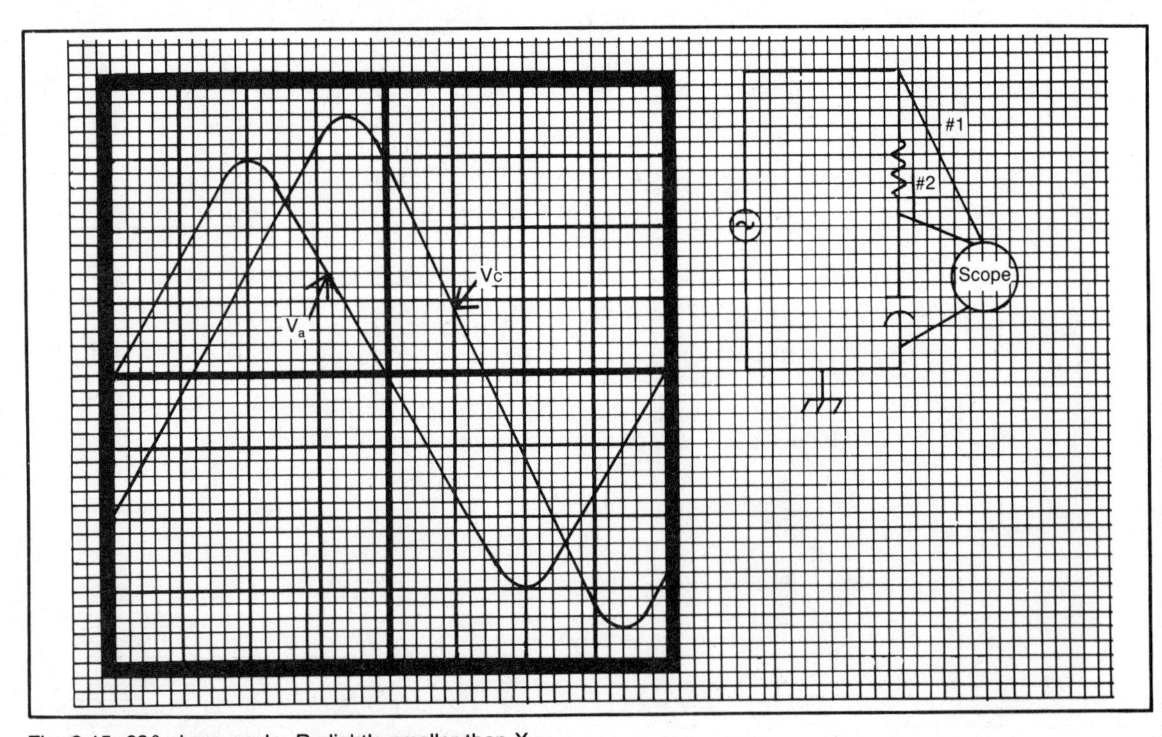

Fig. 9-15. 63° phase angle. R slightly smaller than X_C.

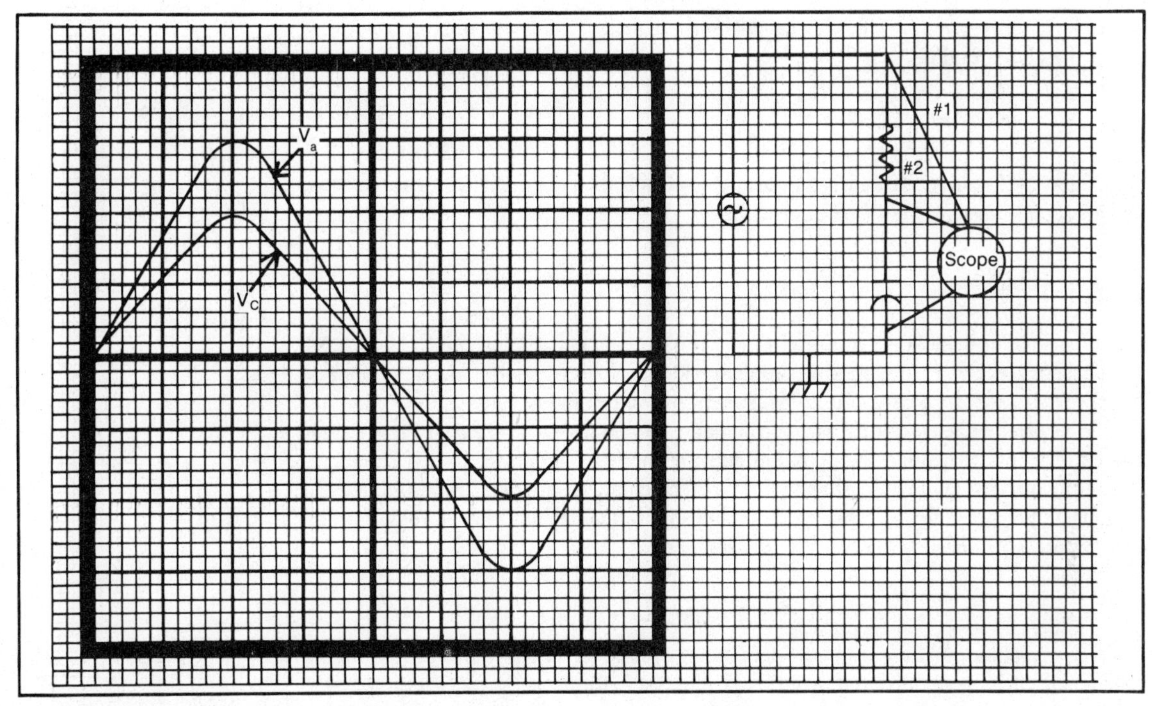

Fig. 9-16. 0° phase angle. R very large compared to X_C.

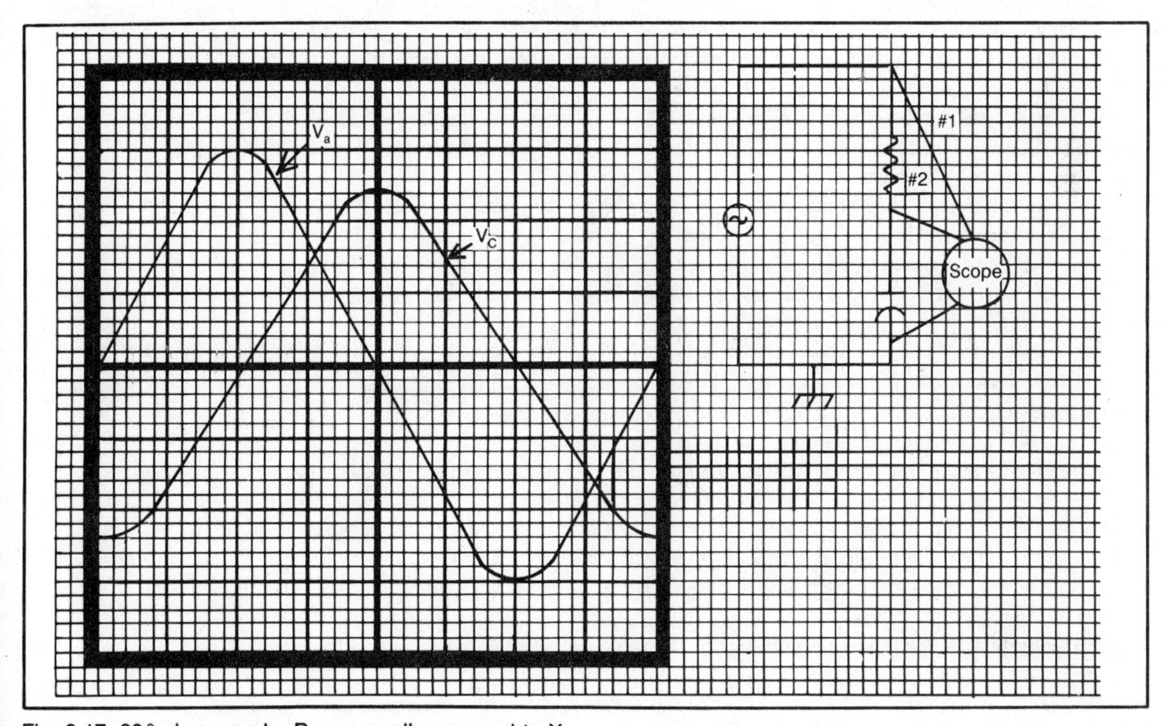

Fig. 9-17. 90° phase angle. R very small compared to X_C.

a very high frequency. The following calculations are made for a phase angle of 5 degrees, using the above outline.

X_C = 8.8 ohms   f = 18,000 Hz

Figure 9-17 shows a 90 degree phase angle, the results for the calculations are shown for an actual phase angle of 88 degrees.

X_C = 2860 ohms   f = 56 Hz

It was pointed out several times in the discussion on calculating the frequency for the phase angle, above, that these are approximations. Not only are the numbers rounded off, the student needs to be aware of the inherent tolerances and other circuit inaccuracies.

Sometimes, when working with a calculator and numbers, it is easy to forget that the components that build the actual circuit are far from accurate, by the standards of the calculations made beforehand. However, calculations are very important for the technician because they are used to predict circuit outcomes and therefore determine if a circuit is performing the way it was intended.

CHAPTER SUMMARY

Capacitive reactance, along with inductive reactance are the two forms of ac resistance. When capacitors are connected in series or parallel, with no resistance in the circuit, the individual reactances can be mathematically combined using the same formulas as are used for pure resistance.

When a capacitor is connected in series with a resistor, there will be a resulting phase shift between the voltage drops across each component. The phase shift will show the capacitive voltage to lag the resistive voltage.

When a capacitor and resistor are connected in parallel, the currents to the individual branches will have a resultant phase angle. The current through the capacitor will lead the current through the resistor.

Power in a reactive circuit must reflect the phase angle, or operating angle of the circuit. There are three types of power; true power, reactive power, apparent power. Apparent will be the combination of both the true and apparent powers, and therefore, the apparent power will always be the largest value. Power Factor is the ratio of the resistive power to the total power of the circuit.

- Capacitive reactance, X_C, is indirectly related to frequency and capacitance.
- In a series circuit, I_C leads V_C, V_R leads V_C always by 90 degrees.
- The voltage vectors for inductance and capacitance are opposite.
- The operating angle of a circuit with a capacitor in series with a resistor will be between 0 degrees and −90 degrees, depending on circuit values.
- In a parallel circuit, I_C leads I_R by 90 degrees.
- Real power is the power dissipated in pure resistance, unit is W (watts).
- Reactive power is the power dissipated in pure reactance, units is VARS (volt-ampere-reactive).
- Apparent power is the power of a circuit containing both resistance and reactance, unit is VA (volt-ampere).
- Power factor is the ratio of the real power to the apparent power. It is a pure number, with no units, between 0 and 1.

Summary of Formulas

capacitive reactance $X_C = \frac{1}{2\pi fC}$ **Formula 9-1**

series reactances $X_{CS} = X_{C1} + X_{C2} + X_{C3} + \ldots$ **Formula 9-2**

parallel reactances $\frac{1}{X_{CP}} = \frac{1}{X_{C1}} + \frac{1}{X_{C2}} + \frac{1}{X_{C3}} + \ldots$ **Formula 9-3**

total voltage in a series circuit $V_T = \sqrt{V_R^2 + V_C^2}$ **Formula 9-4**

operating angle using voltages $\Theta = \tan^{-1} \frac{-V_C}{V_R}$ **Formula 9-5**

impedance in a series circuit $Z = \sqrt{R^2 + X_C^2}$ **Formula 9-6**

operating angle using resistance $\Theta = \tan^{-1} \frac{-X_C}{R}$ **Formula 9-7**

total current in a parallel circuit $I_T = \sqrt{I_R^2 + I_C^2}$ **Formula 9-8**

phase angle of a parallel circuit $\Theta = \tan^{-1} \frac{I_C}{I_R}$ **Formula 9-9**

impedance of a parallel circuit $Z = \frac{V}{I_T}$ **Formula 9-10**

Formulas 9-11 through 9-14 are used when making measurements on the oscilloscope:

$\frac{\text{degrees}}{\text{division}} = \frac{360°}{\text{\# of divisions in one cycle}}$ **Formula 9-11**

phase angle = # of divisions $\times \frac{\text{degrees}}{\text{division}}$ **Formula 9-12**

period $= \frac{\text{\# divisions}}{\text{in one cycle}} \times \frac{\text{time per division}}{\text{(scope setting)}}$ **Formula 9-13**

frequency $= \frac{1}{\text{period}}$ **Formula 9-14**

Chapter 10

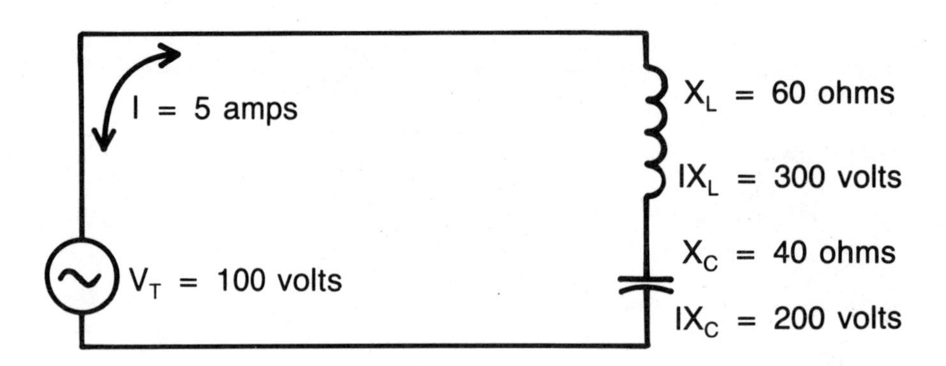

AC Circuits and the J-Operator

This chapter discusses circuits that contain different combinations of resistance, capacitance, and inductance. Using vector addition to combine reactances to determine a net (or total) reactance for the combination is the primary concern of the chapter. The j-operator is a very useful mathematical tool that can be used to allow the calculator to do most of the work in performing calculations on complex circuits.

NET REACTANCE

The net reactance is the reactance of a circuit that has more than one type of reactance; that is to say both inductance and capacitance. In order to best determine the meaning of net reactance, the easiest way is to draw the vectors of inductance and capacitance on the same graph. Notice in Fig. 10-1A that the vector for inductive reactance is plotted vertically up and the vector for capacitive reactance is plotted vertically down. In other words, the vectors for inductive reactance and capacitive reactance are opposite. It can also be demonstrated that all of the vectors dealing with inductance and capacitance are opposite. Figure 10-1A shows a series circuit with inductance and capacitance. The vectors for a series circuit are reactance, with a total impedance and the other vectors are voltage drops, with the total applied voltage. A parallel circuit uses the current triangle (shown in Fig. 10-1B). In all cases, notice the relationship of the inductance vector and the capacitance vector.

- The vectors for inductance and capacitance are opposite.

A very interesting observation can be made from Fig. 10-1. In the series circuit, the opposite reactances partially cancel each other and result in a greatly reduced net reactance. The circuit current is determined by the net reactance, which results in an

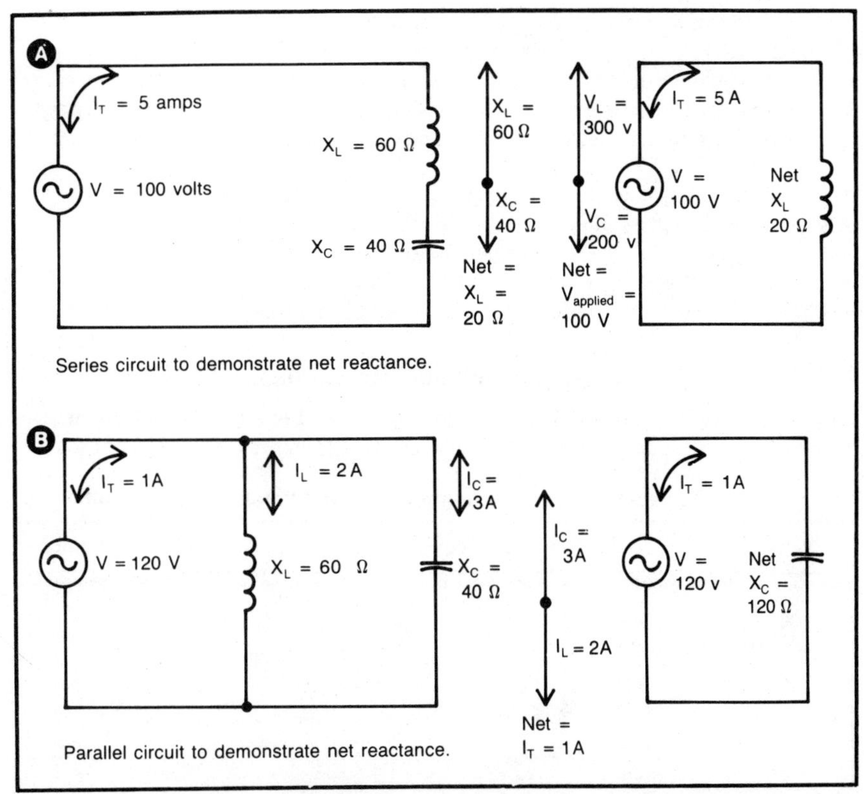

Fig. 10-1. Net reactance is demonstrated with a series circuit and a parallel circuit. A. Series circuit demonstrates net reactance. B. Parallel circuit to demonstrate net reactance.

unusually high current. The individual voltage drops across each component are determined by using Ohm's law with the circuit current and the individual reactances. The result is the voltage drop across the inductor and capacitor, each results in a voltage higher than the applied voltage.

The parallel circuit uses a current vector relationship. The total current is less than the individual branch currents. This is due to the fact that the individual branch currents are calculated using the applied voltage and the branch reactance. The branch currents are 180 degrees out of phase and will result in some cancellation in the total current.

In Fig. 10-1, on the right-hand side, an equivalent circuit showing the net reactance of the circuits is shown. This net reactance is based on the reactance vectors for the series circuit, and based on the total current in the parallel circuit.

The main purpose for using an equivalent circuit when dealing with ac circuits is the fact that the net reactance can be shown as being either an inductor or a capacitor. The fact of the circuit having a net capacitance or inductance becomes very important when determining the characteristics of the circuit.

The above discussion, concerning Fig. 10-1, was all based on circuits that are only inductance and capacitance. In an actual circuit it is also necessary to consider the effects of resistance. Chapters 8 and 9 showed how to deal with the triangles for each of the individual components. In this chapter, the three components will be combined into one circuit. The rules will still be the same except that inductive reactance and capacitive reactance are opposite and must result in a net reactance to be used in the vector triangles.

It is possible for the inductive and capacitive reactances to completely cancel and leave the circuit with only a net resistance. This principle is used in resonant circuits.

Table 10-1 shows the relationship of the vectors used in ac circuits. To summarize:

- In a series circuit, use an impedance triangle and a voltage triangle.
- In a series circuit, X_L and V_L are $+90°$ and X_C and V_C are $-90°$.
- In a parallel circuit, use a current triangle.
- In a parallel circuit, I_C is $+90°$ and I_L is $-90°$.
- Phase angle will be positive or negative depending on the net reactance in the circuit.

Analysis of a Series R-L-C Circuit

Refer to Fig. 10-2. There are six parts to this figure. Part A is the original circuit, containing resistance, capacitance and inductance. Part B shows the vectors of the reac-

Table 10-1. Vector Relationships in a Series and Parallel R-L-C Circuit.

Vector	Series R-L-C circuit	Parallel R-L-C circuit
X_L	+90° (up) ↑	
X_C	−90° (down) ↓	
R	0° (right) →	
Z	Hypotenuse of impedance triangle ◺ or ◹	$Z = \frac{V_a}{I_T}$
V_L	+90° (up) ↑	Same as applied voltage
V_C	−90° (down) ↓	Same as applied voltage
V_R	0° (right) →	Same as applied voltage
V_T	Hypotenuse of voltage triangle ◺ or ◹	Voltage is the same across all branches.
I_L	Same as total current	−90° (down) ↓
I_C	Same as total current	+90° (up) ↑
I_R	Same as total current	0° (right) →
I_T	Current is the same throughout a series circuit.	Hypotenuse of current triangle ◹ or ◺

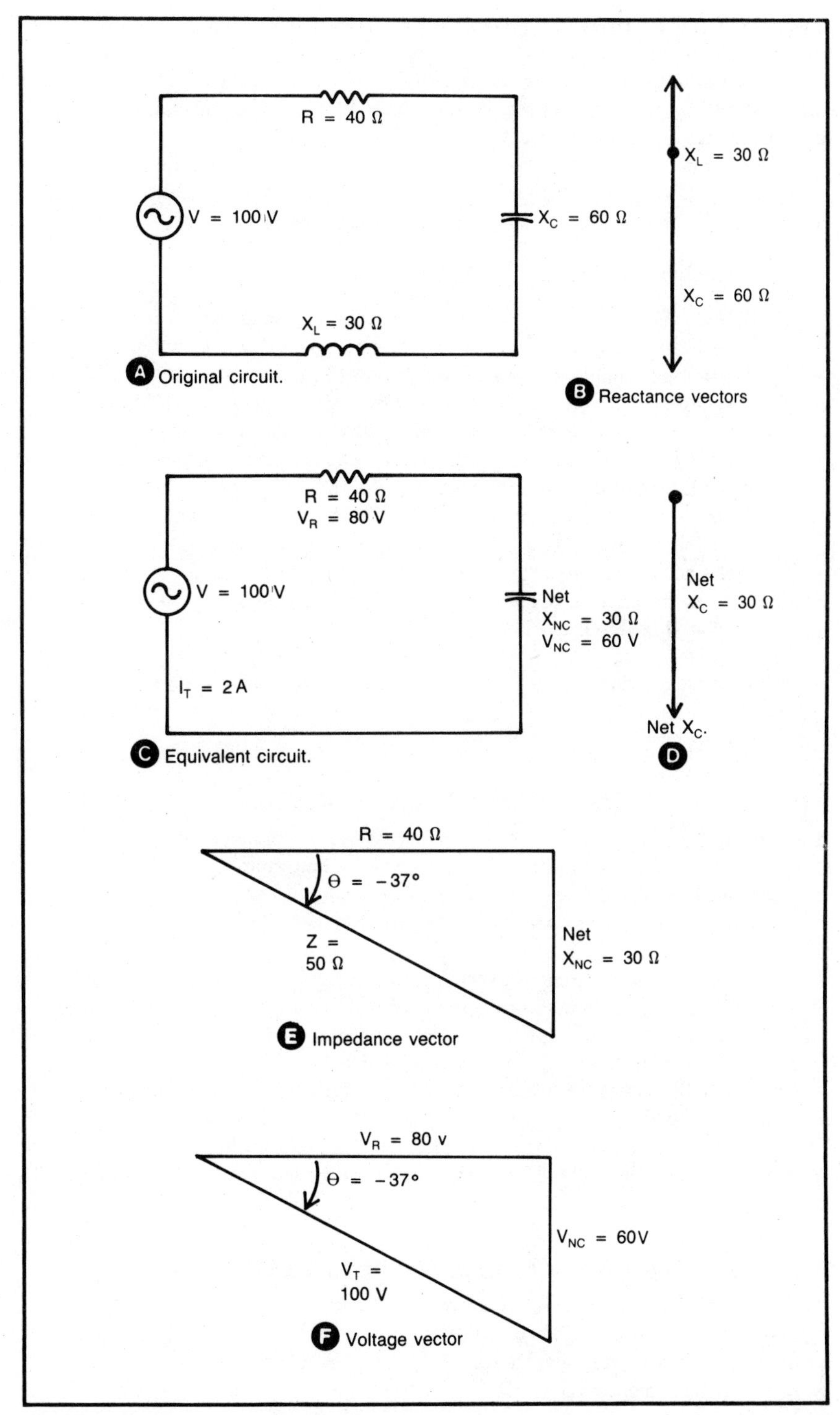

Fig. 10-2. Analysis of a series R-L-C circuit.

tance, not including the resistance vector. This is drawn in this manner in order to impress the idea that X_L and X_C will partially cancel each other. Part C shows an equivalent circuit, resulting from the net reactance. Notice it is drawn as a capacitor, to show a net X_C. Part D is the vector of the net reactance. Part E is the impedance vector, using the resistance and the net reactance. Part F is the voltage vector showing the voltage drop across the resistor and the net reactance. The labeling on the net reactance is shown as X_{NC} and the voltage as V_{NC}. This is to be sure it is clear that these values are for the net and not for the original value of capacitance.

Calculate the Net reactance vector.

Step 1 Draw the vectors of X_L and X_C together with a connecting point to show opposite directions, as in Fig. 10-2B. Because the vectors go in opposite directions, use simple arithmetic to subtract the magnitudes (sizes) of the two vectors. X_L = 30 ohms and X_C = 60 ohms. Using simple arithmetic, subtract the smaller from the larger. 60 − 30 = 30 ohms net. Figure 10-2A shows the resultant net X_C. Notice that the direction of the resultant vector is from the larger of the two original vectors.

Step 2 Draw the equivalent circuit, showing the net reactance. The equivalent circuit shown in Fig. 10-2 is classified as a net capacitive circuit by the fact that the net reactance is capacitive.

Step 3 Draw the impedance and voltage triangles. Current for a series circuit is calculated using Ohm's law.

- Series circuits are classified as inductive or capacitive by whichever is the larger reactance or voltage drop.

Analysis of a Parallel R-L-C Circuit

Refer to Fig. 10-3 for the drawings used in the analysis of a parallel circuit. Figure 10-3A is the original parallel circuit. In order to keep everything simple here, the circuit used has only one component in each branch. Keep in mind, it is possible for the branches to have more than one component forming series circuits, additional parallel or any combination.

Figure 10-3B shows the current vectors for the capacitive and the inductive branches. The resistive current is now shown here in order to show the relationship of these two vectors. Figure 10-3C is the equivalent circuit, showing the resistive branch unchanged and the second branch having a net capacitive reactance.

Figure 10-3D shows the vector of the net reactance. The magnitude, or length, of this vector is found by using simple subtraction of the inductive and capacitive CURRENTS. Direction is determined by the larger of the two vectors. Notice, the larger vector is the larger current, which is the smaller of the two reactances. Figure 10-3E is the current triangle.

Summary for Analysis of a Parallel R-L-C Circuit

- Plot the vectors for the reactive currents.
- Subtract the reactive currents to determine the size and direction (inductive or capacitive) of the Net current and therefore the Net reactance.
- Draw the equivalent circuit.
- Draw the current triangle.

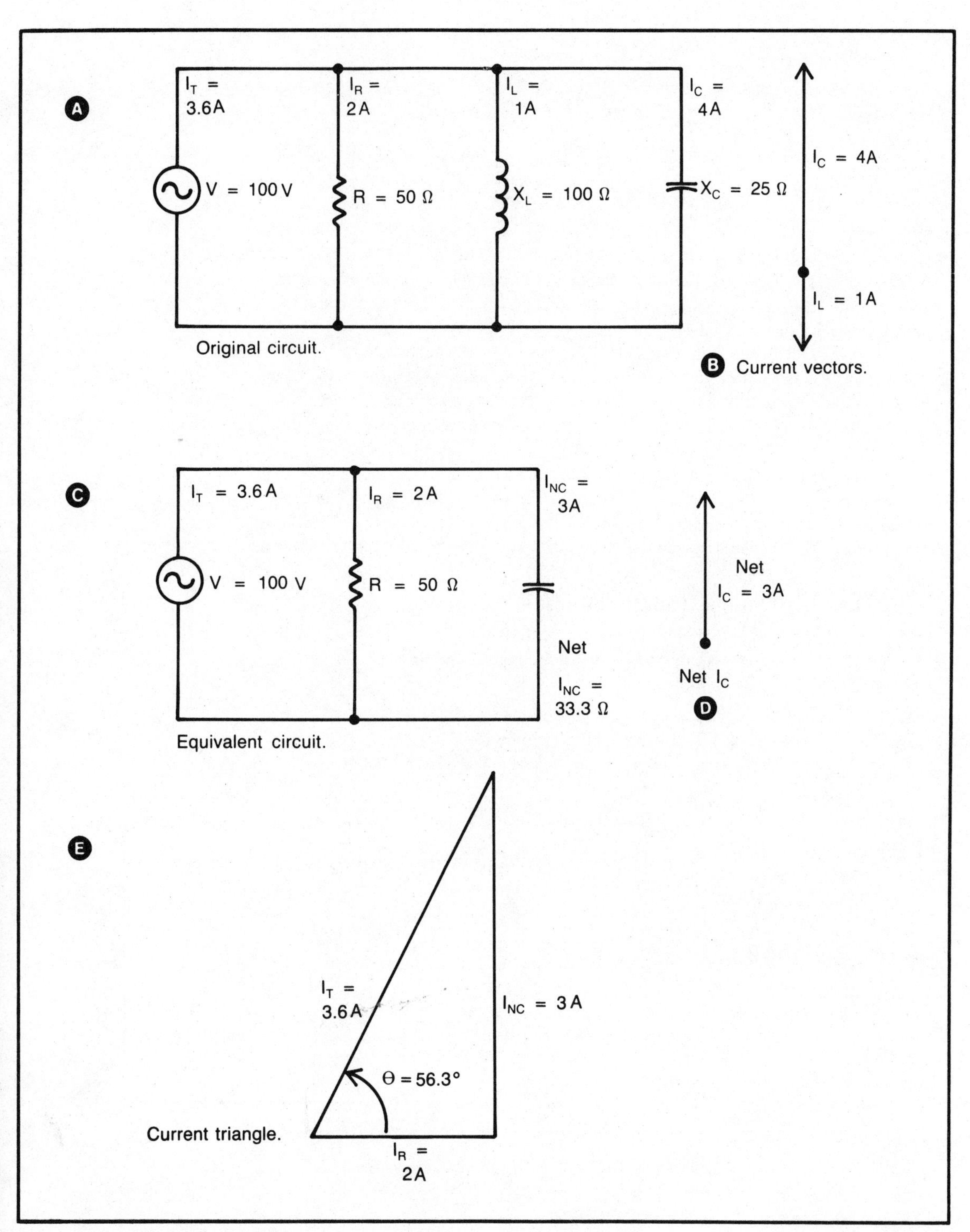

Fig. 10-3. Analysis of a parallel R-L-C circuit.

The following key points should be noted.

- Parallel circuits are classified as inductive or capacitive by whichever has the largest current. The largest current results from the smallest reactance.
- A circuit is classified as resistive only if the Net reactance is zero.

POWER IN AN R-L-C CIRCUIT

In Chapter 8 the subject of power in inductive circuits was discussed. In Chapter 9 the subject of power in capacitive circuit was discussed. This chapter deals with circuits that have both capacitance and inductance. Power is considered separately for each of the separate components. The points to remember are the definitions of the three types of power in an ac circuit.

True power is the power dissipated by the resistance of the circuit, unit is watts. Reactive power is the power dissipated by a reactive component, unit is VARS (volt-ampere-reactive). Apparent power is the power dissipated by a complete ac circuit, unit is VA (volt-ampere). Power factor is the ratio of true power to the total power of the circuit, no unit, it is a pure number.

Due to the fact that the voltage of a series circuit or the current of a parallel circuit will have a phase angle, if a circuit contains both inductance and capacitance, the powers for these two components, since they are both in VARS, will be opposite and will have a cancellation effect, the same as would be seen with either the impedance or the current.

Practice Problems

Use the schematics shown to calculate the answers to the problems. For each problem, find: I_T, Z, and state if the circuit is a net C or L. Find angle theta.

1. X_{C1} = 25 ohms, R_1 = 25 ohms, X_{L1} = 100 ohms, R_2 = 25 ohms
2. X_{C1} = 200 ohms, R_1 = 50 ohms, X_{L1} = 50 ohms, R_2 = 50 ohms
3. X_{C1} = 10 kilohms, R_1 = 1 kilohm, X_{L1} = 5 kilohms, R_2 = 5 kilohms

Problems 1, 2, and 3

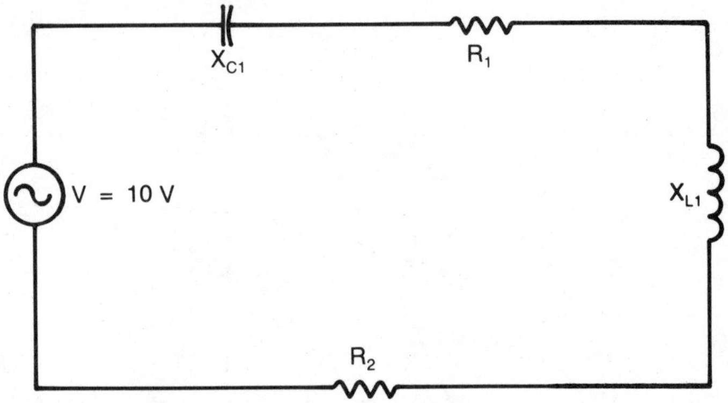

4. X_{L1} = 10 ohms, X_{C1} = 10 ohms, R_1 = 10 ohms, X_{L2} = 20 ohms, X_{C2} = 30 ohms, R_2 = 5 ohms, R_3 = 5 ohms
5. X_{L1} = 200 ohms, X_{C1} = 100 ohms, R_1 = 50 ohms, X_{L2} = 200 ohms, X_{C2} = 100 ohms, R_2 = 75 ohms, R_3 = 75 ohms

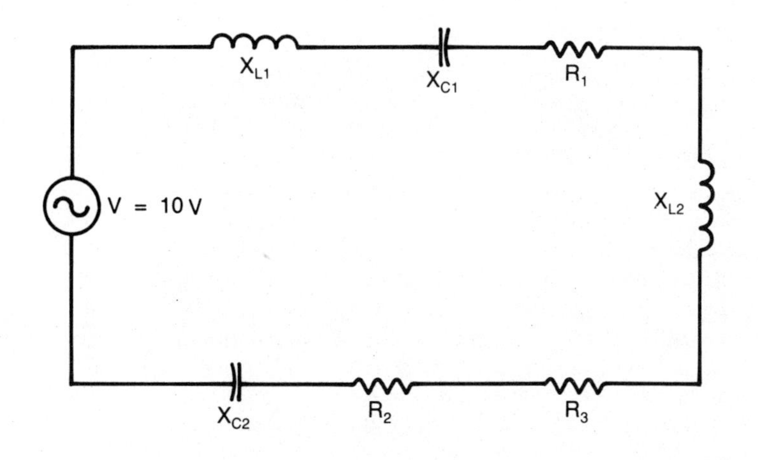

Problems 4 and 5

6. R_1 = 10 ohms, X_{C1} = 10 ohms, X_{L1} = 20 ohms
7. R_1 = 100 ohms, X_{C1} = 200 ohms, X_{L1} = 100 ohms
8. R_1 = 1 kilohm, X_{C1} = 500 ohms, X_{L1} = 2 kilohms

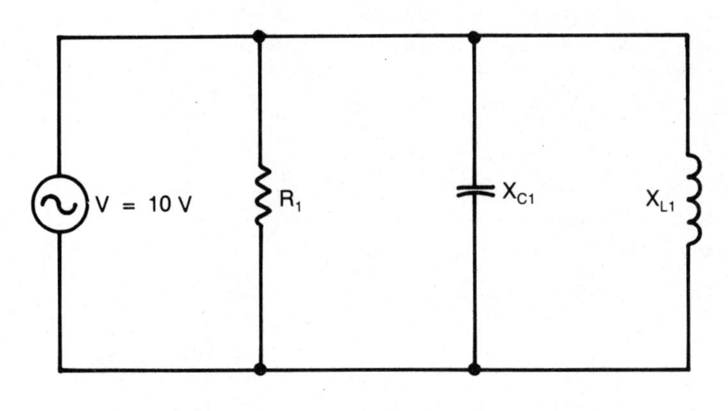

Problems 6, 7 and 8

9. X_{C1} = 20 ohms, R_1 = 10 ohms, X_{C2} = 20 ohms, X_{L1} = 40 ohms, X_{L2} = 50 ohms
10. X_{C1} = 50 ohms, R_1 = 20 ohms, X_{C2} = 40 ohms, X_{L1} = 20 ohms, X_{L2} = 10 ohms

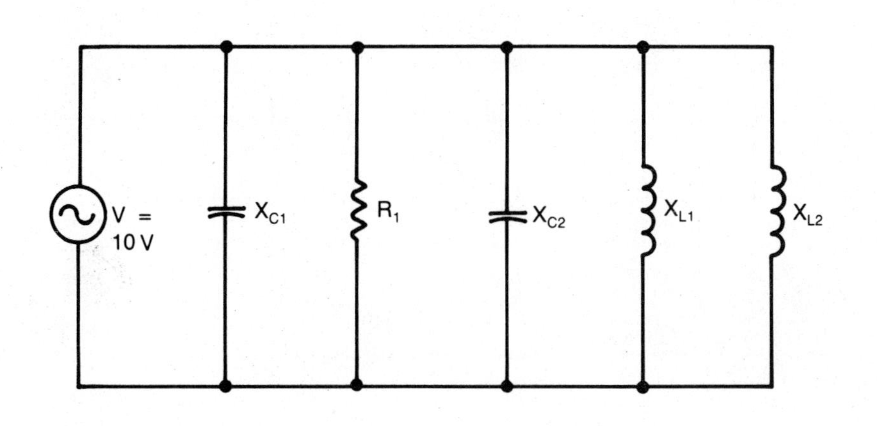

Problems 9 and 10

THE J-OPERATOR

The j-operator is a mathematical tool used to simplify working with complex circuits. The j-operator concept can actually be used under any set of conditions dealing with vectors. Most scientific calculators are equipped with keys to allow the calculator to deal directly with numbers written in the j-operator format. When several vectors are to be combined, the use of trigonometric methods is somewhat time consuming.

The standard, geometric x,y coordinate axis system is relabeled for use with the j-operator. Figure 10-4 shows how the coordinate system is relabeled. The horizontal axis, normally called the x axis, is now called the real axis. The vertical axis, normally called the y axis, is now labeled with +j going up and −j going down.

The real axis is labeled at both 0 degrees and 180 degrees. In electronics, the real axis is used to show the resistance of the circuit. Any point on the right of the center point is positive and any point on the left of the center point is negative. Therefore, in electronics, points to the left cannot be considered for circuits with just resistance, capacitance, and inductance.

The j axis, both positive and negative, is of very important consideration in circuits. It is in these two directions, +90 degrees and −90 degrees, that we find the vectors for inductors and capacitors. It depends on whether the circuit is a series circuit or a parallel circuit that determines the plotting of the vectors, but for pure inductance or capacitance, it will always be 90 degrees in relation to the resistor vector. When dealing with j-operators and ac circuits, reactance is usually what is worked with. Even when the circuit is a parallel circuit, the reactance can be used with j-operators because it is possible to deal with it in terms of using the reciprocal formula necessary for parallel resistances.

Therefore, inductive reactance is called a +j and capacitive reactance is called −j. It is also possible to deal with currents in a parallel circuit, in which case, the inductance would be negative and the capacitance would be positive.

It is interesting to note that the j axis is often called the imaginary axis.

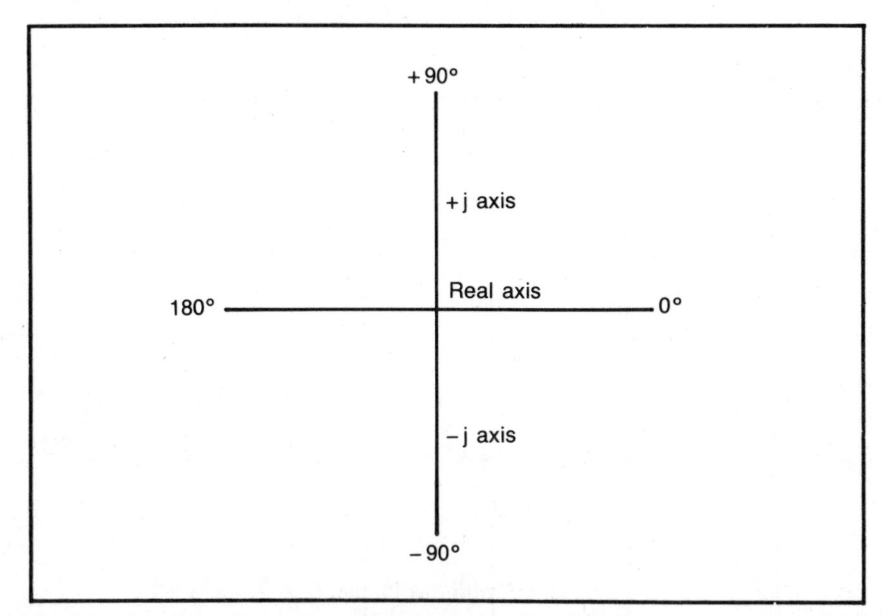

Fig. 10-4. The j-operator shown on the coordinate axis system.

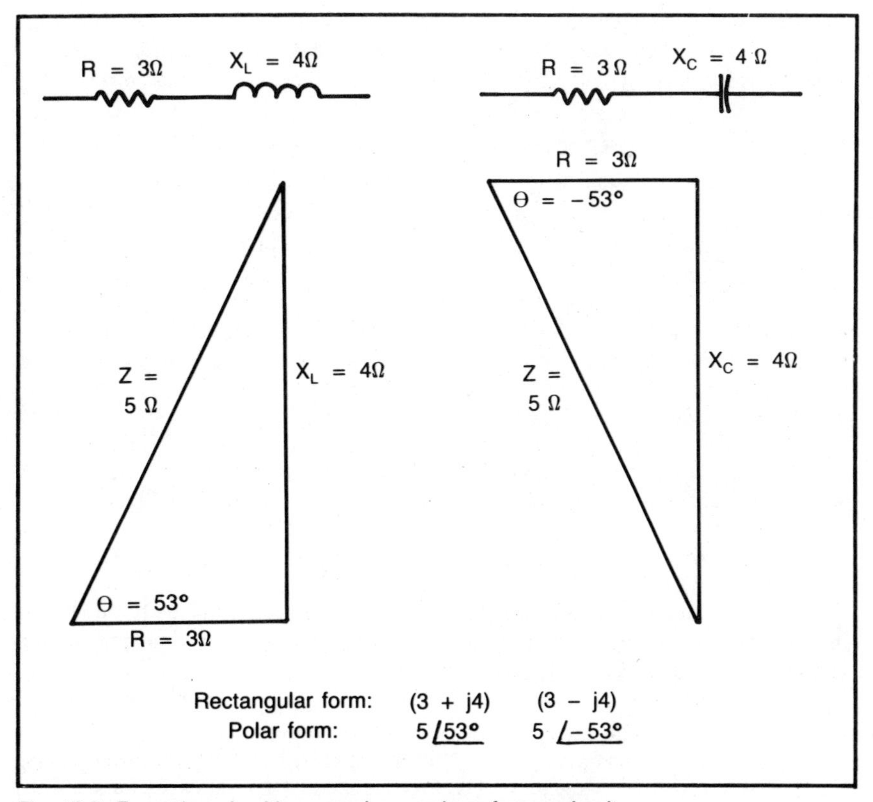

Fig. 10-5. Examples of writing complex numbers from a circuit.

COMPLEX NUMBERS IN RECTANGULAR FORM

A complex number is a number that contains both a real term and a j term. Complex numbers can be written in two forms; rectangular and polar.

The rectangular form of a complex number states the actual quantity of the real term and the actual quantity of the j term. In other words, the rectangular form tells the amount of resistance and the amount of net reactance. The vectors of these two quantities are plotted at right angles to each other, either up or down (+90 degrees or −90 degrees). A complex number, when dealing with resistance and reactance, states the amount of impedance in the circuit, in rectangular form. When working the mathematics of a circuit, sometimes it is necessary to use rectangular form, sometimes polar form.

- The rectangular form of a complex number states the actual quantities of the real and j terms.

When numbers are to be written in rectangular form, the real term is written first, followed by the j term. Because the j term can be either positive or negative, the plus or minus sign can be written as an addition or subtraction sign connecting the real term with the j term.

- The rectangular form is written with this format; (real $\pm$ j term).

Figure 10-5 shows two simple series circuits, one with an inductor and one with a capacitor. With each circuit is the impedance triangle. Shown below each triangle is the impedance written out in both rectangular and polar forms. The rectangular form is simply the two sides of the triangle and the polar form is the hypotenuse.

COMPLEX NUMBERS IN POLAR FORM

The polar form of a complex number is a vector at some angle between 0 degrees and $\pm$ 90 degrees. In terms of an electrical circuit, that means the hypotenuse of the triangle is used to find either impedance, voltage, or current. The polar form also contains the operating angle, theta.

- The polar form of a complex number is written in this format: $Z \angle \Theta$.
- When a number is written in polar form, the symbol $\angle$ means at an angle of.
- Rectangular form shows the horizontal (real) and vertical (imaginary) components.
- Polar form shows the resultant vector and its angle.

It will be necessary to consult the owners manual of the calculator in use to determine the steps involved in converting from rectangular to polar and vice versa.

CALCULATIONS WITH COMPLEX NUMBERS

Calculations with complex numbers have a set of rules that are unique to this type of mathematics. When working with circuits it is sometimes necessary to add or subtract and other times it is necessary to multiply or divide.

To show when it might be necessary to use more than one operation consider an example of a series parallel circuit. When working the parallel portion, total resistance is found by using the reciprocal formula, which involves both addition and division. Then the equivalent resistance of the parallel branches is added to the series resistance. Then, to find the total current, or voltage drops in the circuit, Ohm's law is necessary.

Rules of Calculations with Complex Numbers

Rule 1 Add or subtract in rectangular form.

Rule 2 When adding or subtracting in rectangular form, add/subtract the real terms together and add/subtract the j terms together.

Rule 3 Multiply or divide in polar form.

Rule 4 When multiplying in polar form, multiply the magnitudes (the Z term) and add the angles.

Rule 5 When dividing in polar form, divide the magnitudes (the Z term) and subtract the angles.

Rule 6 When taking the reciprocal of a number, it is to be done in polar form, because it is division. Take the reciprocal of the magnitude and change the sign of the angle.

Series AC Circuits

Whenever it is necessary to solve any series circuit, keep in mind the basic rules of series circuits. In particular, current is the same throughout a series circuit and the voltage drops across each component will add to the applied voltage.

When dealing with ac circuits, the reactive components, inductance and capacitance, should always be changed to their reactances, if not already given in that form. Using the rules for j-operators, inductors have an X_L of $+j$ and capacitors have an X_C of $-j$.

In a series circuit, resistances and reactances, are added to find the total resistance of the series circuit. Notice that the +j for X_L and the −j for X_C will take care of the fact that the reactances are opposite and in one step produce the total resistance and the net reactance of the circuit. This addition is performed in rectangular form and the final result will be in the form of: R ± jX.

The calculations shown below are for the series circuit shown in Fig. 10-6. To solve a series circuit, using complex numbers, it is usually enough to find the total impedance, total current and the voltage drops of each component.

Because this is a series circuit, the individual resistances will all add to form the total resistance. The reactive components will combine to produce a net reactance.

Step 1 Add the resistances for the real term of Z_T.

R_1 = 6 ohms
R_2 = 10 ohms
$R_1 + R_2 = 6 + 10 = 16$ ohms real term

Step 2 Add the reactances, algebraicly (subtract) to produce the net reactance or the j term.

X_C = −j3 ohms
X_L = +j8 ohms
$X_C + X_L = -j3 + +j8 = +j5$ ohms net reactance

Step 3 Combine the real term and the j term to form the complete total impedance, written in rectangular form.

$Z_T = R \pm jX$
$Z_T = (16 + j5)$ ohms

Step 4 Change the Z_T in step 3 to polar form to make it ready to use to find I_T.

$Z_T = (16 + j5)$ rectangular form
$Z_T = 16.76 \angle 17.35$ ohms polar form

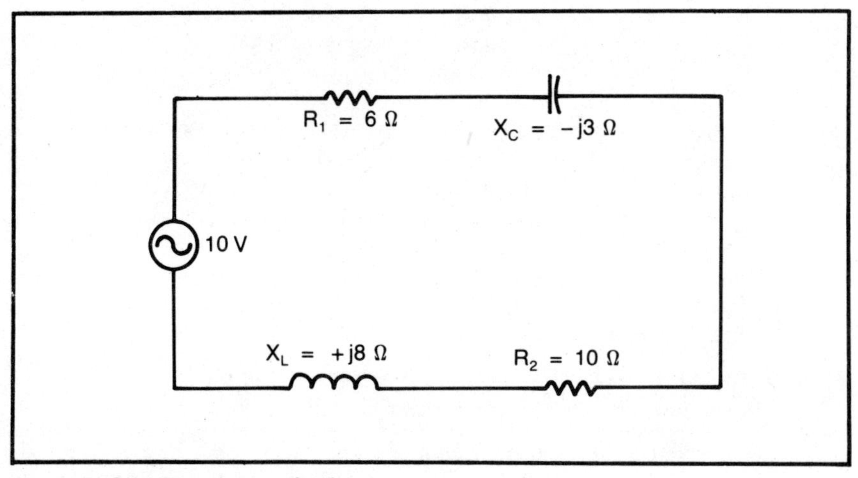

Fig. 10-6. Sample series ac circuit.

Step 5 Use the applied voltage and total impedance with Ohm's law to find the total current. Because this is a division problem, it must be worked using the numbers in polar form. The applied voltage is shown with no specific angle. Because it needs an angle for the calculations, an angle of zero degrees will be assigned.

$I_T = \frac{V}{Z_T}$ Ohm's law formula to be used

$I_T = \frac{10 \angle 0 \text{ volts}}{16.76 \angle 17.35 \text{ ohms}}$ substitute values

$I_T = .596 \angle -17.35$ amps solve by dividing the magnitudes and subtracting the angles, bottom from top.

Step 6 The voltage drops across each component can be found by using Ohm's law and the total current. Because Ohm's law is a multiplication problem, each component must be shown in polar form. This is best done by first writing in rectangular form and then changing to polar form.

$R_1 = (6 + j0)$ rectangular form (zero shows no reactance)
$R_1 = 6 \angle 0$ polar form
$X_C = (0 - j3)$ rectangular form (zero shows no resistance)
$X_C = 3 \angle -90$ polar form
$R_2 = (10 + j0)$ rectangular form
$R_2 = 10 \angle 0$ polar form
$X_L = (0 + j8)$ rectangular form
$X_L = 8 \angle 90$ polar form

Step 7 Find the voltage drops of each component using the polar form of each of the values found in Step 6, with the total current found in Step 5. E = IR

$E_{R1} = (.596 \angle -17.35 \text{ amps})(6 \angle 0 \text{ ohms}) = 3.576 \angle -17.35$ volts
$E_{XC} = (.596 \angle -17.35 \text{ amps})(3 \angle -90 \text{ ohms}) = 1.788 \angle -107.35$ volts
$E_{R2} = (.596 \angle -17.35 \text{ amps})(10 \angle 0 \text{ ohms}) = 5.96 \angle -17.35$ volts
$E_{XL} = (.596 \angle -17.35 \text{ amps})(8 \angle 90 \text{ ohms}) = 4.768 \angle 72.65$ volts

Step 8 Notice that simply trying to add the voltage drops to be equal to the applied voltage does not work. The reason is that the voltage drops must be added in rectangular form. Change each of the voltage drops to rectangular form.

$E_{R1} = 3.4 - j1.07$
$E_{XC} = -.53 - j1.7$
$E_{R2} = 5.68 - j1.78$
$E_{XL} = 1.42 + j4.55$

Step 9 To add the rectangular numbers from Step 8, add the column of j terms separately, then the column of real term separately.

$V_T = 9.97 + j0$ rectangular form
$V_T = 9.97 \angle 0$ volts polar form for comparison with the applied voltage

Power can also be found for each component by using the power formula, $P = I \times E$. Power would be apparent power if it is used with the total voltage and total current. It would produce true power when total current is used with the voltage drop across a resistor and it would be reactive power if total current is used with the voltage drop of a reactive component.

Parallel AC Circuits

When dealing with a parallel circuit, remember that voltage is the same across all parallel branches and the current divides to each branch.

Figure 10-7 shows a parallel circuit with two branches. Recall from working with dc circuits that there are several ways to work out a parallel circuit. Notice in this sample circuit, each branch is composed of a series circuit. It is probably best to work each branch individually as a series circuit, trying to find branch currents. The individual branch currents can then be added to find the total current and that will produce the total circuit impedance.

Step 1 Find each branch current. $I = \dfrac{E}{R}$

$$I_A = \frac{10 \angle 0 \text{ volts}}{(10 + j15) \text{ ohms}}$$ branch impedance needs to be changed to polar form

$$I_A = \frac{10 \angle 0 \text{ volts}}{18 \angle 56.3 \text{ ohms}} = .556 \angle -56.3 \text{ amps}$$

$$I_B = \frac{10 \angle 0 \text{ volts}}{(10 - j5) \text{ ohms}}$$ branch impedance needs to be changed to polar form

$$I_B = \frac{10 \angle 0 \text{ volts}}{11.2 \angle -26.6 \text{ ohms}} = .893 \angle 26.6 \text{ amps}$$

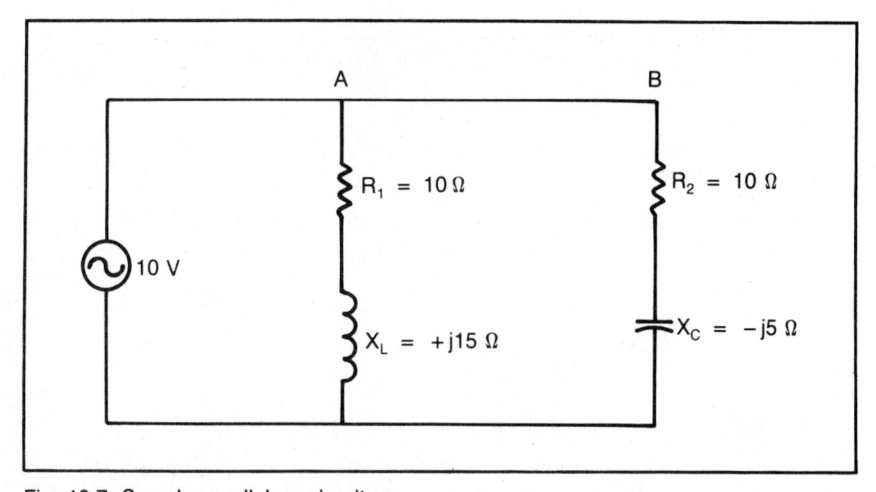

Fig. 10-7. Sample parallel ac circuit.

Step 2 In order to add the individual branch currents to find the total current, the currents must be changed from polar form to rectangular form.

I_A = (.308 – j.46) amps polar form
I_B = (.798 + j.399) amps polar form

Step 3 Add the j terms separately, then add the real terms separately to find the total current.

I_T = (1.106 – j.061) amps

Step 4 Change to the polar form of the total current from the rectangular form for use in finding the total impedance.

$I_T = 1.11 \angle -3.15$ amps

Step 5 Find total circuit impedance using the applied voltage and the total current.

$$R = \frac{E}{I}$$

$$Z_T = \frac{10 \angle 0 \text{ volts}}{1.11 \angle -3.15 \text{ amps}} = 9 \angle 3.15 \text{ ohms}$$

Complex AC Circuits

The process of solving complex ac circuits using the j-operator, uses the same basic thought process involved in solving complex dc circuits. The parallel branches are solved first for the total impedance, and then the series resistance is added. This combination of parallel, then series, results in the total circuit resistance of the circuit. That is the same basic procedure that will be used with the following sample problem.

Refer to Fig. 10-8. There are three parallel branches and a complex series impedance. The method of finding the equivalent parallel impedance will be the reciprocal method.

Step 1 Write the impedances for each individual branch in rectangular form, then change to polar form.

Branch A. (10 + j15) ohms = $18 \angle 56.3$ ohms
Branch B. (15 – j10) ohms = $18 \angle -33.6$ ohms
Branch C. (5 + j5) ohms = $7 \angle 45$ ohms

Step 2 Use the polar form of each branch resistance in the reciprocal formula to solve for total parallel impedance.

$$\frac{1}{R_T} = \frac{1}{R_1} + \frac{1}{R_2} + \frac{1}{R_3} \quad \text{formula}$$

$$\frac{1}{Z_{Parallel}} = \frac{1}{18 \angle 56.3} + \frac{1}{18 \angle -33.6} + \frac{1}{7 \angle 45} \quad \text{substitute values}$$

To solve this type of equation, it is necessary to first take the reciprocal of the magnitudes (the numbers) and change the sign of the angles. For this step, it is probably best to write down each reciprocal, as a decimal, as it is taken, to save on confusion.

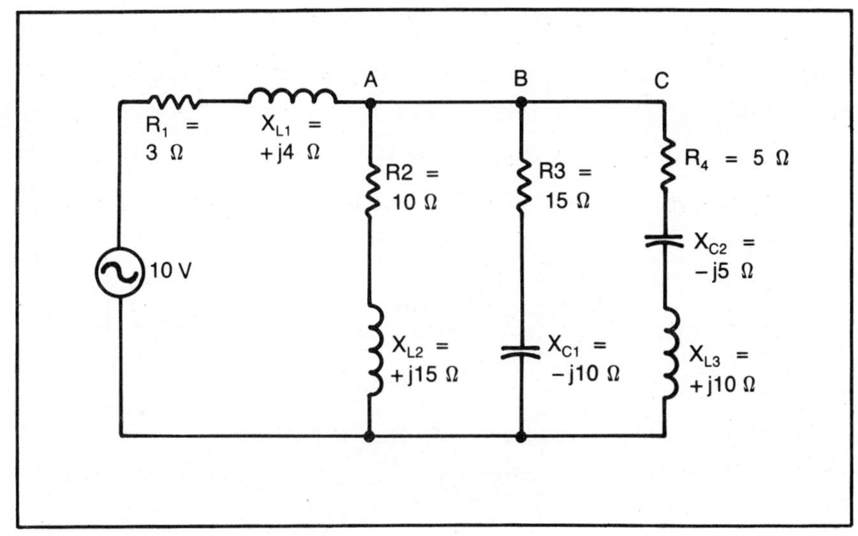

Fig. 10-8. Sample complex ac circuit.

$$\frac{1}{Z_P} = .05556\,\underline{/-56.3} + .05556\,\underline{/33.6} + .1428\,\underline{/-45}$$

The formula now says to add these numbers. Remember, it is necessary to add in rectangular form. Therefore, change each of the polar forms to rectangular form.

$$\frac{1}{Z_P} = (.0308 - j.0462) + (.0462 + j.0307) + (.1009 - j.1009)$$

Now it is possible to perform the indicated addition. Add the real terms together, then add the j terms together.

$$\frac{1}{Z_P} = (.1779 - j.1164) \text{ rectangular form}$$

Now the equation indicates taking the reciprocal of both sides in order to change to Z_P rather than $1/Z_P$. In order to take the reciprocal, it is necessary to perform this in polar form. Therefore, change to polar form.

$$\frac{1}{Z_P} = .2125\,\underline{/-33.19} \text{ polar form}$$

Now it is possible to take the reciprocal and end up with the value of the impedance of the parallel branches.

$$Z_P = 4.7\,\underline{/33.19} \text{ parallel equivalent impedance}$$

Step 3 It is necessary to combine the parallel impedance with the series impedance. Before doing this, it is necessary to change the parallel impedance to rectangular form.

$$Z_P = (3.9 + j2.57) \text{ rectangular form}$$

Write the series impedance in rectangular form so it can be added to the parallel impedance.

$$Z_S = (3 + j4) \text{ ohms}$$

Now, add the parallel impedance with the series impedance. This will be the total circuit impedance in rectangular form.

$$Z_T = Z_P + Z_S = (3.9 + j2.57) + (3 + j4)$$
$$Z_T = (6.9 + j6.57) \text{ ohms rectangular form}$$
$$Z_T = 9.52 \angle 43.59 \text{ ohms polar form}$$

Practice Problems

Calculate the answers to the problems using the j-operator and the circuit diagrams shown.

1. In the circuit shown, find the total impedance and the total current.

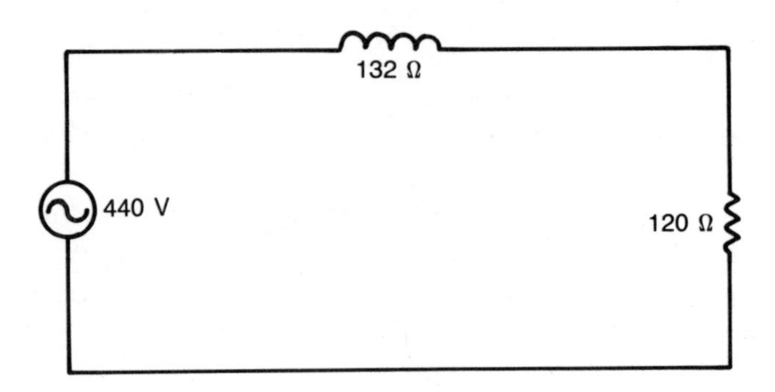

2. In the circuit shown, find the voltage drops across the resistor and capacitor, total current, and total impedance.

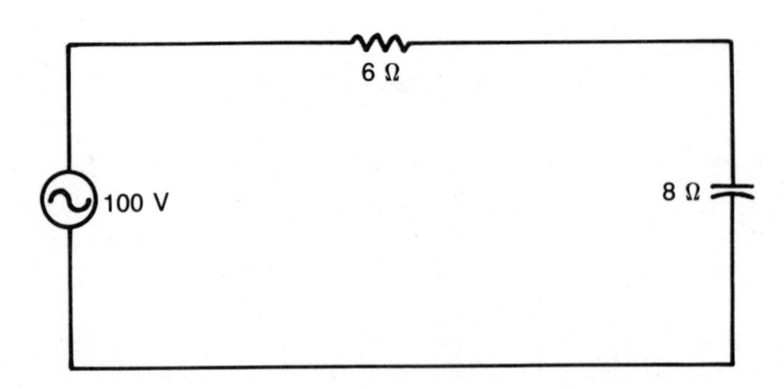

3. A circuit consisting of 175 ohms resistance in series with a capacitor of 5.0 μF is connected across a source of 150 volts, 120 hertz. Determine the impedance of the circuit and the current through the circuit.
4. A circuit consisting of 100 ohms resistance, 0.35 H inductance, 13 μF capacitance is connected across a 220 volt, 60 Hz voltage source. Find the voltage drops across each component, total current and total impedance.
5. In the circuit shown, find the total impedance and the total current.

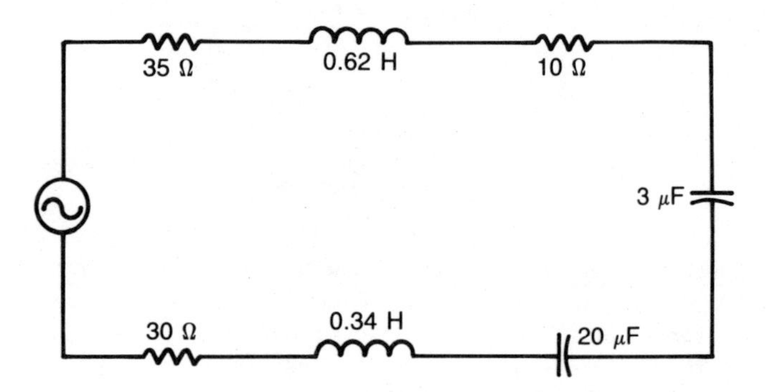

6. In the circuit shown, find the current through the resistor and inductor, total current and total impedance.

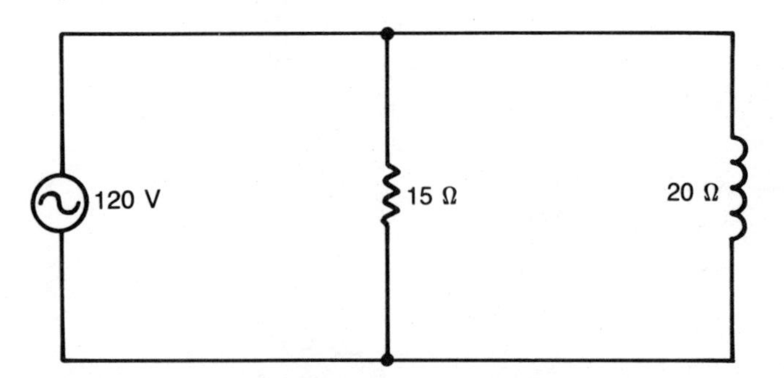

7. In the circuit shown, find the current through the resistor and capacitor, total current and total impedance.

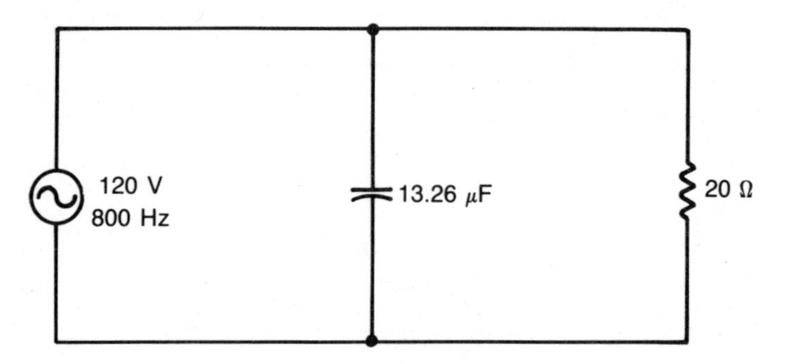

8. In the circuit shown, find the series equivalent impedance from point A to B with a frequency of 5 MHz.

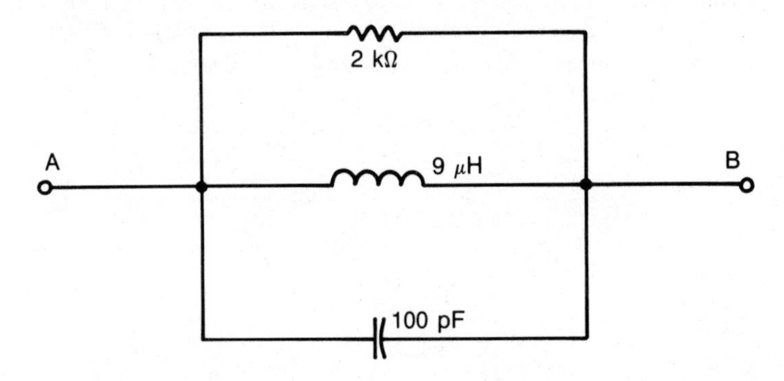

9. A parallel circuit contains two branches. The first branch has a 75 ohm resistor in series with a capacitive reactance of 30 ohms. The second branch has a 35 ohms resistor in series with an inductive reactance of 50 ohms. Find the total impedance.
10. A parallel circuit contains two branches. The first branch has an 80 ohm resistor in series with an inductive reactance of 26 ohms. The second branch has only a capacitive reactance of 100 ohms. Find the total impedance.
11. In the circuit shown, find the total impedance.

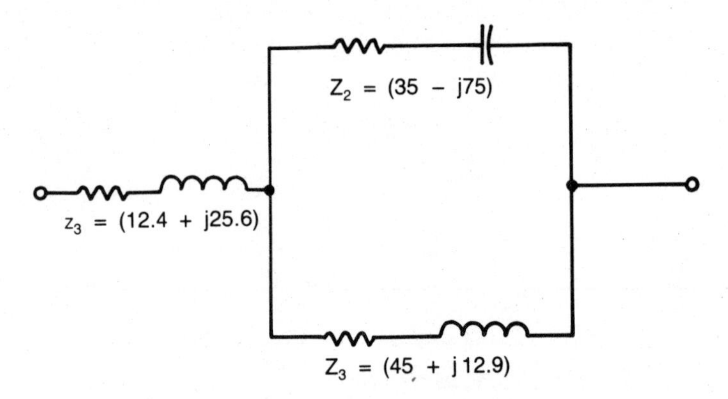

12. In the circuit shown, Z_1 = (27.7 – j50), Z_2 = (150 + j76.2), $Z_3 = 111.5 \angle 21$. Find the total impedance.

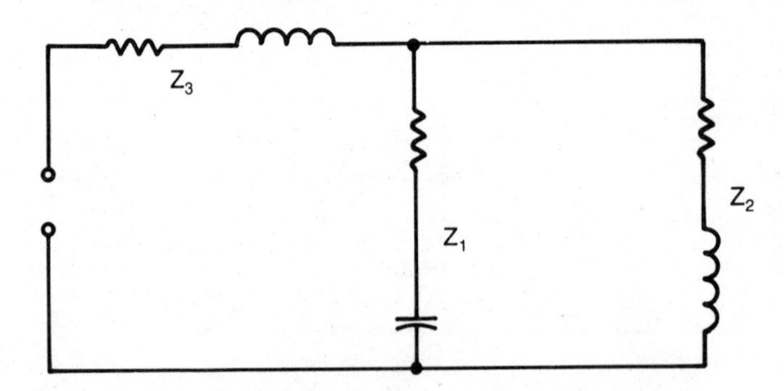

13. What is the equivalent impedance of two impedances $Z_1 = 151 \angle 4.07$ and $Z_2 = 50 \angle 53.1$ connected in parallel?
14. What is the equivalent impedance of two impedances $Z_1 = (73.8 - j34.4)$ and $Z_2 = (30 + j40)$ connected in parallel?
15. What is the equivalent impedance of two impedances $Z_1 = 60.5 \angle 20$ and $Z_2 = (100 + j0)$ connected in parallel?

CHAPTER SUMMARY

Alternate current circuits that contain both inductance and capacitance, along with resistance are dealt with in a manner very similar to circuits with only one reactive component. The key to remember is the fact that the vectors representing capacitive reactance and inductive reactance are opposite in their direction. This results in at least a partial cancellation of the vectors. The result is called the net reactance.

Sometimes it is necessary to classify a circuit as being either inductive or capacitive. The reason for the classification is to be able to predict how the circuit will react in terms of a leading or lagging voltage or current.

The j-operator is a mathematical tool used to simplify the mathematics involved in solving complex ac circuits. When the j-operator is written in a number, the number then is called a complex number. Complex numbers can be written in either rectangular or polar form.

The following are considered the key points of the chapter:

- The vectors for inductance and capacitance are opposite.
- Series circuits are classified as inductive or capacitive by whichever is the larger reactance or voltage drop.
- Parallel circuits are classified as inductive or capacitive by whichever has the largest current. The largest current results from the smallest reactance.
- A circuit is classified as resistive only if the net reactance is zero.
- The rectangular form of a complex number states the actual quantities of the real and j terms.
- The rectangular form is written with this format; (real $\pm$ j-term).
- The polar form of a complex number is written in this format; $Z \angle \Theta$.

Rules of Calculations with Complex Numbers

Rule 1 Add or subtract in rectangular form.

Rule 2 When adding or subtracting in rectangular form, add/subtract the real terms together and add/subtract the j terms together.

Rule 3 Multiply or divide in polar form.

Rule 4 When multiplying in polar form, multiply the magnitudes together (the Z term) and add the angles.

Rule 5 When dividing in polar form, divide the magnitudes (the Z term) and subtract the angles.

Rule 6 When taking the reciprocal of a number, it is to be done in polar form, since it is division. Take the reciprocal of the magnitude and change the sign of the angle.

Chapter 11

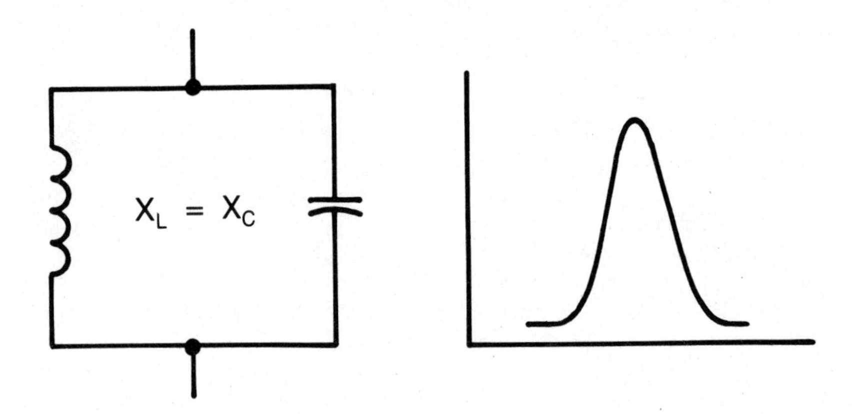

Resonance

Resonance is a very common use for an inductor and capacitor in the same circuit. A resonant circuit can be either series or parallel and the characteristics of each are completely different. Some of the uses of a resonant circuit include the tuning of an antenna, tuning of a radio receiver, and filtering circuits. Resonant circuits can accept only certain frequencies, a group of frequencies or it can reject certain frequencies, all depending on the application of the resonant circuit in the total circuit.

THE RESONANCE EFFECT

In Chapters 8 and 9, inductive reactance and capacitive reactance are discussed. Remember, inductive reactance will increase with an increase in frequency and capacitive reactance will decrease with an increase in frequency. Not only is the effect of frequency opposite, but all characteristics of the two reactances are opposite.

Once a set of values for inductance and capacitance have been selected, it is possible to calculate the frequency where both inductive reactance and capacitive reactance are equal. It is this point, where the reactances are equal that is called the resonant frequency.

- At the resonant frequency, $X_L = X_C$.

Chapter 10 mentions that a circuit could be described as being either a capacitive circuit or an inductive circuit, depending on the net reactance. In a resonant circuit, the net reactance is zero.

CALCULATING THE RESONANT FREQUENCY

By the definition given above, the resonant frequency can be calculated by determining the frequency where the inductive reactance equals the capacitive reactance. Therefore:

Step 1 $X_L = 2\pi fL$ and $X_C = \dfrac{1}{2\pi fC}$ reactance formulas

Step 2 $X_L = X_C$ definition of resonance

Step 3 $2\pi fL = \dfrac{1}{2\pi fC}$ Substitute reactance formulas

Step 4 $f^2 = \dfrac{1}{2^2\pi^2 LC}$ rearrange to solve for f, take the square root of both sides

frequency at resonance $f_r = \dfrac{1}{2\pi\sqrt{LC}}$ **Formula 11-2**

When the resonant frequency is calculated, it does not depend on it being a series circuit or a parallel circuit, the formula is still the same. The frequency of resonance depends only on the combined values of L and C.

Table 11-1 shows that the values of inductance and capacitance can be selected over quite a wide range of values. Once one of the two is selected, the other value is calculated so the reactance of the two is equal, and therefore, resonant at the desired frequency. The table is calculated for a resonant frequency of 1000 kHz, of 1 MHz.

The five different combinations shown in the table all have $X_L = X_C$ but, there is a wide range of values that the reactance can have. Because there is such a wide range, what is the best way to select the values of the components? As a general rule, the inductor will have the deciding factor. Usually, with small values of inductance, the resistance of the wire used to make the inductor, becomes smaller. The ratio of X_L to its coil resistance, R_s (or r_s) is called Q, or the quality of the coil. We will investigate the effects of Q later in this chapter. To conclude this discussion in simplified terms, for normal circuits, X_L is usually selected to be approximately 1500 ohms. A lower value means the resonant circuit does not have as sharp of a bell curve (the curve of the output) and a higher value of X_L results in a circuit with a very sharp bell curve.

Calculating the Value of C when Resonant Frequency Is Known

The sample shown here is selected from one of the points in Table 11-1.

Step 1 $f_r = \dfrac{1}{2\pi\sqrt{LC}}$ resonant frequency formula

Table 11-1. L-C Combinations Resonant at 1000 kHz.

L (μH)	C (pF)	Actual Value $X_L = X_C$ (ohms)
23.9	1060	150
119.5	212	750
239	106	1500
478	53	3000
2390	10.6	15,000

Step 2 $f_r^2 = \dfrac{1}{4\pi^2 LC}$ square both sides to remove the square root sign

Step 3 $C = \dfrac{1}{4\pi^2 L f_r^2}$ equation to solve for C

Find the value of the capacitor to be used if the value of inductor is selected to be 239 μH, with a resonant frequency of 1000 kHz.

Step 1 $C = \dfrac{1}{4\pi^2 L\, f_r^2}$ formula

Step 2 $C = \dfrac{1}{4\pi^2 \times 239\ \mu H \times (1000\ kHz)^2}$ substitute values

Step 3 C = 106 pF value of C

Calculating the Value of L when Resonant Frequency Is Known

The sample shown here is selected from one of the points in Table 11-1.

Step 1 $f_r = \dfrac{1}{2\pi\sqrt{LC}}$ resonant frequency formula

Step 2 $f_r^2 = \dfrac{1}{4\pi^2 LC}$ square both sides to remove the square root sign

Step 3 $L = \dfrac{1}{4\pi^2 C\, f_r^2}$ equation to solve for L

Find the value of inductance to be used in a resonant circuit where the value of capacitance is 212 pF and the frequency of resonance is 1000 kHz.

Step 1 $L = \dfrac{1}{4\pi^2\, C\, f_r^2}$ formula

Step 2 $L = \dfrac{1}{4\pi^2 \times 212\ pF \times (1000\ kHz)^2}$ substitute values

Step 3 L = 119.5 μH (Value of L)

Effect Above and Below Resonant Frequency

When a circuit is at the resonant frequency, the effects of inductive reactance and capacitive reactance cancel each other, leaving a net reactance of zero, since they are equal and opposite. However, when the frequency drifts from resonance, the circuit will start to display characteristics of either an inductive circuit or a capacitive circuit because not all of the reactance will be canceled. The characteristics of inductive or capacitive depend on the circuit being either a series circuit, with the effects being determined by the voltage drops, or a parallel circuit, with the effects being determined by the current through each branch.

Even though the circuit configuration (series or parallel) determines the characteristics, it is possible to predict the actual values of inductive reactance and capacitive reactance, above and below the resonance frequency.

Table 11-2, shown below, has a resonant frequency of 1000 kHz. Reactance is calculated below resonant frequency and above resonant frequency. The values of reactance calculated in this table will apply to both series and parallel circuits.

Notice, below resonant frequency, the net reactance is larger in the capacitive side. Above the resonant frequency, the net reactance is larger on the inductive side. However, keep in mind that the effects of net reactance on the circuit depend on the circuit being series or parallel.

SERIES RESONANCE

Figure 11-1 shows a series resonant circuit. Notice, there is a resistor in series with the inductor and capacitor. This resistor is labeled r_s and is intended to represent the resistance of the inductor, resistance of the wires and any other resistance that may be present in the circuit. This r_s should be less than 10 ohms or the effect of the resonant circuit may be minimized to a point of uselessness.

By the definition of resonance, if $X_L = X_C$ and they cancel each other, the only remaining component in the circuit is the series resistor. It should make sense, then that if the reactances cancel and the only thing left is the resistor, the impedance of the circuit is very small. In fact, the impedance of the circuit is the remaining resistance.

The impedance is considered to be a minimum. If the impedance is minimum, the current flowing through a series resonant circuit would be large, restricted only by the series resistor.

- In a series resonant circuit, impedance is minimum and current is maximum.

The curve shown with the circuit in Fig. 11-1 is called the bell curve. Across the horizontal axis is the frequency and the vertical axis is the current. Right at the resonant frequency is the peak of the curve. It is at this point that the current through the circuit is the maximum value. Notice, the curve drops off very steeply on either side of the resonant frequency. This drop off in current can also be seen in Table 11-2 by looking at the net reactance.

In a series circuit, the circuit is classified as either inductive or capacitive depending on the voltage drops of the individual components. The voltage drops depend on the relative sizes of the two components, that is to say the reactances.

Table 11-2. Reactance at Frequencies Near 1000 kHz.

Frequency kHz	X_L ohms	X_C ohms	Net Reactance ohms
600	900	2500	1600
800	1200	1875	675
1000	1500	1500	0 (Resonance)
1200	1800	1250	550
1400	2100	1070	1030

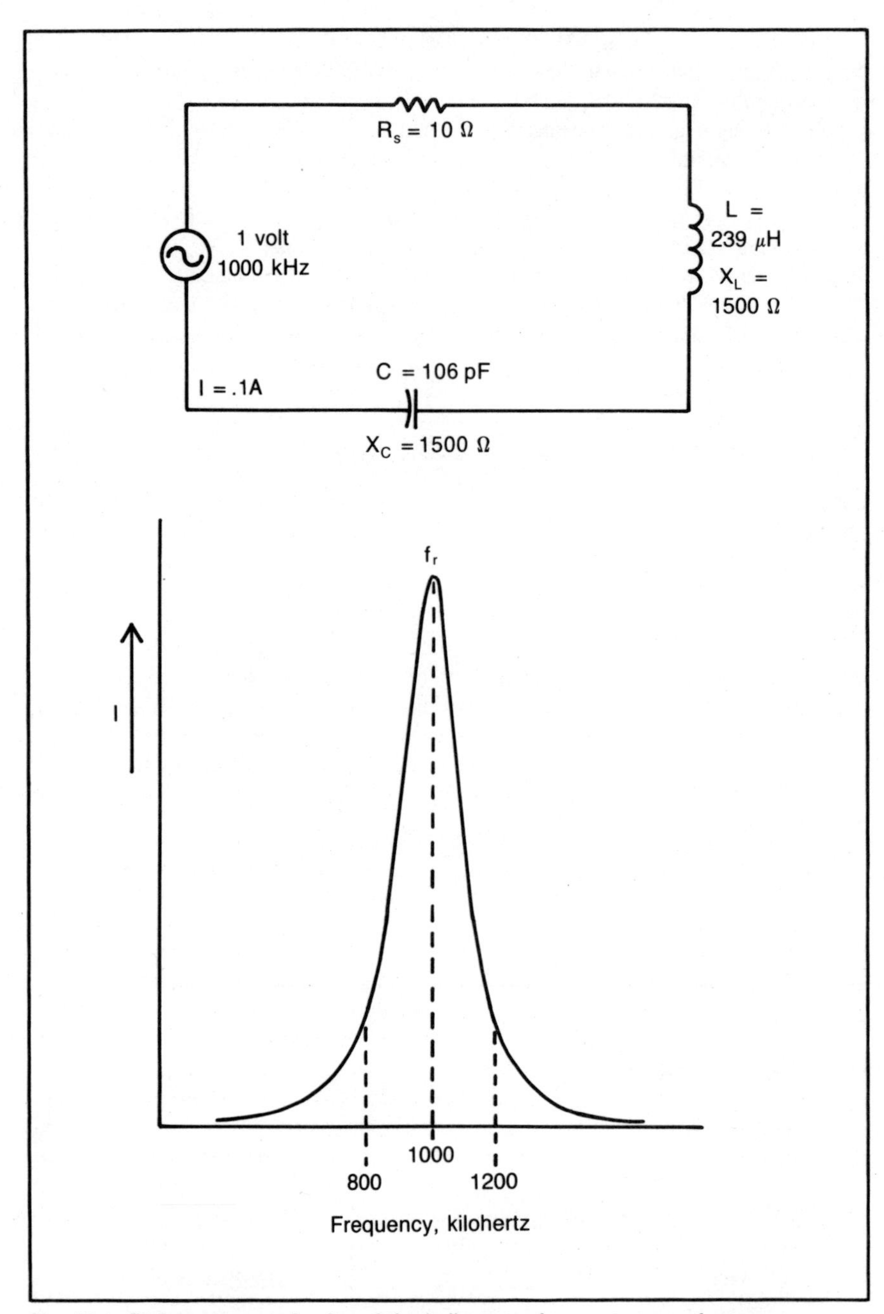

Fig. 11-1. Series resonant circuit and the bell curve of current versus frequency.

Referring to Table 11-2, the capacitive reactance below resonant frequency is higher than the inductive reactance. Above resonant frequency, the inductive reactance is higher.

- In a series circuit, the circuit is net capacitive below resonant frequency and net inductive above resonant frequency.

PARALLEL RESONANCE

Many of the items discussed earlier will still hold true in a parallel circuit because the circuit configuration does not effect such things as the resonant frequency, and the values of inductive or capacitive reactance.

The main change between the series circuit and the parallel circuit is the circuit conditions at resonance. In a series circuit, $X_L = X_C$ and cancel each other to leave the series resistance as the only component to limit current. In a parallel circuit, $X_L = X_C$ but it is the branch currents that cancel, leaving a path of no current flow. In a series circuit, the maximum current resulted in a minimum impedance. In a parallel circuit, the canceled branch currents make the circuit appear to be a very high impedance. In the parallel circuit, the higher reactance above or below frequency results in the lower value of branch current. The higher branch current determines the circuit characteristics. Therefore, below resonant frequency, the capacitive reactance is higher than the

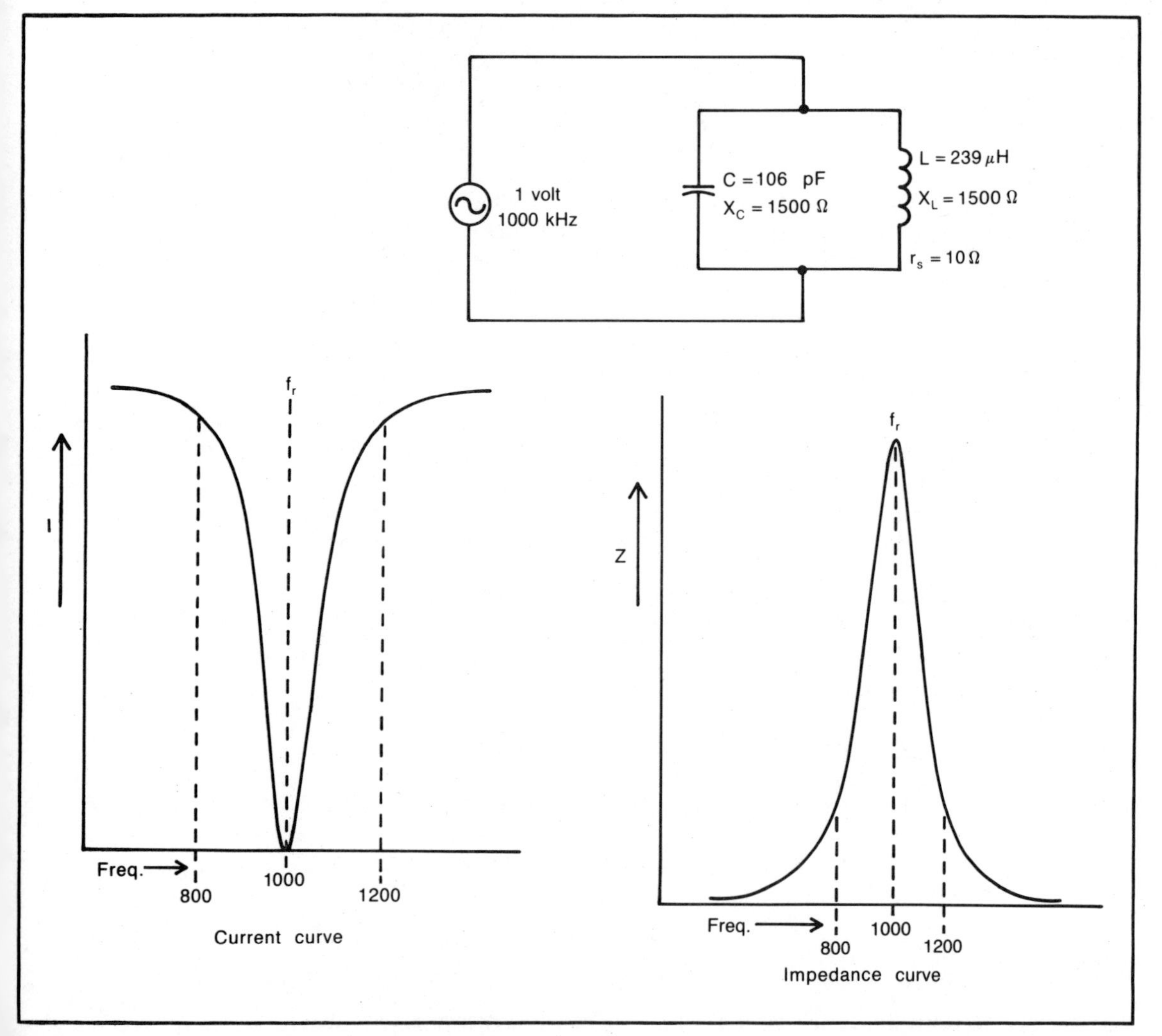

Fig. 11-2. Parallel resonant circuit and the bell curves of current versus frequency and impedance versus frequency.

inductive reactance, this produces more current in the inductance branch. Therefore, the circuit is classified as an inductive circuit.

- In a parallel resonant circuit, impedance is maximum and current is minimum.
- In a parallel circuit, the circuit is net inductive below resonant frequency and net capacitive above resonant frequency.

Figure 11-2 shows a parallel resonant circuit with its respective bell curves. Notice in the circuit, the resistor, r_s is placed in series with the inductor. This is to represent the resistance of the inductor windings. As long as the value is quite small, there is little or no effect on the resonant circuit.

The two bell curves show the comparison between the current vs frequency and the impedance versus frequency. The current curve is to show the sharp drop in current at the resonant frequency. The impedance curve shows a very sharp increase at the resonant frequency. If this parallel circuit is used in conjunction with another circuit, the impedance curve could also be used to represent the voltage dropped across the resonant circuit in comparison to another section of a circuit this might be connected to. For example, this could be connected as a filter in parallel with the applied voltage and the remainder of the circuit. The resonant frequency would see the filter as an open circuit. All frequencies that are not the resonant frequency would see the filter as a path to ground.

Using the same idea as an example for the series resonant circuit, if the series circuit were used as a filter, it would be connected in series with the load and voltage source. At the resonant frequency, the resonant circuit would appear as only a piece of wire. At any other frequency, the circuit would have high impedance and would stop a signal from passing through by having too large of a voltage drop.

Series and parallel resonant circuits can be used in combination with each other to produce filters that fit specialized needs. Although filters are not the only application of resonant circuits, it is an easy one to describe.

Q OF A RESONANT CIRCUIT

The Q of a resonant is a measure of the quality or figure of merit of the circuit. In general, the higher the circuit Q, the sharper the resonance effect, that is to say, a sharper bell curve.

The Q of a resonant circuit, either parallel or series can be calculated by:

Q of a resonant circuit.
Q is a r_s ratio; it has no units.

$$Q = \frac{X_L}{r_s}$$

Formula 11-2

X_L in the formula is the inductive reactance, measured in ohms, and r_s is the resistance of the coil windings. This formula assumes r_s is small in comparison to X_L and the capacitor is assumed not to have any leakage. If r_s becomes larger, the resonant effect becomes less sharp. The bell curve flattens out.

In a parallel circuit, the Q can be used to determine the Z of the circuit. The Z of a parallel circuit is defined as maximum.

impedance of a parallel resonant circuit $Z_p = Q \times X_L$

Formula 11-3

The Z of a parallel circuit is expected to be high. The higher Q produces a sharper resonant effect, therefore, a higher impedance.

BANDWIDTH OF A RESONANT CIRCUIT

The bell curve that has been discussed throughout this chapter is shown as having sloped sides to the curve. The resonant frequency curve allows calculations of the resonant frequency. When the resonant frequency is shown on the bell curve, however, it is seen that there are frequencies that are close that have almost the same amplitude as the resonant frequency.

In order to make calculations concerning bandwidth, a standard has been set.

- Bandwidth is defined as the range of frequencies with a response of 70.7 percent (or more) of the maximum response. 70.7 percent of current (series) or 70.7 percent of voltage (parallel). Measured in frequency units.

The bandwidth is also known as the half-power points, or the −3dB points. Half-power and −3dB are equal to 70.7 percent of the maximum current or voltage.

Bandwidth can be represented by Δ f or by BW. The first symbol best describes exactly what the bandwidth is. The triangle in the first symbol is actually the Greek letter Delta. It is used here to mean change in or range of and the small letter f stands for frequencies. When the two symbols are put together, Δf, it means change in frequencies or range of frequencies.

Refer to Fig. 11-3. This drawing is a simplified drawing of a bell curve. Its purpose is to demonstrate the bandwidth in terms of the 70.7 percent of the maximum amplitude and to show the bandwidth in terms of the Δf. Notice, there are points labeled f_1 and f_2. These two points are the frequencies at the 70.7 percent point. f_1 is the lower frequency and f_2 is the higher frequency. If f_1 is subtracted from f_2, the difference will be the range of frequencies between these two points, also called the bandwidth.

If the bell curve is perfectly symmetrical, the resonant frequency is found exactly in the middle, at the highest point on the curve. Making the assumption that the bell curve is perfectly symmetrical, the bandwidth can be divided in half. Half will be above resonant frequency and half will be below resonant frequency.

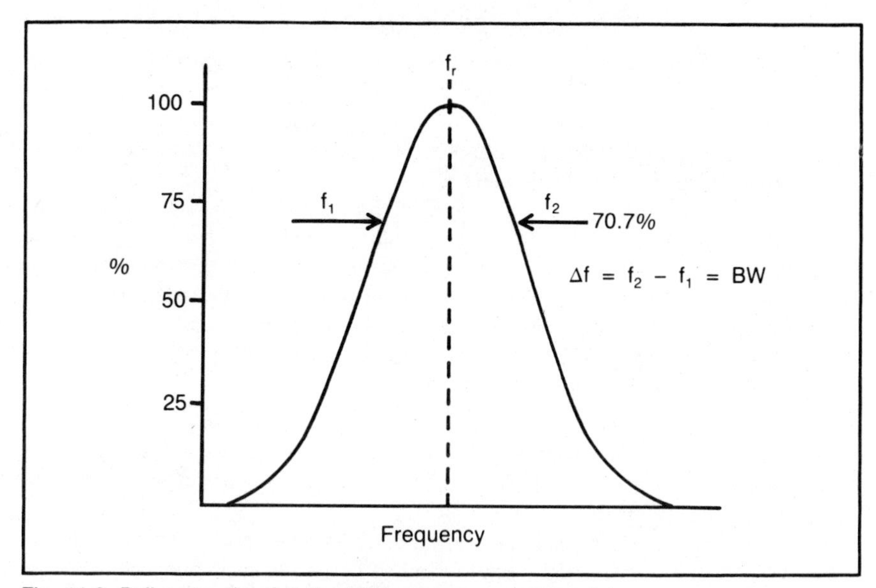

Fig. 11-3. Bell curve showing the 70.7 percent response point.

For example, in Fig. 11-3, if the resonant frequency, f_r, is 100 kHz and the bandwidth is 20 kHz, then the points at the 70.7 percent level could be identified as 90 kHz and 110 kHz.

$$f_1 = f_r - \frac{BW}{2} \quad \text{substitute } f_1 = 100 \text{ kHz} - \frac{20 \text{ kHz}}{2} = 90 \text{ kHz}$$

$$f_2 = f_r + \frac{BW}{2} \quad \text{substitute } f_2 = 100 \text{ kHz} + \frac{20 \text{ kHz}}{2} = 110 \text{ kHz}$$

It is possible to calculate the bandwidth of a circuit in terms of the circuit values. The bandwidth is affected by the Q of the circuit. A circuit with a very high Q will have a very small bandwidth, steep bell curve. A circuit with a low Q will produce a very wide bandwidth or shallow bell curve.

bandwidth in terms of circuit Q. $BW = \frac{f_r}{Q}$ **Formula 11-4**

bandwidth in terms of frequency. $BW = \Delta f = f_2 - f_1$ **Formula 11-5**

Refer to Fig. 11-4. There are three bell curves drawn. Each curve has the same resonant frequency. The three curves show the effect of circuit Q on the bandwidth. The curve with the Q of 80 has a steep bell curve, therefore, at the 70.7 percent points, the bandwidth is quite narrow. The middle curve, with a Q of 40 is not as steep of a bell curve because the resonant effect is not as great and the bandwidth is twice as wide as the top curve. The lowest curve on the graph shows a Q of 10. This curve has very little resonant effect and a very wide bandwidth. The curve for a pure resistance would be a flat line.

CHAPTER SUMMARY

The resonant circuit is useful for accepting, or rejecting certain frequencies. The main key to any resonant circuit is the fact that the inductive reactance and capacitive reactances are equal, at resonant frequency.

The Q of a resonant circuit is an important factor, because it will determine the sharpness, or bandwidth of the resonant effect.

Bandwidth is defined as the 70.7 percent of maximum response. The points that have less than this response are considered too small to be of any practical use. If the circuit has a narrow bandwidth, the sides of the curve are so steep that the frequencies below the 70.7 percent points would be very hard to find with measurements.

If the bell curve is perfectly symmetrical, the bandwidth can be divided in half and the upper and lower frequencies (the 70.7 percent points) can be calculated.

- At the resonant frequency, $X_L = X_C$.
- In a series resonant circuit, impedance is minimum and current is maximum.
- In a series, the circuit is net capacitive below resonant frequency and net inductive above resonant frequency.
- In a parallel resonant circuit, impedance is maximum and current is minimum.
- In a parallel circuit, the circuit is net inductive below resonant frequency and net capacitive above resonant frequency.

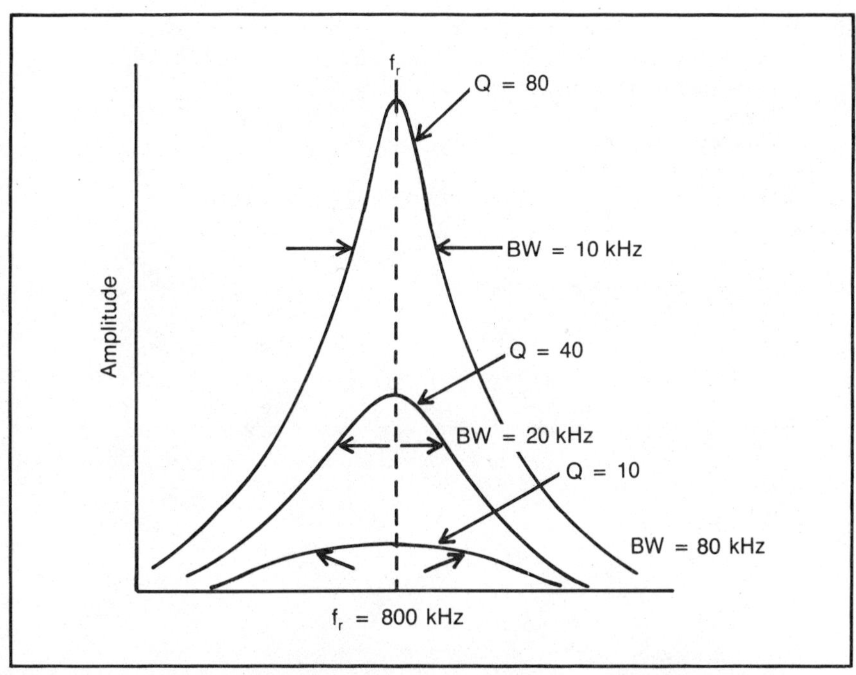

Fig. 11-4. Three bell curves, each with the same resonant frequency and different Q.

- Bandwidth is defined as the range of frequencies with a response of 70.7 percent (or more) of the maximum response.

Summary of Formulas

resonant frequency $f_r = \dfrac{1}{2\pi\sqrt{LC}}$ **Formula 11-1**

Q of a resonant circuit (no units) $Q = \dfrac{X_L}{r_s}$ **Formula 11-2**

impedance of a parallel resonant circuit $Z_P = Q \times X_L$ **Formula 11-3**

bandwidth in terms of circuit Q $BW = \dfrac{f_r}{Q}$ **Formula 11-4**

bandwidth in terms of frequency $BW = \Delta f = f_2 - f_1$ **Formula 11-5**

Practice Problems

Calculate the answers to each problem.

1. Find the f_r for a series circuit with a 10 μF capacitor and an inductor of 16 H, with 5 ohms internal resistance.

2. Find the f_r for a parallel circuit with an L of 80 μH and C of 120 pF.
3. In a series resonant circuit, X_L is 1500 ohms and the internal coil resistance is 15 ohms. At the resonant frequency determine:

    a. How much is the Q of the circuit?
    b. How much is X_C?
    c. With a generator voltage of 15 mV, how much is I?
    d. How much is the voltage across X_C?

4. What value of L is necessary with a C of 100 pF for a resonant frequency of:

    a. 1 MHz
    b. 4 MHz

5. Calculate the lowest and highest values of a variable capacitor needed with a 0.1 μH inductor to tune through the commercial FM broadcast band of 88 MHz to 108 MHz.

Chapter 12

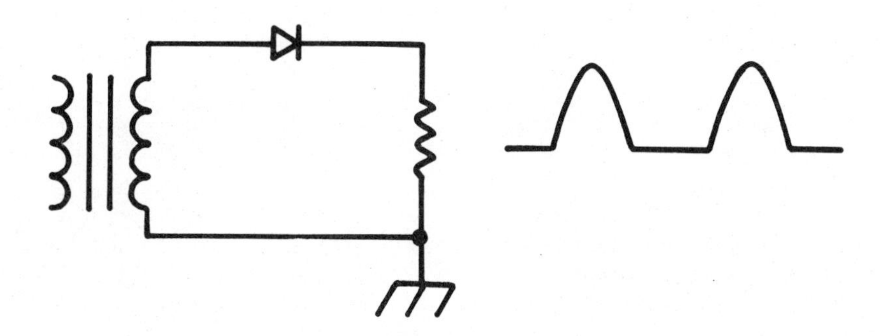

Power Supplies

Power supplies are a very important part of any electronic system. After all, without a power supply of some sort, electrical circuits cannot operate. As an electronic technician, it is equally important to determine if a power supply is operating properly.

A power supply is simply a means of changing ac power to dc for use in the circuits. Very few circuits can operate without dc. The primary device used for converting ac to dc is the diode.

THE P-N JUNCTION DIODE

Diodes that are used for the purposes of rectifying ac to dc are of the P-N junction type of diode. The P-N junction is two pieces of semiconductor material joined together.

Semiconductor materials are doped to become either P-type or N-type materials. The doping is an impurity atom that is added to the pure silicon or germanium crystals. It causes the semiconductor material to have either one extra electron in orbit, or be missing one electron in its outer orbit. The extra electron, or missing electron is what makes the semiconductor material useful in the field of electronics.

- A semiconductor with an extra electron is called an N-type semiconductor.
- A semiconductor missing an electron is called a P-type semiconductor. The atom with a missing electron is referred to as a hole.

Figure 12-1A shows a P-N junction diode. The point where the two pieces of semiconductor material are joined is called the junction. On the P-type side, there is a large number of holes. These are the majority carriers of a P-type material. In the P-type material, there will be a very small number of minority carriers, which are electrons in the P material. Holes in the P-type material are represented as empty circles, to show they are an atom that is void one electron. They can be said to carry a positive charge.

- P-type semiconductors have a majority of holes, and a minority of electrons. Holes carry a positive charge.

Figure 12-1A also shows an N-type of material. The N-type has a majority of electrons and a minority of holes. The electrons of N-type material and holes of P-type material are considered to be current carriers of the semiconductor. The circles with a minus sign represent the electrons.

- N-type semiconductors have a majority of electrons, and a minority of holes. Electrons carry a negative charge.

Figure 12-1B shows the P-N junction diode with a depletion zone. As soon as the P material and the N material are joined together, some of the electrons will join with some of the holes in the vicinity of the junction. The region is then called the depletion zone because it becomes void of any mobile carrier because when the holes and electrons join, the atom is no longer missing or extra an electron.

- The depletion zone is depleted of any mobile carriers and acts like an insulation region.

In the area of the junction, the depletion zone will increase only to a certain point. It cannot expand any further because the attraction between the positive and negative charges is not strong enough without an outside voltage being applied to the semiconductor.

Figure 12-1C shows the P-N junction diode with a forward bias applied to the semiconductor. A forward bias is a voltage that will cause the semiconductor to conduct current flow. Current flow is always assumed to be electron flow. However, keep in mind, inside a semiconductor material, the current flow is the majority carrier.

In order to have current flow, there must be a complete circuit, which means that for every electron that leaves the negative terminal of the battery, an equal number of electrons must return to the positive terminal of the battery.

In Fig. 12-1C, the electrons leave the negative terminal of the battery, travel to the N-type semiconductor material. These electrons enter the N-type material and become majority carriers. The negative force of the battery repels the electrons in the N material to travel towards the junction. The depletion zone will then start to collapse. While this is happening at the N material, the positive battery is attracting electrons from the P-type material. This attraction removes the electrons from the P-type material and causes them to travel to the battery. The positive force of the battery will also repel the positive holes, causing them to move toward the junction, collapsing the depletion zone.

With the depletion zone of the forward biased diode collapsed, the electrons from the N-type material will enter the P-type material and recombine with the holes and be swept toward the positive terminal. The holes of the P-type material will enter the N-type and be recombined with the electrons and be swept toward the negative terminal. The result is, in the forward direction, the diode conducts current easily.

Even though the forward biased diode conducts easily, the junction still acts like a region of resistance. This is called the barrier potential which is .1 volt to .3 volt for germanium and .5 volt to .7 volt for silicon. This small voltage drop must be considered in all semiconductor circuits.

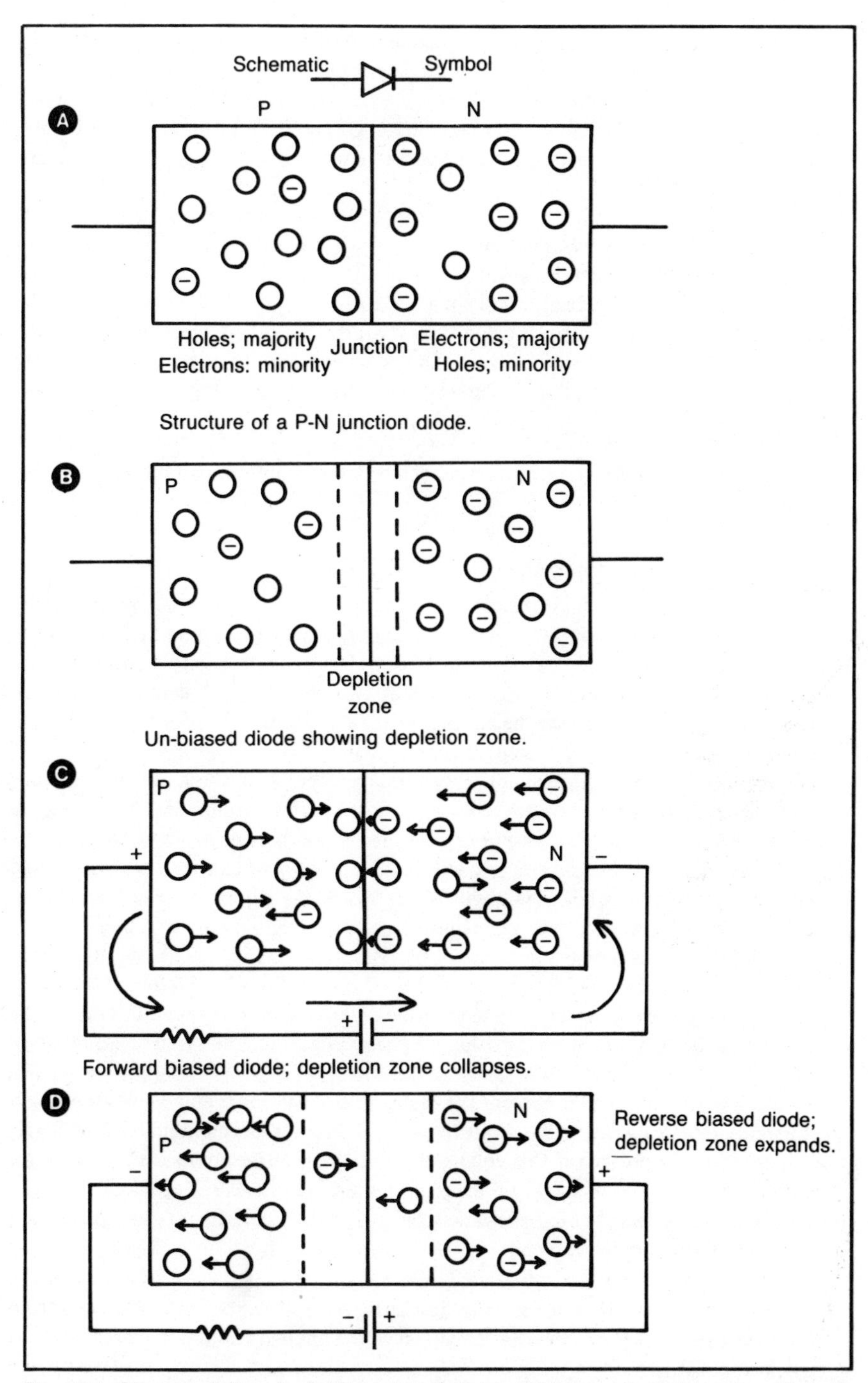

Fig. 12-1. Characteristics of a P-N junction diode. A. Structure of a P-N junction diode. B. Un-biased diode showing depletion zone. C. Forward biased diode; depletion zone collapses. D. Reverse biased diode; depletion zone expands.

- A forward biased diode has negative connected to the N material and positive to the P material.

Figure 12-1D shows a reverse biased diode. The positive terminal of the battery is connected to the N-type material and the negative terminal of the battery is connected to the P-type material. The positive on the N material attracts the electrons toward the battery, away from the junction. This causes the depletion zone to widen. The negative terminal of the battery attracts the holes toward the battery, away from the junction. This causes the depletion zone to widen. The depletion zone acts like a region of insulation and when it widens, it will not allow current to flow. Even though the depletion zone is quite wide, some electrons from the P material and some holes from the N material will travel across the junction. This is called minority current flow. Minority current flow is very small and is the only type of current in a reverse biased P-N junction diode.

- A reverse biased diode has positive connected to the N material and negative to the P material.

HALF-WAVE RECTIFIER

The half-wave rectifier is an ideal example of an application for the P-N junction diode. The half-wave rectifier is a power supply circuit made by using one diode.

Figure 12-2 shows a half-wave rectifier drawn to produce a positive output in Fig. 12-2A and a negative output in Fig. 12-2B. The two circuits are essentially the same with the exception of the diodes being in opposite directions. In order to simplify the discussion, Fig. 12-2A, the positive output is the one that will be discussed.

In Fig. 12-2 the diode is connected to a transformer. With the half-wave rectifier circuit, this is really not necessary. However, most rectifier circuits will be connected to a transformer because that is an easy way to step down the ac input voltage.

The secondary of the transformer will have a sine wave exactly the same shape as the primary, with a different voltage. When discussing any rectifier circuit, the input sine wave is discussed in terms of positive half-cycle and negative half-cycle.

When the input sine wave is in the positive half-cycle, Fig. 12-2A, the diode will have a positive polarity on its P-type material. At the same time, the bottom of the transformer will be negative, which puts a negative polarity on the N-type material of the diode. The conditions for a forward biased junction are satisfied and the diode will conduct.

When the diode conducts the positive half-cycle of the input sine wave, there will be forward diode drop across the diode of either .3 volt for Germanium (Ge) or .7 volt for Silicon (Si). The polarity of this voltage drop will be positive on the P side and negative on the N side. The resistor will drop the remainder of the voltage applied to the circuit. The voltage across the resistor will have a positive polarity on the top side and negative on the bottom side.

- When a rectifier diode is conducting, in the forward direction, the diode will have a voltage drop of .3 V or .7 V, the remainder of the applied voltage is dropped across the load resistor.

When the input sine wave swings into the negative half-cycle, the P material will have a negative voltage applied to it and the N material will have a positive voltage applied to it. It should be noted that the input sine wave polarity is always made in

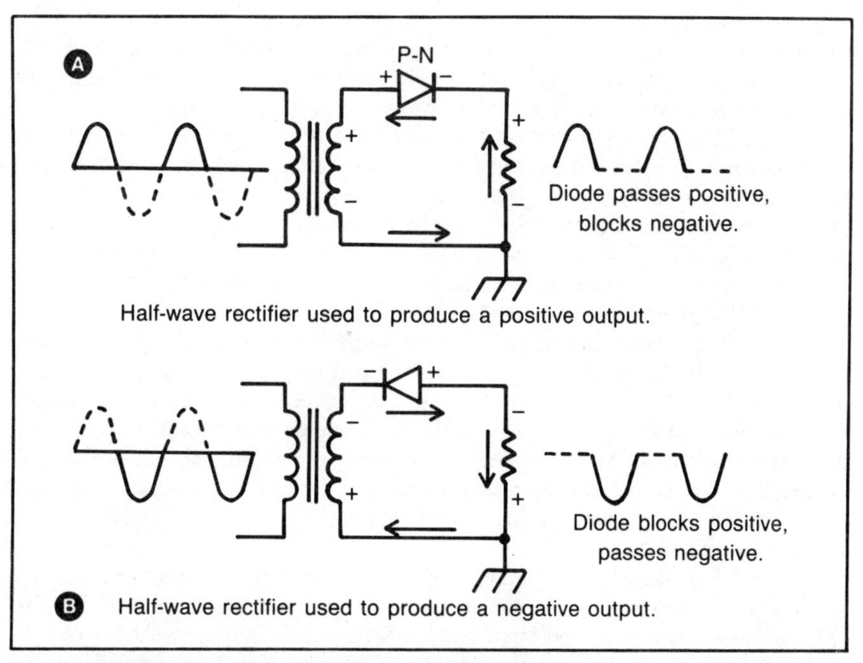

Fig. 12-2. Half-wave rectifier. A. Half-wave rectifier used to produce a positive output. B. Half-wave rectifier used to produce a negative output.

reference to the circuit ground, which is usually shown at the bottom of the transformer. Therefore, the negative half-cycle gives a transformer polarity of negative on the top and positive on the bottom of the transformer secondary. When the input sine wave is in its negative half-cycle, the diode is reverse biased and no current can flow. Minority current that can flow in the reverse direction is so small that it is considered zero.

- When the rectifier diode is reverse biased, there is no current flow. All of the applied voltage will be dropped across the diode and no voltage will be dropped across the load resistor.

When the two conditions of forward and reverse bias are both considered, we see that the arrows of both circuits (Figs. 12-2A and 12-2B) represent the current flow. Current flows only in one direction, which means the output voltage is considered dc. The name of this dc is pulsating-dc. The dashed lines in the output waveform represent a period of time when there is no output. The input voltage is blocked during this time.

FULL-WAVE RECTIFIER

The full-wave rectifier circuit is shown in Fig. 12-3. The circuit uses two diodes instead of one, like the half-wave circuit. The primary advantage is the fact that the full-wave circuit will reproduce both halves of the input sine wave.

The full-wave rectifier circuit requires the use of a center tapped transformer. The function of the center tapped transformer is to produce two sine waves, 180 degrees apart, from the one input sine wave. The center tapped transformer can also be used as a step up or as a step down transformer. The center tap is connected to ground reference, which means the center tap is always zero volts. Figure 12-3B shows the secondary sine wave split into two sine waves 180 degrees apart.

When the input sine wave is in its positive half cycle, as it is in Fig. 12-3B, the top part of the secondary becomes positive, in relation to the center tap. The bottom half of the secondary becomes negative, in relation to the center tap.

In Fig. 12-3C, the input sine wave is in its negative half cycle and the secondary produces two sine waves 180 degrees apart, and the portion of the sine wave of interest is the second part, shown by the solid line. The polarity of the transformer reverses, putting positive at the bottom, with respect to the grounded center tap. The waveform looks like it is upside down. However, the wave is drawn correctly because it is in reference to the grounded center tap (imagine the bottom of the transformer to be drawn over the half of the circuit connected to the top of the transformer).

Figure 12-3A shows a complete full wave rectifier circuit. The solid lines drawn on the waveforms of the secondary show which part of the wave is in use with each diode. The dashed lines show the portion not in use at the time. Notice that each diode is responsible to produce only one half of the input sine wave. The center tap is a common return for current to the transformer. The arrows indicate only one direction of current flow. The output waveform drawn for Fig. 12-3A shows both halves of the sine wave, with the negative portion made positive. The labeling of D1 and D2 indicate which diode produced the output.

Figure 12-3B shows the operation of diode D1 in the circuit. When the input sine wave is in its positive half cycle, the secondary has the polarity shown, with positive to D1 and negative to the center tap. The diode is forward biased and will allow current to flow to the load resistor. The diode will drop .3 V or .7 V and the remainder of the voltage will be dropped across the load resistor. During this portion of the sine wave, diode D2 is reverse biased because the bottom of the transformer is negative. Current flow follows the arrows, from the center tap, through the load, through the diode D1 to the transformer.

Figure 12-3C shows the operation of diode D2. When the input sine wave is in its negative half cycle, it produces the polarity across the transformer as shown, with a positive to diode D2 and a negative to the center tap. Diode D2 is forward biased and will allow current to flow to the load. The diode will have a voltage drop of .3 V or .7 V and the load will drop the remainder of the applied voltage. Notice the connection of diode D2 to the top side of the load resistor. It is this connection that causes the output to have both sides of the sine wave reproduced as positive. The polarity of the output is in relation to the ground reference point.

The full-wave rectifier circuit has the advantage over the half-wave circuit because it produces both halves of the input sine wave. This eliminates the period of time when there is an off condition produced with the half-wave rectifier circuit. The disadvantage is the fact that it does require a center-tapped transformer. Also, the center-tapped transformer produces one-half of its full secondary voltage between the tap and either side. This means the output voltage is only one-half of the voltage possible.

BRIDGE RECTIFIER

The bridge rectifier circuit is the most popular with modern power supplies. The reason is the fact that the circuit has some protection against failure. Also, bridge rectifiers are being manufactured as a single component for use in most low power circuits. The single unit makes the cost of building power supplies cheaper than other methods.

Figure 12-4 shows a bridge rectifier. There are four diodes used in this type of circuit. Two diodes take turns operating at a time.

Figure 12-4A shows the normal complete schematic. The output shown is the wave-

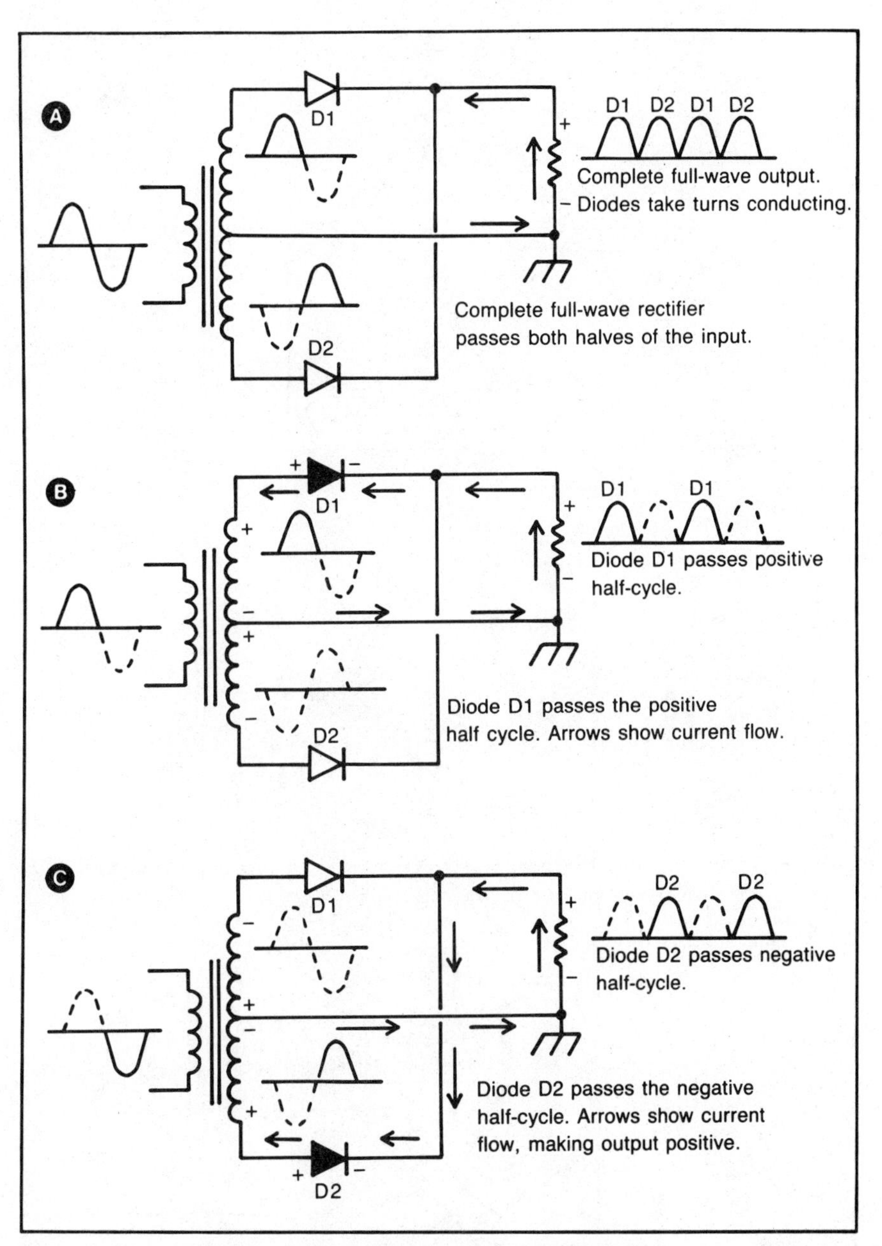

Fig. 12-3. Full wave rectifier. A. Complete full wave rectifier passes both halves of the input. B. Diode D1 passes the positive half-cycle. Arrows show current flow. C. Diode D2 passes the negative half-cycle. Arrows show current flow, output is positive.

form of a full wave rectifier, with both half-cycles being reproduced.

Figure 12-4B shows the bridge rectifier operating when the input sine wave is in its positive half cycle. The positive will forward bias D2 and the negative of the transformer is correct to forward bias D3. To follow current flow, start from the bottom side of the transformer, which is now marked negative, to the junction labeled ac. Pass through

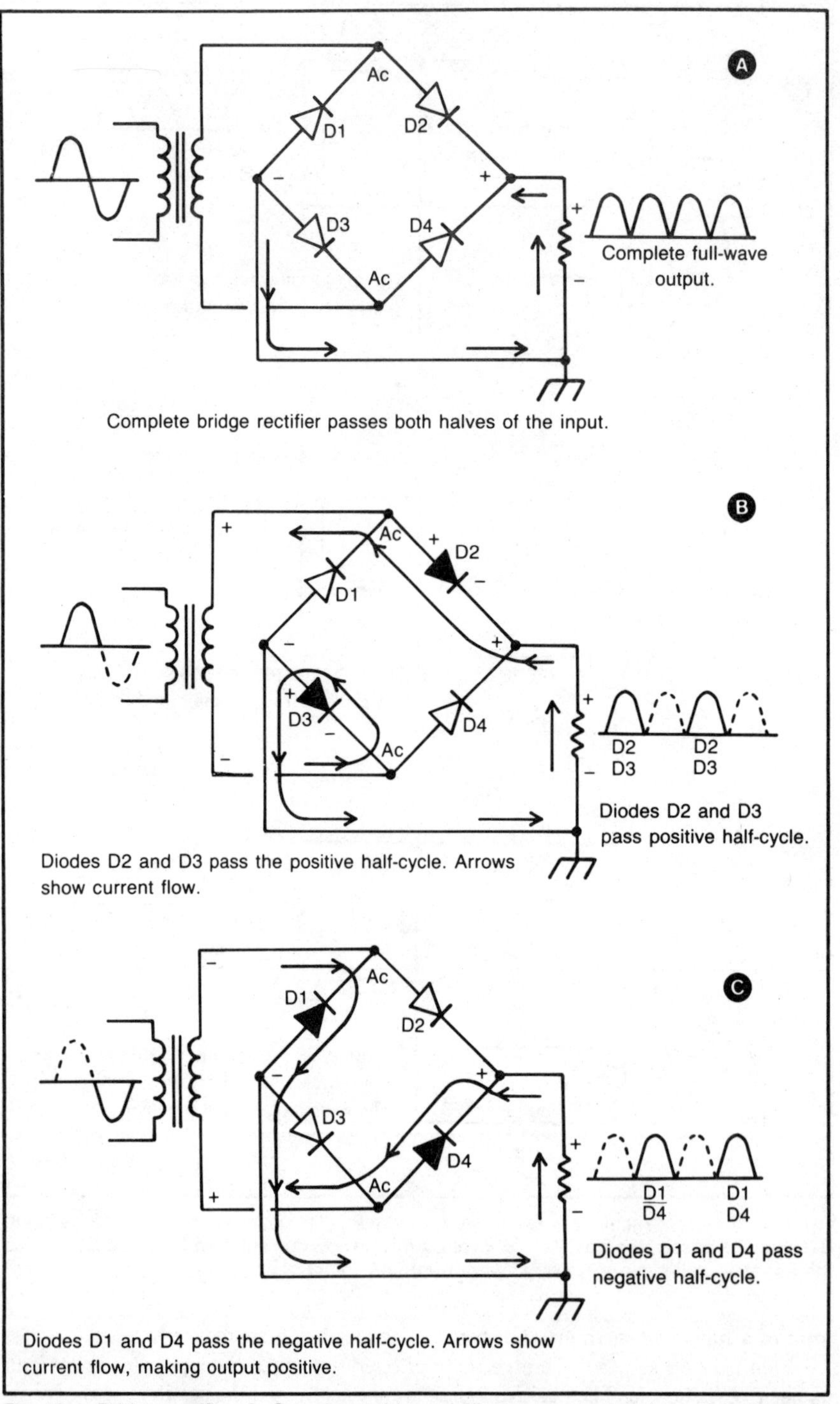

Fig. 12-4. Bridge rectifier. A. Complete bridge rectifier passes both halves of the input. B. Diodes D2 and D3 pass the positive half-cycle. Arrows show current flow. C. Diodes D1 and D4 pass the negative half-cycle. Arrows show current flow, output is positive.

diode D3, dropping .7 volt to allow for diode drop. Continue to follow the current arrows to the junction marked negative, then proceed to the load. The bottom of the load will be negative and the load will drop the majority of the voltage available to the circuit (all except 1.4 V diode drops). The large voltage drop across the load results in the load having the polarity as shown. Continue to trace current flow through the load, to the junction marked positive. Notice that diode D2 is forward biased, pass through D2 with a .7 volt diode drop, to the junction marked ac and return to the transformer.

From the discussion of the previous paragraph, it is seen that the output voltage for the positive half cycle of the input is from diodes D2 and D3.

Figure 12-4C shows the bridge rectifier as it operates on the negative half cycle. Diodes D1 and D4 are now forward biased. Follow the arrows shown in the drawing to trace the path of current to prove how the load once again will be positive.

The full wave pattern is the result of the bridge rectifier. Because it uses two diodes at any one time, the reverse ratings of the diodes can be smaller than the conventional full wave rectifier.

CALCULATING THE RECTIFIER OUTPUT

The output of a rectifier circuit, without filtering, is called pulsating dc. It does qualify as a dc because it does not change direction. Therefore, if it was necessary to measure the output voltage, a dc voltmeter would be the proper measuring instrument to use.

A dc voltmeter will measure the average of the voltage applied to it. Even though a sine wave has its average located at 63.6 percent of peak, if a dc voltmeter were to be used with a sine wave, the result would be to measure zero.

Figure 12-5 compares the outputs of a half-wave, full-wave, and the input sine wave. In all three cases, the peak value is the same. The peak is measured from the zero reference line to the very peak of the wave. The values of RMS and average apply only to the sine wave and not to the rectified outputs.

In Fig. 12-5, each of the rectified outputs shows a value called dc average. The average referred to there is the value of voltage that would be read on a dc voltmeter. The dc value is much lower on the half-wave output than it is on the full wave output. In fact, the dc average value of the full wave output is the same as the ac average of an equivalent sine wave. In other words, the dc average and the ac average both equal 63.6 percent of peak. Since the half-wave rectifier output has only one-half as many half cycles as the full wave circuit, the output must be one half the voltage. These two facts can be expressed in formulas.

formula to find dc output voltage of a full wave or bridge rectifier circuit

$$V_{dc} = .636 \times \text{Peak}$$ **Formula 12-1**

formula to find dc output voltage of a half wave circuit

$$V_{dc} = \frac{.636 \times \text{Peak}}{2}$$ **Formula 12-2**

When dealing in terms of the peak, average and RMS, it is often more convenient to deal in terms of RMS because that is how most ac voltages are expressed. The following will demonstrate a simple conversion from RMS to Average of a sine wave.

Step 1 Average = .636 × Peak starting formula
Step 2 RMS = .707 × Peak formula for RMS

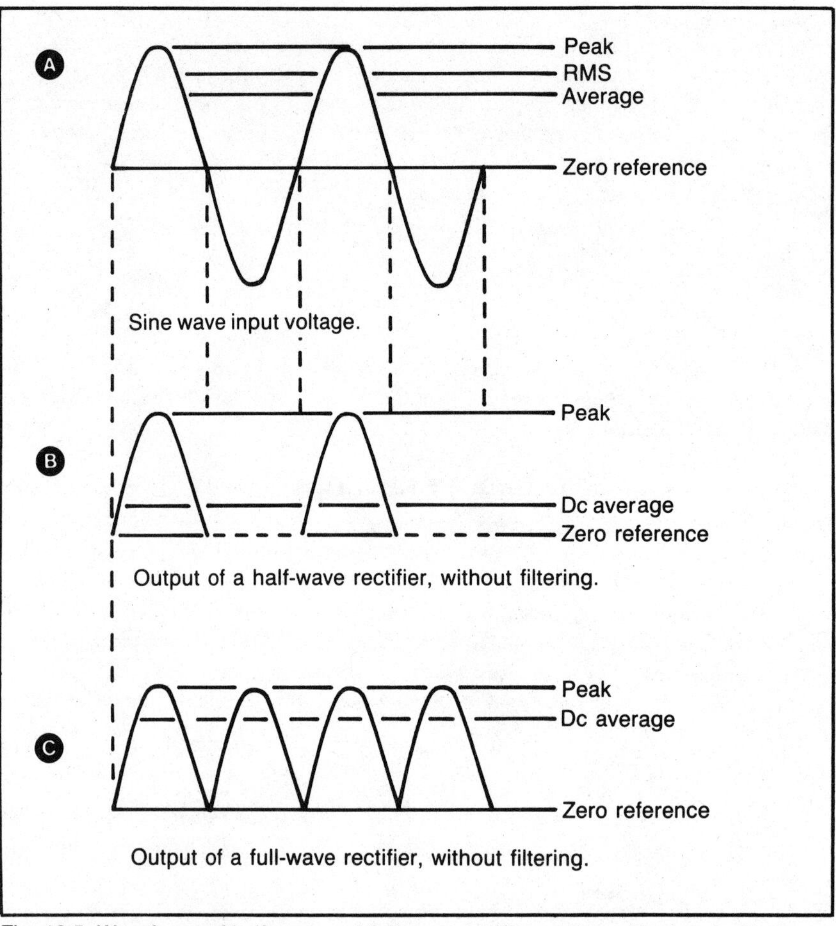

Fig. 12-5. Waveforms of half-wave and full wave rectifiers, without filtering. A. Sine wave input voltage. B. Output of a half-wave rectifier, without filtering. C. Output of a full wave rectifier, without filtering.

Step 3 Peak = $\frac{\text{RMS}}{.707}$ rearrange the formula

Step 4 Average = $.636 \times \frac{\text{RMS}}{.707}$ substitute for Peak

Step 5 Average = $.9 \times$ RMS simplify

The discussion shown above leads to the following formulas:

formula to find dc output voltage of a full wave or bridge rectifier $\quad V_{dc} = .9 \times \text{RMS}$ **Formula 12-3**

formula to find dc output voltage of a half wave rectifier circuit $\quad V_{dc} = \frac{.9 \times \text{RMS}}{2}$ **Formula 12-4**

The next step in making calculations for the output of a rectifier circuit is to determine the frequency of the output waveform. Regardless of the frequency of the input sine wave, the output always follows the input exactly. The half-wave circuit will reproduce either the positive or negative half cycles. The full wave, or bridge, reproduces both half cycles.

Frequency of a waveform is a measure of how often the waveform repeats itself. Notice from Fig. 12-5, the half wave rectifier repeats itself after completing the positive voltage, then the off time, then it is back to the positive voltage. This is exactly the same period of time as the input sine wave. The full wave waveform repeats itself each half cycle of the input, therefore its frequency is twice as much as the input sine wave. The output frequency is called the ripple frequency and is important when considering the characteristics of a power supply.

Ripple Freq. Half Wave = Input Frequency **Formula 12-5**
Ripple Freq. Full Wave = 2 times Input Frequency **Formula 12-6**

With the simple rectifier circuits that have been dealt with so far, there are only two calculations necessary; output voltage and output frequency. One consideration that is sometimes taken into account is the reverse voltage rating of the diodes, this is often called the peak inverse voltage. It is not unusual to find diodes with a PIV of 500 or even 1000 volts. For this reason, the PIV is usually not an important consideration with modern circuits using relatively low voltages, usually under 100 volts.

FILTERING A RECTIFIER CIRCUIT

The purpose of a filter on a rectifier circuit is to smooth the tremendous fluctuations in voltage. Many circuits cannot operate properly if there is any change in voltage. The filter is one step in making the dc voltage output a smooth, constant voltage to supply power to a circuit.

The filter that will be discussed here is the simple capacitor filter. There are many kinds of filters available for power supplies, but the principle of operation is the same for all the filters.

With the simple capacitor filter, the principle characteristic of the capacitor is relied upon. That characteristic is the fact that a capacitor will charge and store a voltage. The stored voltage can then be discharged through a different path than it was charged from.

Figure 12-6 shows the three basic rectifier circuits, each with a simple capacitive filter connected. The capacitor is connected in parallel with the circuit, especially the load resistor. The rectifier diodes will charge the capacitor in the positive direction, then when the diode turns off, the capacitor will discharge through the load. Even though the load is represented by a resistor here, it could be any circuit that needs to be powered by a dc source.

Figure 12-7 shows what takes place with the filter in the circuit. Figure 12-7A is the input sine wave shown here as a figure of reference. Figure 12-7B is the waveform of a half wave rectifier and Fig. 12-7C is the waveform of a full wave and bridge circuit.

Figure 12-7B, with the half wave circuit waveform presents the best picture for an explanation of how the filter capacitor does its job. The dotted lines are for reference only, to show what the output would look like if there were no filter. The solid lines represent the actual voltage that is applied to the output of the rectifier circuit. The output voltage is a combination of voltage from the diode and voltage from the capacitor. The figure assumes the circuit has been in operation for some period of time.

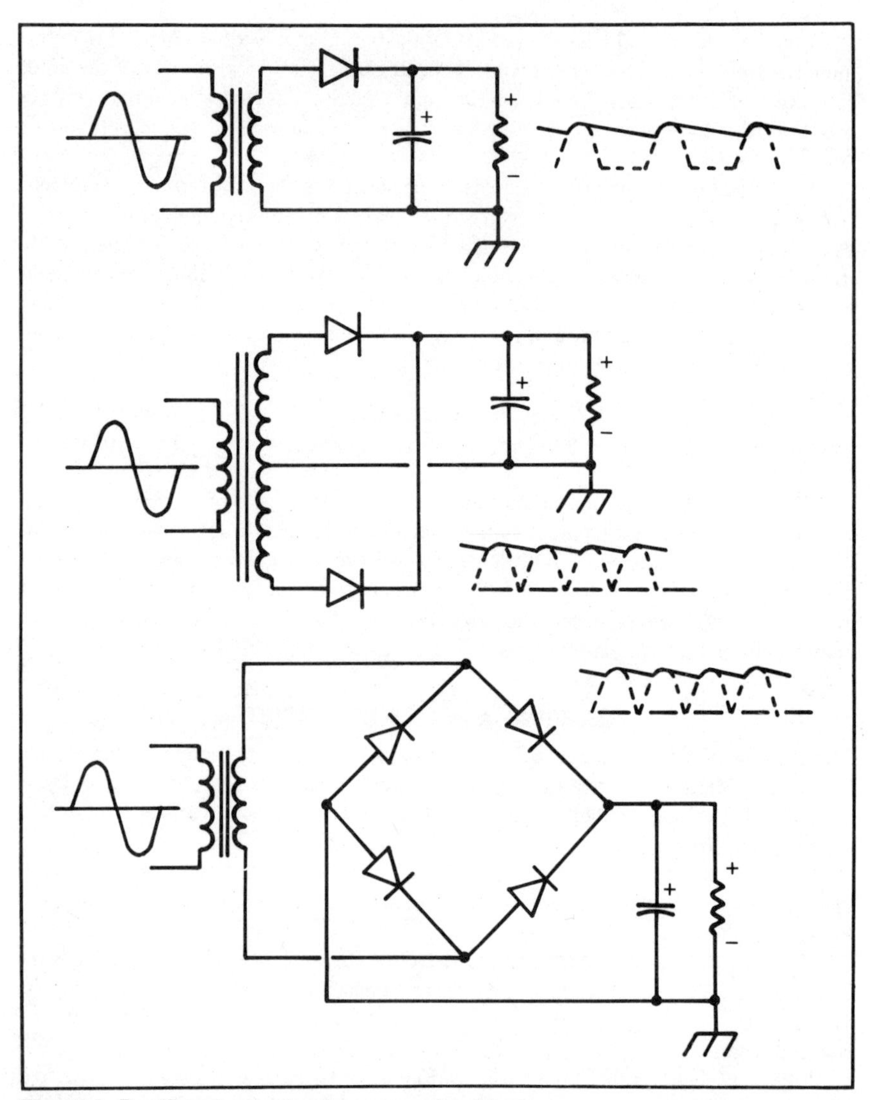

Fig. 12-6. Rectifier circuits showing a capacitor filter.

Looking at the left hand side of Fig. 12-7B, notice that the on time is slightly delayed from the beginning of the waveform. This is also true for each of the other on times shown. It is during this on time that the diode is supplying current to the load. During the off times, the capacitor is supplying current to the load.

The only time the diode can turn on is when the diode has the proper forward bias applied. This will occur when the input sine wave is greater than the voltage across the capacitor. Once the diode turns on, it will follow the input voltage to peak. While the diode is on, it is supplying current to the load and, because the capacitor is in parallel, it will also charge the capacitor. Once the input sine wave reaches peak, the input voltage starts to decrease. At this moment, the capacitor has peak voltage across it. When the input voltage drops, the capacitor voltage is higher than the input and the diode no longer has the required .7 volts forward bias to hold it on. Therefore, when the input sine wave voltage starts to decrease at peak, the diode turns off. The capacitor will then start to discharge through the load, maintaining the voltage at a constant, but decaying voltage. The rate at which the capacitor will discharge depends on how much current the load draws. The capacitor will continue to discharge until the input sine wave increases in voltage enough to become larger than the capacitor voltage.

The exact amount that the capacitor discharges can be calculated, if the value of the capacitor is known and the value of the load resistor is known. However, in an actual power supply, very seldom is the load resistor value known. The way it would be

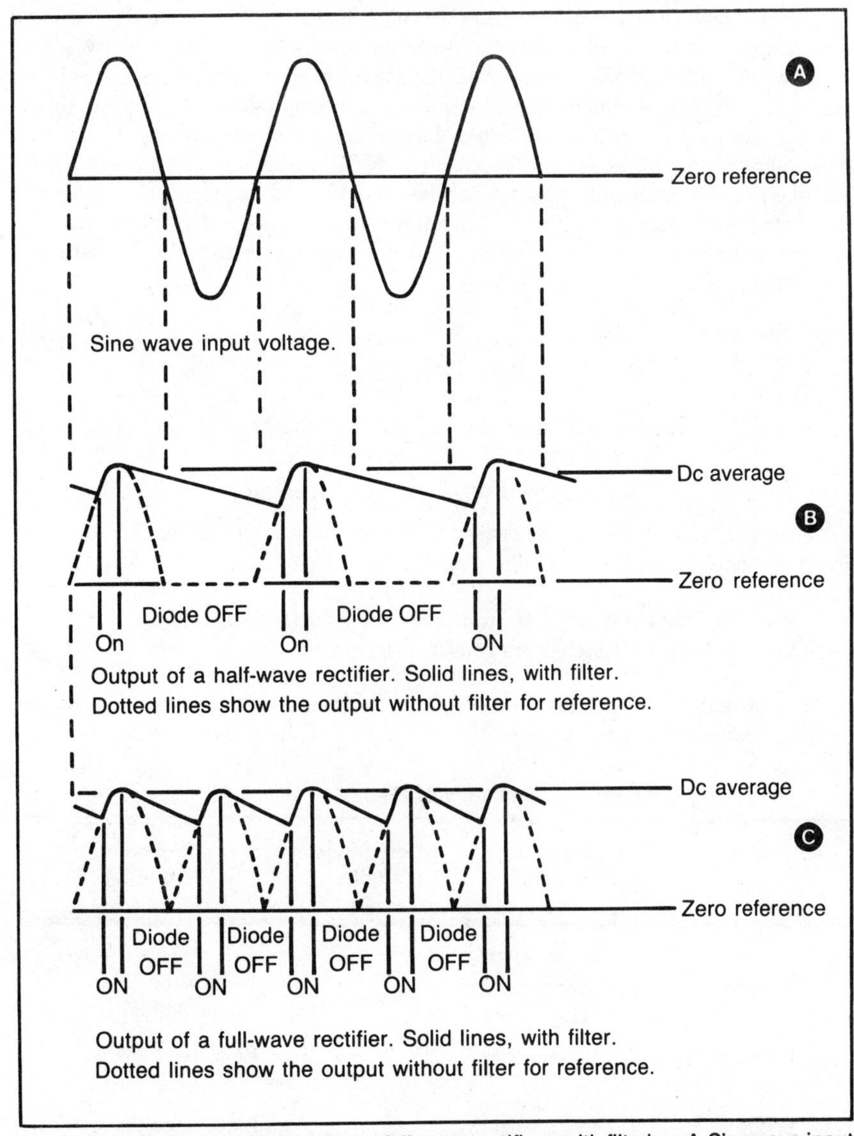

Fig. 12-7. Waveforms of half-wave and full wave rectifiers, with filtering. A. Sine wave input voltage. B. Output of a half-wave rectifier. Solid lines, with filter. Dotted lines show the output without filtering for reference. C. Output of a full wave rectifier. Solid lines, with filter. Dotted lines show the output without filtering for reference.

calculated would be using the R-C time constant and determining exactly how much time is allowed for discharge, based on the frequency of the input sine wave.

Compare Fig. 12-7B to Fig. 17-7C. Notice that the slope of the capacitor discharge is the same for both figures, which indicates the same value of load resistor and filter capacitor. Notice, however, that even though the slope is the same, the length of discharge is greatly reduced by the fact that the waveform occurs twice as often. Notice, on the right hand side of the drawing, there is the labeling of dc average. The half wave circuit should have a little less dc voltage than the full wave circuit due to the fact that the full wave does not discharge the capacitor as much as the half wave circuit.

Even though there is a difference between the half wave circuit and the full wave circuit waveforms, it is still possible to determine the value of voltage that would be read on a dc voltmeter. Keep in mind that the actual value of load resistance is seldom, if ever known. The reading on the dc voltmeter can be approximated. An approximate reading on a voltmeter is often enough to determine if the circuit is working correctly. The following formula is a figure between the dc average of a circuit without a filter and the peak value, because that's what the capacitor charges to. This formula will calculate the dc voltage of a circuit with a filter within 10 percent of the actual value that a voltmeter will read. This formula can be used for any circuit with simple capacitor filtering, full wave or half wave.

approximation for dc voltage for a rectifier with a filter $V_{dc} = 1.2 \times RMS$ **Formula 12-7**

The RMS value in the formula shown above is from the ac input voltage.

Sample Half-Wave Rectifier Circuits

With the following conditions, determine the dc output voltage and ripple frequency. Refer to Fig. 12-2 for the schematic diagram.

Alternating current input to transformer: 120 V/60 Hz (line voltage).
Transformer is a 5:1 step-down transformer.

Calculate the ac applied to the rectifier. The rectifier is connected to the secondary of the transformer, not the primary.

Table 12-1. Summary of Rectifier Formulas.

	Dc voltage without filter	Dc voltage with filter	Ripple frequency
half-wave	$V_{dc} = \frac{.9 \times RMS}{2}$	$V_{dc} = 1.2 \times RMS$	same as ac input freq.
full wave	$V_{dc} = .9 \times RMS$	$V_{dc} = 1.2 \times RMS$	twice the ac input freq.
bridge	$V_{dc} = .9 \times RMS$	$V_{dc} = 1.2 \times RMS$	twice the ac input freq.

Step 1 $\frac{\text{turns primary}}{\text{turns secondary}} = \frac{\text{voltage primary}}{\text{voltage secondary}}$ formula

Step 2 $\frac{5}{1} = \frac{120\text{ V}}{x}$ substitute values

Step 3 voltage secondary $= \frac{120\text{ V}}{5}$ rearrange formula

Step 4 ac voltage to rectifier = 24 Vac RMS

Use the V_{dc} formula for no filter to determine the output voltage.

Step 1 $V_{dc} = \frac{.9 \times \text{RMS}}{2}$ formula

Step 2 $V_{dc} = \frac{.9 \times 24\text{ V}}{2}$ substitute RMS value

Step 3 $V_{dc} =$ 10.8 Vdc dc voltage without a filter

Calculate the dc voltage if the circuit had a simple capacitor filter.

Step 1 $V_{dc} = 1.2 \times \text{RMS}$ formula for circuits with a filter
Step 2 $V_{dc} = 1.2 \times 24$ Vac substitute RMS value
Step 3 $V_{dc} =$ 28.8 Vdc dc voltage with a filter

Calculate the ripple frequency.

Step 1 Ripple frequency of a half wave rectifier circuit is the same as the input frequency. Ripple Frequency = 60 hertz

Sample Full Wave Rectifier Circuits

Refer to Fig. 12-3. With 120 V/60 Hz applied to the primary of a transformer with a turns ratio of 2:1, step down, secondary has a center tap, determine the dc output voltage with and without a filter. Also determine the ripple frequency.

When determining the ac voltage applied to the rectifier circuit, a certain amount of caution must be exercised. The center tapped transformer splits the voltage of the secondary in half. Therefore, only one-half of the secondary voltage is actually applied to the rectifier.

First using the turns ratio, determine the total secondary voltage, then divide that voltage in half.

Step 1 $\frac{\text{turns primary}}{\text{turns secondary}} = \frac{\text{voltage primary}}{\text{voltage secondary}}$ formula

Step 2 $\frac{2}{1} = \frac{120\text{ V}}{x}$ substitute values

Step 3 total secondary voltage = 60 Vac

With the total secondary voltage calculated, divide this figure in half to find the voltage to the rectifier.

Step 1 $\text{Vac to rectifier} = \dfrac{\text{total secondary voltage}}{2}$

Step 2 $\text{Vac} = \dfrac{60}{2} = 30$ Vac RMS voltage to rectifier

Use the V_{dc} formula to calculate the output voltage of this full wave rectifier without a filter.

Step 1 $V_{dc} = .9 \times \text{RMS}$ formula to be used
Step 2 $V_{dc} = .9 \times 30$ Vac substitute
Step 3 $V_{dc} = 27$ Vdc dc voltage without a filter

Using the same value of RMS applied to the rectifier, determine the value of the dc voltage with a simple capacitor filter.

Step 1 $V_{dc} = 1.2 \times \text{RMS}$ formula
Step 2 $V_{dc} = 1.2 \times 30$ Vac substitute
Step 3 $V_{dc} = 36$ Vdc dc voltage with a filter

Determine the ac voltage applied to the rectifier circuit. Because the entire secondary is used, the secondary voltage is simply calculated from the turns ratio.

Step 1 Ripple frequency $= 2 \times 60$ Hz $= 120$ hertz

Sample Bridge Rectifier Circuits

Refer to Fig. 12-4. Determine the dc voltage with and without a filter. Also determine the ripple frequency. The ac input to the transformer is 120 V/60 Hz, and the transformer has a 8:1 turns ratio, step down.

Determine the ac voltage applied to the rectifier circuit. Because the entire secondary is used, the secondary voltage is simply calculated from the turns ratio.

Step 1 $\dfrac{\text{turns of primary}}{\text{turns of secondary}} = \dfrac{\text{volts primary}}{\text{volts secondary}}$

Step 2 $\dfrac{8}{1} = \dfrac{120 \text{ Vac}}{x}$ substitute

Step 3 Secondary volts $= 15$ Vac RMS to rectifier

Calculate the dc voltage without a filter.

Step 1 $V_{dc} = .9 \times \text{RMS}$ formula
Step 2 $V_{dc} = .9 \times 15$ Vac substitute
Step 3 $V_{dc} = 13.5$ Vdc output without a filter

Calculate the dc voltage with a filter.

Step 1 V_{dc} = 1.2 × RMS formula
Step 2 V_{dc} = 1.2 × 15 Vac substitute
Step 3 V_{dc} = 18 Vdc dc voltage with a filter

ZENER REGULATOR

The voltage regulator is the section of a power supply circuit that regulates the voltage to compensate for any changes in load current or input voltage. Regulators range from the simplest being a zener diode to very complex circuits used to regulate the voltage in a complex computer.

This section will deal with only the zener diode in order to stay within the boundaries of the text.

Figure 12-8 is a schematic diagram of a simple zener regulator circuit. A zener regulator circuit contains the zener diode, connected in reverse, and a series dropping resistor.

Figure 12-9 is the characteristic curve of a zener diode. In the forward direction, the zener acts like any rectifier diode. Caution should be used in the forward direction, however, because the zener was not designed for forward operation. Its current capability in the forward direction is quite limited. Notice that the diode drop in the forward direction is .7 V because most zeners are silicon.

The zener diode in the reverse direction has a very unique characteristic curve. Starting from zero reverse volts, and increasing, there will be very small current flow, so small as to be considered zero. Each zener diode has a zener breakdown rating. This rating is the zener voltage in the reverse direction. When increasing the voltage, from zero, the reverse current will start to increase near the zener knee. Then, breakdown will be reached and there will be a large increase in current flow and the voltage across the zener will remain constant.

The reason there must be a series dropping resistor is the fact that when the voltage is being increased in the reverse direction, during zener breakdown, the voltage not across the zener will be dropped across the resistor.

When the zener is operating in the zener region, there is a large amount of current through the zener, limited only by the series resistor and the applied voltage. Since the zener is conducting a significant amount of current and there is a voltage drop, there will be a considerable amount of power dissipated by the zener. For this reason, all zeners will have a zener voltage rating and a maximum wattage.

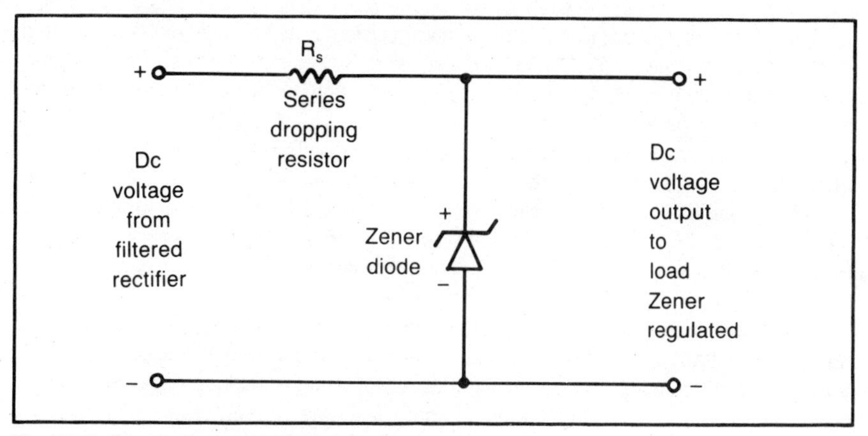

Fig. 12-8. Simple zener regulator circuit.

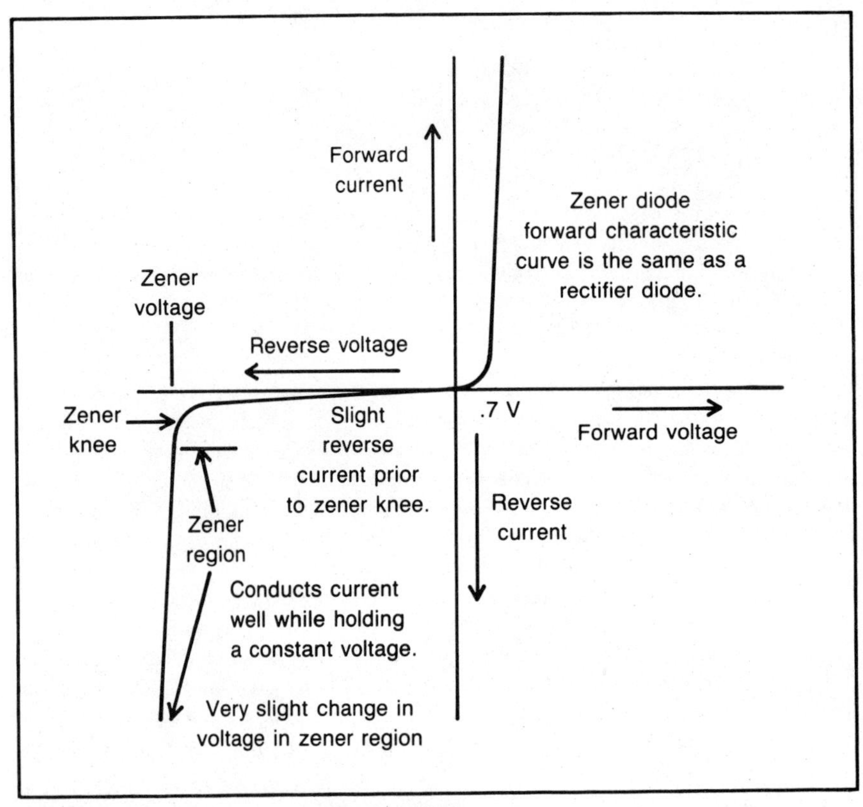

Fig. 12-9. Characteristic curve of a zener diode.

Zener diodes can have voltage ratings ranging from 2 V or so to quite high voltages. Wattage ratings can be from a few milliwatts to several watts. The higher the voltage rating, the more heat the zener must dissipate.

Block Diagram of a Power Supply

Figure 12-10 shows a simple block diagram of a power supply. The block diagram is an easy tool to use because it shows the names of the circuits involved but is not limited to any one circuit inside the blocks. Almost every power supply uses a transformer. However, it really is not necessary with the half wave and the bridge circuits. The transformer is shown because it is so common.

The first block of the diagram is the rectifier circuit. This could be any type of rectifier. Its purpose is to change the ac input to pulsating dc.

The second block of the diagram is the filter. Even though only the simple capacitor filter was discussed, there are many types of filters used with power supplies. All filters have the same function. Their function is to smooth out the fluctuating dc and to make it into as constant a voltage as possible.

The third block of the diagram is the regulator circuit. There are many types of regulators available, even though only the zener regulator will be discussed here. The function of the regulator circuit is to hold the output voltage constant under any changing load conditions or input conditions. Some regulators contain circuitry to turn off the power if the output voltage changes due to a failed regulator.

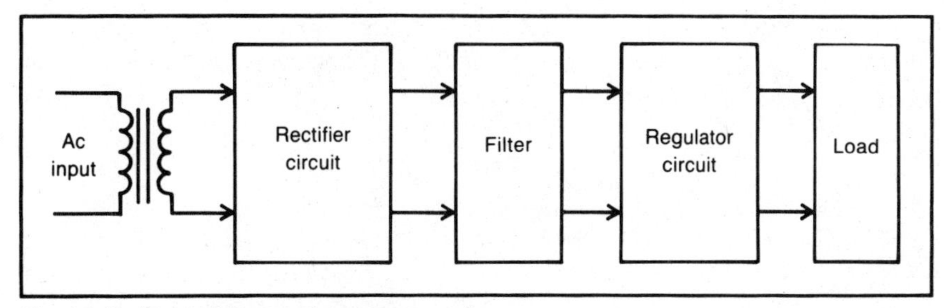

Fig. 12-10. Block diagram of a power supply.

The last block of the diagram is the load. The load is the device that requires the dc power supply. The load could be a resistor, such as a heating element, it could be a radio, a tape player, record player, amplifier, a computer, or anything else that requires dc power to operate. It is quite normal for the load to vary its current requirements while maintaining the same voltage requirements. An example of this might be a tape player that is switched from the play mode to fast forward. The changing speeds of the motor will require different amounts of current.

Calculating a Zener Regulator

Usually when calculating the requirements of the regulator circuit, the load is the first thing that is known. The reason for knowing the load first is the fact that the power supply would not even be needed if there wasn't a load to connect to it. Once the load is known, it is usually a simple matter of working the calculations backward in order.

Figure 12-11 shows the inside of the block diagram and is very useful when making the calculations for the zener regulator. The zener is in parallel with the load. That means that whatever voltage is required by the load, is also the zener voltage. Select the zener to match the load.

- Whatever voltage that is required by the load, is also the zener voltage. Select the zener to match the load.

R_s will be calculated, so we will come back to this. Keep in mind, R_s must drop whatever voltage the zener does not drop, coming from the filtered rectifier. A good rule to follow when determining the output voltage of the filter is to make it approxi-

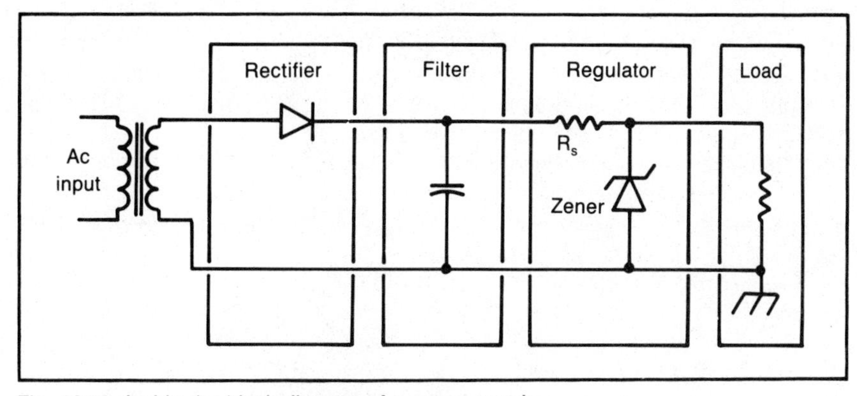

Fig. 12-11. Inside the block diagram of a power supply.

mately double the zener voltage. That will give plenty of room for any fluctuations. Once the filter voltage has been approximated, then the output of the transformer can be approximated.

Sample Zener Regulator Calculation

The load requires 12 volts, maximum current of 1.5 amps.

With the information given, the zener is rated for 12 volts. The zener is in parallel with the load. It must have the same voltage as the load. The next step is to determine the total current for the circuit. Notice how the total current is made up of both zener current and load current. Because the zener must always be operated beyond the zener knee, a good rule to follow is to allow a zener current of 10 percent of the maximum load current.

total current in a zener regulator $I_T = I_Z + I_L$ **Formula 12-8**

Using the formula for total current, Formula 12-8, calculate the total current, allowing full load current and 10 percent of the maximum load current for the zener.

Step 1 $I_T = I_Z + I_L$ formula
$I_L = 1.5$ A given in original problem
$I_Z = 10\%$ of 1.5 A minimum zener current
$I_Z = .15$ A zener current when load is drawing full amount
Step 2 $I_T = .15\text{ A} + 1.5\text{ A}$ substitute values
Step 3 $I_T = 1.65$ A total circuit current

The load will seldom, if ever draw the full rated value of current. Whatever the load does not use, the zener will take the remainder. This way, the total current remains constant as long as the input voltage is a constant.

Next determine the voltage from the filtered rectifier. It is a good rule to use double the zener voltage, approximately, to allow for fluctuations in the input voltage.

Zener voltage = 12 V, rectifier output = 24 V

With the value of voltage known across the zener, and the value of voltage across the filter known, the difference of these two values must be dropped across the series resistor, R_s.

series voltage drop in a zener regulator $V_{RS} = V_{dc} - V_Z$ **Formula 12-9**

Step 1 $V_{RS} = V_{dc} - V_Z$ formula
Step 2 $V_{RS} = 24 - 12$ substitute values
Step 3 $V_{RS} = 12$ V voltage across R_S

Calculate the ohmic value of R_S. Total current will flow through R_S, therefore, the value of the resistor can be calculated using total current and the voltage dropped across the resistor.

Step 1 $R = \frac{E}{I}$ formula

Step 2 $R_s = \dfrac{12\ V}{1.65\ A}$ substitute values

Step 3 $R_s = 7.27$ ohms calculated value of R_s

Note The calculated value of R_s will have to be adjusted to a value that is possible to purchase at a parts store.

Using the calculated value of R_s, the calculated voltage drop across R_s and the calculated total current; calculate the wattage rating of the resistor.

Step 1 $P = I \times E$ formula to be used
Step 2 $P = 1.65\ A \times 12\ V$ substitute values
Step 3 $P_{RS} = 19.8\ W$ calculated power of R_s

Using the calculated values, calculate the value of the zener wattage. Use total current because the worst condition for the zener is when the load is not using any current.

Step 1 $P = I \times E$ formula
Step 1 $P_Z = 1.65\ A \times 12\ V$ substitute values
Step 3 $P_Z = 19.8$ watts calculated power of the zener

Next in designing the zener regulator circuit is to determine the value of secondary voltage that will supply the calculated dc voltage across the filter. Remember, the filter voltage was calculated using the V_{dc} formula:

Step 1 $V_{dc} = 1.2 \times RMS$ starting formula

Step 2 $RMS = \dfrac{V_{dc}}{1.2}$ rearrange formula

Step 3 $RMS = \dfrac{24\ V}{1.2}$ substitute values

Step 4 $RMS = 20$ Vac RMS secondary voltage

Note Primary of the transformer will be as needed, usually 120 V/60 Hz.

Selecting the Actual Zener Circuit Values

Even though the zener regulator circuit is such a simple circuit, it demonstrates the problems that are encountered when designing a working circuit.

When the calculations for a circuit are made, it is most common to start off at the load. The calculated values give a starting point, but that is not the final value. Once the calculations have been made, the easiest place to start with in adjusting the values is the transformer and work towards the load.

Transformers come in almost any value of secondary voltage, if round numbers are selected. The hard part in selecting a transformer is the fact that larger values of secondary current will result in larger transformers, physically. Most transformers will come with a center-tap so any type of rectifier circuit can be used. The actual output voltage of the transformer will not be critical in the final circuit, provided there is enough voltage.

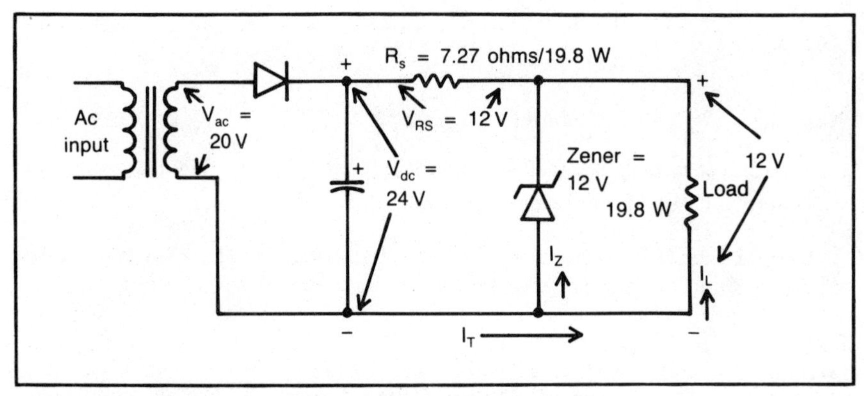

Fig. 12-12. Complete zener circuit, showing calculated values.

The next step in the design is the rectifier circuit. The bridge rectifier has many advantages over the other types, and is usually selected for low power applications.

The filter capacitor can be selected over a wide range of values. It should be an electrolytic capacitor. It will have a voltage rating and a capacitor rating. The voltage rating must be a minimum of the peak value of the output of the rectifier. A good rule to follow is to double the necessary voltage rating. As far as the capacitor rating is concerned, 100 μF is a satisfactory value, but the higher, the better.

The next step in adjusting the values of the zener circuit is to select the zener. The load is usually selected first and will therefore determine the actual value of the zener. However, if that value cannot be found, it is possible to place zeners in series to arrive at a new value. For example, two 6 volts in series equals 12 volts. Do not try to connect them in parallel; it does not work that way.

The only component left in the circuit is the hardest one to choose. R_s determines the total current of the circuit. It must be selected in such a way as to ensure enough current will flow to the load and still have a small amount of current to keep the zener in the zener region. If the value of R_s is increased beyond the calculated value, there may not be enough current for all conditions. If the value of R_s is made smaller than the calculated value, it will increase the total current of the circuit and will therefore demand the zener handle more current. The zener must be able to handle all of the total current when the load is disconnected. Usually, the zener current rating will be selected high enough to allow for extra current. So, R_s should be selected slightly lower in value than the figure actually calculated. As far as the wattage rating is concerned, make sure the wattage rating is higher than needed. Resistors can be placed in series or parallel to make up any combination of resistance or power.

CHAPTER SUMMARY

The power supply is the part of an electronic system that is often forgotten about, but yet is a very important part when it is considered to be the thing that makes the electronic device operate.

Changing ac power to dc is what the power supply is used for. Its primary component for this purpose is the rectifier diode. The rectifier diode is made up of two pieces of semiconductor material, one N-type and the other P-type. The two types of semiconductor materials joined together form the P-N junction. It is this P-N junction that allows the diode to conduct only in one direction.

The three types of rectifier circuits discussed in this chapter are the half wave, full wave, and bridge circuits. These three circuits are the circuits used in most modern power supplies, with the bridge being the most favorite. The rectifier circuit is the circuit responsible for the actual changing of ac to dc. The different configurations of circuitry allow the use of diodes in different manners to accomplish the task.

The second major part of a power supply is the filter. This chapter discussed only the capacitor filter because it demonstrates all of the principles of any filter available. The function of the filter is to smooth the pulsating dc that comes from the rectifier.

The next step in a power supply is the regulator. The regulator is responsible for holding a constant output voltage to the load. A power supply without a regulator would vary in voltage whenever the load conditions of current were to change. Some devices would be destroyed by a supply voltage that is too low or too high. This chapter discusses the zener regulator and demonstrates the calculations that are involved in the zener circuit. Even though it is such a simple regulator circuit, the calculations can become quite complex. Zener circuits are only suitable for low power applications. Other circuits can be chosen for loads using over 5 amps.

The following have been chosen as the key points of this chapter:

- A semiconductor with an extra electron is called an N-type semiconductor.
- A semiconductor missing an electron is called a P-type semiconductor. The atom missing an electron is called a hole.
- P-type semiconductors have a majority of holes, and a minority of electrons. Holes carry a positive charge.
- N-type semiconductors have a majority of electrons, and a minority of holes. Electrons carry a negative charge.
- The depletion zone is depleted of any mobile carriers and acts like an insulation region.
- A forward biased diode has negative connected to the N material and positive connected to the P material.
- A reverse biased diode has positive connected to the N material and negative to the P material.
- When a rectifier diode is conducting, in the forward direction, the diode will have a voltage drop of .3 V or .7 V, the remainder of the applied voltage will be dropped across the load resistor.
- When the rectifier diode is reverse biased, there is no current flow. All of the applied voltage will be dropped across the diode and no voltage will be dropped across the load resistor.
- Whatever voltage is required by the load in a simple zener regulator, is also the zener voltage. Select the zener to match the load.

Summary of Formulas

output voltage of a full wave or bridge rectifier circuit	$V_{dc} = .636 \times \text{peak}$	**Formula 12-1**
output voltage of a half-wave rectifier circuit	$V_{dc} = \frac{.636 \times \text{peak}}{2}$	**Formula 12-2**
output voltage of a full wave or bridge rectifier circuit	$V_{dc} = .9 \times \text{RMS}$	**Formula 12-3**
output voltage of a half-wave rectifier circuit	$V_{dc} = \frac{.9 \times \text{RMS}}{2}$	**Formula 12-4**

ripple frequency half-wave = input frequency		**Formula 12-5**
ripple frequency full wave = 2 × input frequency		**Formula 12-6**
approximation for dc output voltage for a rectifier with a filter, all configurations	$V_{dc} = 1.2 \times RMS$	**Formula 12-7**
total current in a zener regulator circuit	$I_T = I_Z + I_L$	**Formula 12-8**
voltage drop across the series resistor of a zener regulator	$V_{RS} = V_{dc} - V_Z$	**Formula 12-9**

Practice Problems

The problems have the load conditions given and the RMS value of the transformer secondary. Use a bridge rectifier circuit with simple capacitive filtering. In all problems, allow a minimum zener current of 10 percent of the full load condition.

Find for each:
V_{dc} from filter (input to zener regulator), zener voltage, zener current (maximum when the load is disconnected), zener wattage, voltage drop across R_s, value of R_s in ohms and value of R_s in wattage, total circuit current.

1. Load; 12 volts/250 mA — Input ac; 24 volts RMS
2. Load; 9 volts/50 mA — Input ac; 18 volts RMS
3. Load; 6 volts/750 mA — Input ac; 12 volts RMS
4. Load; 15 volts/1.5 A — Input ac; 30 volts RMS
5. Load; 24 volts/150 mA — Input ac; 60 volts RMS

Appendix

Answers to Practice Problems

CHAPTER 1

page 3	page 4	page 5	page 7
1. 5610	1. −2	1. 6	1. 8.76×10^{5}
2. 5610	2. 10	2. 20	2. 1.03×10^{9}
3. 5600	3. 6	3. −15	3. 3.2×10^{4}
4. 29900	4. 4	4. −18	4. 2.5×10^{1}
5. 360,000	5. −4	5. −6	5. 5.8×10^{0}
6. 569	6. 16	6. 2	6. 3.0×10^{-2}
7. 573	7. 6	7. 0	7. 5.6×10^{-4}
8. 333	8. −7	8. .2	8. 4.05×10^{-3}
9. 12.1	9. −7	9. −.18	9. 1.0×10^{-7}
10. 23.2	10. −2	10. 4	10. 2.0×10^{-1}
11. 54.5	11. 0	11. −8	11. 1.3×10^{9}
12. 57.0	12. −1	12. 5	12. 5.2×10^{6}
13. 8.2	13. −11	13. −5	13. 4.5×10^{4}
14. 8.43	14. 20	14. −5	14. 3.9×10^{-2}
15. 9.00	15. .5	15. 5	15. 5.6×10^{-3}
16. .00333	16. 5.3	16. −4	16. 5.2×10^{1}
17. .900	17. 7.96	17. −1/6	17. 3.2×10^{-3}
18. .000932	18. −10.76	18. 6	18. 4.6×10^{-9}
19. .0100	19. −9.11	19. 3	19. 7.05×10^{-5}
20. .200	20. − 1 3/8	20. 27	20. 4.0×10^{-14}
21. .667			
22. .00556			
23. .745			
24. .0909			
25. .0125			

page 7	page 9		page 12	page 13
1. 480,000	1. 2,400 kΩ	2.4 MΩ	1. 100 (10^{0})	1. 44 kilo (10^{3})
2. 785	2. 353 kHz	.353 MHz	2. 60 kilo (10^{3})	2. 914 (basic units, 10^{0})
3. 8.9	3. 2,500,000 mW	2.5 kW	3. 15 kilo (10^{3})	3. 76.2 kilo (10^{3})
4. 346	4. 25,000 mV	.025 kV	4. 8 milli (10^{-3})	4. 1.0 mega (10^{6})
5. 457,000	5. 10,000 μH	10 mH	5. 30 nano (10^{-9})	5. 725 kilo (10^{3})
6. .1	6. 1500 Ω	.0015 MΩ	6. 5 milli (10^{-3})	6. 1350 (basic units, 10^{0})
7. 0.00201	7. 25,000 kHz	.025 GHz	7. 2 kilo (10^{3})	7. 12.5 (basic units, 10^{0})
8. 0.00000003	8. 30 MHz	30,000,000 Hz	8. 4 micro	8. 25 kilo (10^{3})
9. 1	9. 56,000 kV	56,000,000 V	9. 5 kilo (10^{3})	9. 86 milli (10^{-3})
10. 0.1	10. 75,000 W	.075 MW	10. 62.8 milli (10^{-3})	10. 1.05 milli (10^{-3})

page 7

11. 35
12. 4.0
13. 0.709
14. 5,600,000
15. 65
16. 98
17. 0.000109
18. 0.00000078
19. 10
20. 10

page 9

11. 25,000 μA .025 A
12. .0015 μA 1.5 mA
13. 1,000,000,000 pF .001 F
14. .000000001 μF .000001nF
15. 25 μV .000025 V
16. 7.5 W 7,500,000 μW
17. .5 kV 500,000 mV
18. 2.4 V .0024 kV
19. .001 kA 1000 mA
20. 10,000 mV .01 kV

page 12

11. 628 kilo (10^3)
12. .159 (10^0)
13. 15.9 mega (10^6)
14. 1 giga (10^9)
15. 5 milli (10^{-3})

page 13

11. 6 milli (10^{-3})
12. 9.99 milli (10^{-3})
13. 10 micro (10^{-6})
14. 251.5 micro (10^{-6})
15. 25.1 nano (10^{-9})
16. .011 micro (10^{-6})
17. 40 pico (10^{-12})
18. .0115 micro (10^{-6}) or 11.5 nano (10^{-9})
19. 1.(basic units, 10^0)
20. 221.1 (basic units, 10^0)

CHAPTER 2

page 16

1. E = 20 volts
2. E = 50 volts
3. E = 20 volts
4. E = 300 volts
5. E = 12 volts
6. I = .1 amps
7. I = 1 amp
8. I = 10 mA
9. I = 10 mA
10. I = 10 mA
11. R = 10 ohms
12. R = 10 ohms
13. R = 2 kohms
14. R = 2 Mohms
15. R = 5 kohms
16. I = .1 amps
17. E = 30 volts
18. I = 1 amp
19. R = 1.2 megohm
20. R = 1 ohm

page 20 (answers are underlined)

	Voltage	Current	Resistance	Power
1.	E = 30 volts	I = 15 amps	R = 2 ohms	P = 450 watts
2.	E = 2.5 volts	I = 2 amps	R = 1.25 ohms	P = 5 watts
3.	E = 100 volts	I = .01 amps	R = 10 kohms	P = 1 watt
4.	E = 250 volts	I = 50 amps	R = 5 ohms	P = 12,500 W
5.	E = 15 volts	I = 6.67 mA	R = 2250 ohms	P = 100 mW
6.	E = 3.16 V	I = 3.16 mA	R = 1 kΩ	P = 10 mW
7.	E = 250 volts	I = 25 mA	R = 10 kΩ	P = 6.25 watts
8.	E = 12.5 V	I = 2 mA	R = 6.25 kΩ	P = 25 mW
9.	E = 50 volts	I = 5 mA	R = 10 kohms	P = 250 mW
10.	E = 5 volts	I = 2 mA	R = 2500 ohms	P = 10 mW
11.	E = 0 volts	I = 0 amps	R = 100 ohms	P = 0 watts
12.	E = 10 volts	I = 1 amp	R = 10 ohms	P = 10 watts
13.	E = .1 volts	I = 10 μA	R = 10 kΩ	P = 1 μW
14.	E = 10 volts	I = 100 mA	R = 100 ohms	P = 1 watt
15.	E = 25 volts	I = 1 amp	R = 25 ohms	P = 25 watts
16.	E = 20 volts	I = 100 mA	R = 200 ohms	P = 2 watts
17.	E = 1 volt	I = 1 amp	R = 1 ohm	P = 1 watt
18.	E = 10 volts	I = 10 mA	R = 1000 ohms	P = .1 watt
19.	E = 10 kV	I = 1000 mA	R = 10,000Ω	P = 10 kW
20.	E = 10 volts	I = 100 mA	R = 100 ohms	P = 1 watt

page 26

1. I_T = .2 amps, R_T = 250 ohms, E_{R1} = 20 volts, E_{R2} = 30 volts, P_{R1} = 4 watts, P_{R2} = 6 watts
2. I_T = 50 mA, R_T = 300 ohms, $E_{R1} = E_{R2} = E_{R3}$ = 5 volts, $P_{R1} = P_{R2} = P_{R3}$ = .25 watts
3. R_T = 280 ohms, I_T = 50 mA
4. I_T = 3 amps, V = 150 volts
5. R_T = 80 ohms, I_T = 1.5 amps, V = 120 volts

6. R_T = 2.5 kilohms, R_1 = 1.5 kilohms
7. R_T = 800 ohms, R_3 = 100 ohms, V = 80 volts
8. I_T = 25 mA, R_2 = 320 ohms, E_{R1} = 4 volts
9. V = 32 volts, I_T = 400 mA
10. V = 120 volts, I_T = 2 amps, R_2 = 20 ohms

Note Due to rounding of numbers and the fact that some calculations are used to make other calculations, your answers may be slightly different.

page 35

1. R_T = 13.6 ohms, I_T = 1.83 amps, I_A = 1 amp, I_B = .5 amps, I_C = .33 amps
2. R_T = 512.8 ohms, I_T = 29 mA
3. R_T = 40 ohms
4. R_T = 125 ohms
5. R_3 = 1500 ohms
6. R_4 = 33.3 ohms
7. R_3 = 1000 ohms, I_T = 100 mA, R_T = 200 ohms
8. R_T = 60 ohms, I_A = .333 A, I_B = .2 A, I_C = .1333 A
9. R_T = 200 ohms, I_A = 50 mA, I_B = 16.7 mA, I_C = I_D = 33.3 mA
10. R_T = 9.1 ohms, I_A = 1 amp, I_B = .364 amps, I_C = .637 amps, I_D = .228 amps, I_E = .409 amps, I_F = .228 amps, I_G = .182 amps

Note Due to rounding of numbers and the fact that some calculations are used for other calculations, your answers may be slightly different.

page 45

1. R_T = 2.2 k, I_T = 20 mA, E_{R1} = 20 V, E_{R2} = 24 V, E_{R3} = 24 V, I_{R2} = 12 mA, I_{R3} = 8 mA
2. R_T = 4 k, I_T = 5 mA, E_{R1} = 5 V, E_{R2} = 1.5 V, E_{R3} = 1.5 V, E_{R4} = 13.5 V, I_{R2} = 2.21 mA, I_{R3} = 2.68 mA
3. R_T = 50 ohms, I_T = 100 mA, E_{R1} = 1 V, E_{R2} = E_{R3} = .5 V, E_{R4} = 1 V, E_{R5} = E_{R6} = .5 V, E_{R7} = E_{R8} = 1 V, I_{R2} = I_{R5} = 50 mA
4. R_T = 314.5 ohms, I_T = 31.8 mA, I_{R1} = 31.8 mA, I_{R2} = 21.2 mA, I_{R3} = 10.6 mA, I_{R4} = 31.8 mA, I_{R5} = 20 mA, I_{R6} = 8 mA, I_{R7} = 4 mA, E_{R1} = 3.18 V, E_{R2} = E_{R3} = 5.72 V, E_{R4} = .7 V, E_{R5} = E_{R6} = E_{R7} = .4 V
5. R_T = 133.3 ohms, I_T = 75 mA, I_{R2} = 50 mA, I_{R3} = 25 mA, E_{R1} = 7.5 V, E_{R2} = 2.5 V
6. R_T = 16.9 ohms
7. R_T = 100 ohms
8. R_T = 1 k
9. R_T = 275 ohms, I_T = 200 mA, E_{R1} = E_{R2} = 20 V, E_{R3} = 15 V, E_{R4} = 2.5 V, E_{R5} = 5 V, E_{R6} = 7.5 V, E_{R7} = E_{R8} = E_{R9} = 2.5 V, I_{R1} = I_{R2} = 200 mA, I_{R3} = I_{R4} = I_{R5} = 100 mA, I_{R6} = I_{R7} = I_{R8} = I_{R9} = 50 mA
10. R_T = 2520 ohms, I_T = 39.7 mA

Note Due to rounding of numbers and the fact that some calculations are used for other calculations, your answers may be slightly different.

page 57

1. R_T = infinity (open circuit), I_T = 0, E_{R1} = E_{R2} = 0
2. R_T = 25 k (notice the short), I_T = 1 mA, E_{R1} = E_{R2} = 0, E_{R3} = 25 V
3. R_T = 1000 ohms, I_T = 30 mA, V_a = 15 V, V_b = 6 V

4. R_T = 667, I_T = 18 mA, I_{R1} = 6 mA, I_{R2} = 12 mA
5. R_T = 25 ohms, I_T = 1 amp, I_{R1} = .5 amp, I_{R2} = .25A, I_{R3} = .25A
6. R_T = 62.5, I_T = 1.92 amp, P_T = 230.4 watts
7. R_T = 55.3 ohms, I_T = 2.17A, P_T = 260 W, R_1 = 144 Ω, R_2 = 192 Ω, R_3 = 576 Ω, R_4 = 240 Ω
8. R_T = 76.7 ohms, I_T = .196 amps, I_{R1} = .196A, I_{R2} = .130A, I_{R3} = .065A, E_{R1} = 11.7 V, E_{R2} = 3.24 V, E_{R3} = 3.24 V % of I_T in R_2 = 66%
9. R_T = 40 ohms, I_T = 1 amp, I_{R1} = 1A, I_{R2} = I_{R3} = I_{R4} = I_{R5} = .5A, I_{R6} = 1A, E_{R1} = 10 V, E_{R2} = 7.5 V, E_{R3} = 2.5 V, E_{R4} = E_{R5} = 5 V, E_{R6} = 20 V
10. I in meter = 0 . . . This is a balanced bridge

CHAPTER 3

page 68

1. a) RMS = 21.2 b) p to p = 60 c) avg = 19.1
2. a) p to p = 200 b) avg = 63.6 c) RMS = 70.7
3. a) RMS = 28.3 b) peak = 40 c) avg = 25.4
4. a) peak = 130 b) avg = 82.7 c) RMS = 91.9
5. a) peak = 99 b) p to p = 198 c) avg = 63
6. a) avg = 10.8 b) p to p = 34 c) peak = 17
7. a) RMS = 17.8 b) peak = 25.2 c) p to p = 50.3
8. a) peak = 189 b) p to p = 377 c) RMS = 133
9. a) peak = 120 b) p to p = 240 c) avg = 76.5
10. a) p to p = 140 b) avg = 44.5 c) RMS = 49.5
11. a) p to p = 40 b) peak = 20 c) RMS = 14.1 d) avg = 12.7
12. a) p to p = 70 b) peak = 35 c) RMS = 24.7 d) avg = 22.3
13. a) p to p = 400 b) peak = 200 c) RMS = 141 d) avg = 127
14. a) p to p = 120 b) peak = 60 c) RMS = 42.4 d) avg = 38.2
15. a) p to p = 80 b) peak = 40 c) RMS = 28.3 d) avg = 25.4

page 70

1. .1 s
2. .01 s
3. 1 ms
4. .1 ms
5. .01 ms
6. 1 μs
7. .1 μs
8. .01 μs
9. .001 ns
10. .1 ns
11. 20 Hz
12. 200 Hz
13. 2 kHz
14. 20 kHz
15. 2 Hz
16. 60,000 cm
17. 6,000 cm
18. 600 cm
19. 6 cm
20. .06 cm

page 73

1. .48 V p to p, period of 3.4 ms, 294 hertz
2. .056 V p to p, period of .10 μs, 10 MHz
3. .18 V p to p, period of .8 μs, .125 MHz
4. 2.22 V p to p, period of 23 μs, 43 kHz
5. 4 V p to p, period of .43 μs, 1.9 MHz
6. .05 V p to p, period of .02 ms, 50 kHz
7. .044 V p to p, period of .16 ms, 6.2 kHz
8. 1 V p to p, period of 64 μs, 15.6 kHz
9. 16 V p to p, period of 8 ms, 167 Hz
10. .14 V p to p, period of .052 ms, 19.2 kHz

Note Due to the fact that these drawings are difficult to read accurately, the answers have been rounded considerably.

CHAPTER 4

page 82 (answers are underlined)

	Turns Ratio (Pri:Sec)	Primary Volts (RMS)	Secondary Volts (RMS)	Primary Current (Amps)	Secondary Current (Amps)
1	10:1	120	12	.1	1
2	6:1	120	20	.1	.6
3	1:5	30	150	1	.2
4	1:4	24	96	4	1
5	3:1	81	27	.2	.6
6	1:7	17	119	.21	.03
7	2:5	10	25	.3	.12
8	3:4	120	160	1.7	1.3
9	4.5:2	120	53.3	.5	1.1
10	7:6	28	24	.18	.21

CHAPTER 5

page 89

1. 1200 μH
2. 90 mH
3. 470 mH
4. 3.65 H
5. 66.7 mH
6. 200 mH
7. 220 mH
8. 180 mH
9. 220 mH
10. 180 mH

page 96

1. C_T = 5 μF
2. C_T = 1015 pF
3. C_T = .231 μF
4. C_T = 141 μF
5. C_T = 500 μF
6. C_T = .667 μF, V_{C1} = 6.67 V, V_{C2} = 3.33 V
7. C_T = 3.33 pF, V_{C1} = 6.67 V, V_{C2} = 3.33 V
8. C_T = .0091 μF, V_{C1} = 4.55 V, V_{C2} = .45 V
9. C_T = 3.33 μF, V_{C1} = 5 V, V_{C2} = 5 V, V_{C3} = 5 V
10. C_T = .067 μF, V_{C1} = 13.3 V, V_{C2} = 6.67 V

CHAPTER 6

page 110

1. Charge time = .02 seconds, full charge current = .15 A
   discharge time = .02s, discharge voltage = 15 V
2. Charge time = 2 ms, charge current = 15 mA
   discharge time = .2s, discharge voltage = .15 V
3. Charge time = .2s, charge current = 1.5 A
   discharge time = .2 ms, discharge voltage = 15,000 V
4. Charge time = 5 μs, charge current = .75 mA
   discharge time = 5 μs, discharge voltage = 15 V
5. Charge time = 30 ms, charge current = .12 A
   discharge time = 6.6 μs, discharge voltage = 54,000 V
6. Charge time = 5 ms, charge voltage = 12 volts
   discharge time = 5 ms, discharge current = .12 A
7. Charge time = 10 ms, charge voltage = 12 V
   discharge time = 25 ms, discharge current = 48 mA
8. Charge time = 375 ms, charge voltage = 15 V
   discharge time = 150 ms, discharge current = 15 mA
9. Charge time = 10 ms, charge voltage = 15 V
   discharge time = 20 μs, discharge current = .75 A

10. Charge time = 1.25s, charge voltage = 1.5 V
discharge time = 500 μs, discharge current = 1.5 A

Note The times shown here for charge and discharge are all for *one* time constant. To show full charge and discharge multiply by 5.

CHAPTER 8

page 125

1. 628 ohms 2. 15.9 mH 3. 3.2 mH 4. 1990 Hz 5. 5970 Hz
6. 62.8 kΩ 7. 6.4 mH 8. 9950 Hz 9. 942 Ω 10. 0 Ω

page 134

1. L_T = 6H, X_{L1} = 377 Ω, X_{L2} = 754 Ω, X_{L3} = 1131 Ω, X_{LT} = 2262 Ω
2. L_T = 140 mH, X_{L1} = 7.54 Ω, X_{L2} = 15.1 Ω, X_{L3} = 30 Ω, X_{LT} = 52.6 Ω
3. Z = 141 Ω, I = .71A, V_R = 70.7 V, V_L = 70.7 V, Θ = 45°

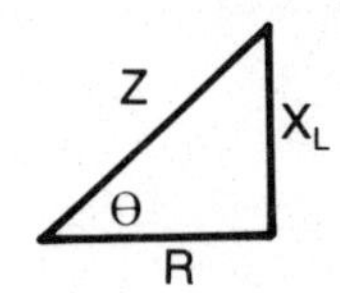

4. Z = 55.9 Ω, I = 1.79A, V_R = 44.7 V, V_L = 89.4 V, Θ = 63.4°

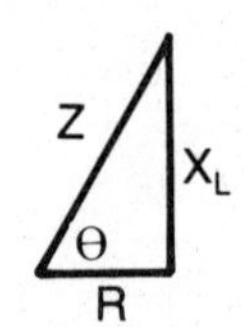

5. Z = 100.5 Ω, I = .99A, V_R = 9.9 V, V_L = 99.5 V, Θ = 84.3°

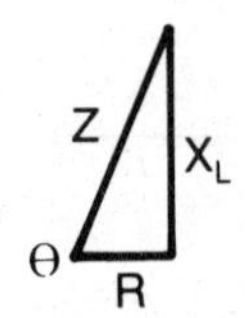

6. Z = 75.2 Ω, I = 1.33A, V_R = 99.8 V, V_L = 6.65 V, Θ = 3.8°

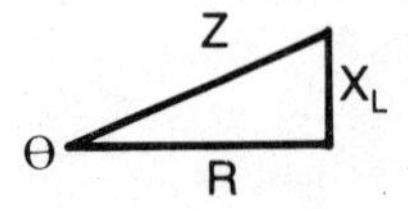

7. Z = 79 Ω, I = 1.26A, V_R = 94.9 V, V_L = 31.6 V, Θ = 18.4°

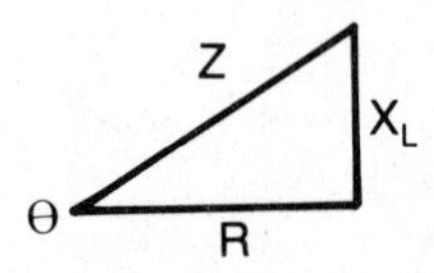

8. Z = 50 Ω, I = 2A, V_R = 0 V, V_L = 100 V, Θ = 90°

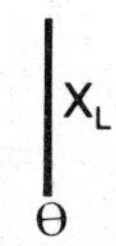

9. $R = 80\ \Omega$, $X_L = 160\ \Omega$, $Z = 179\ \Omega$, $V_a = 44.7$ V, $\Theta = 63.4°$

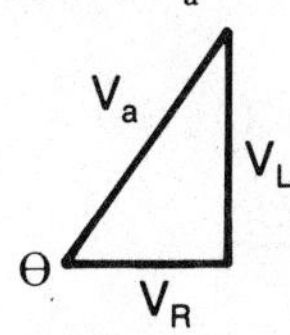

10. $R = 75\ \Omega$, $X_L = 45\ \Omega$, $Z = 87.7\ \Omega$, $V_a = 29.2$ V, $\Theta = 31°$

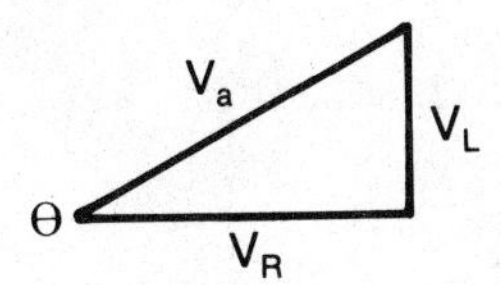

page 140

1. $L_T = .55$ H, $X_{L1} = 314\ \Omega$, $X_{L2} = 628\ \Omega$, $X_{L3} = 942\ \Omega$, $X_{LT} = 171\ \Omega$
2. $L_T = 11.4$ mH, $X_{L1} = 6.28\ \Omega$, $X_{L2} = 12.6\ \Omega$, $X_{L3} = 25\ \Omega$, $X_{LT} = 3.6\ \Omega$
3. $I_R = 1$A, $I_L = 1.4$A, $\Theta = -45°$, $Z = 70.7$ ohms

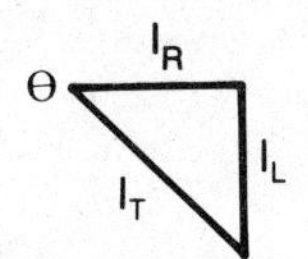

4. $I_R = 4$A, $I_L = 2$ A, $I_T = 4.5$A, $\Theta = -26.6°$, $Z = 220$ ohms

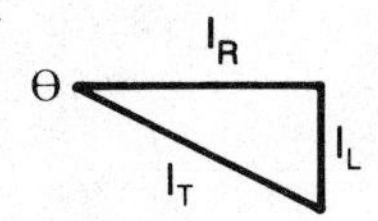

5. $I_R = 10$A, $I_L = 1$A, $I_T = 10.05$A, $\Theta = -5.7°$, $Z = 10$ ohms

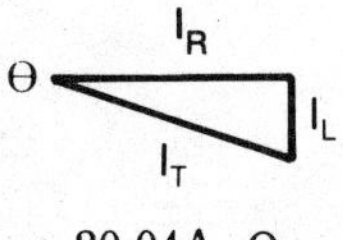

6. $I_R = 1.3$A, $I_L = 20$A, $I_T = 20.04$A, $\Theta = -86.3°$, $Z = 5$ ohms

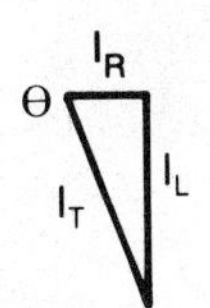

7. $I_R = 1.3$A, $I_L = 4$A, $I_T = 4.2$A, $\Theta = -72°$, $Z = 23.8$ ohms

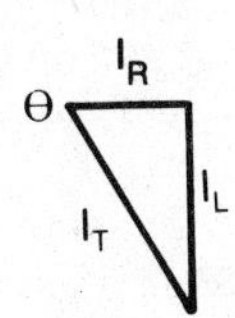

8. I_R = 20A, I_L = 20A, I_T = 28.3A, Θ = −45°, Z = 3.5 ohms

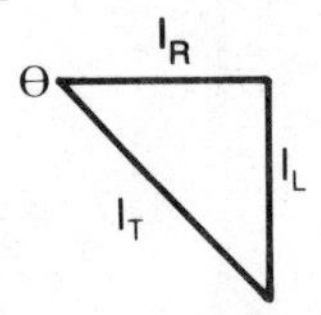

9. R = 100 Ω, X_L = 33.3 Ω, I_T = 3.16A, Θ = −71.6°, Z = 31.6 Ω

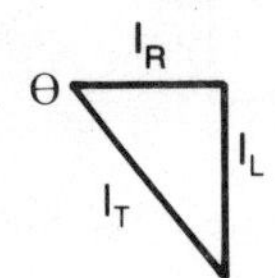

10. R = 4 kΩ, X_L = 10 kΩ, I_T = 27 mA, Θ = −21.8°, Z = 3.7 kΩ

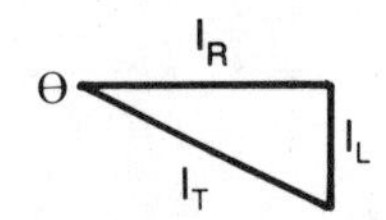

page 144

1. Z = 707 ohms, I = .141 amps, V_R = 70.7 volts, V_L = 70.7 volts, Θ = 45°, P_R = 10 watts, P_X = 10 VARS, P_A = 14.1 VA, PF = .707
2. Z = 269 ohms, I = .372 amps, V_R = 37.2 volts, V_L = 92.9 volts, Θ = 68.2°, P_R = 13.8 watts, P_X = 34.6 VARS, P_A = 37.2 VA, PF = .371
3. Z = 1250 ohms, I = .08 amps, V_R = 80 volts, V_L = 60 volts, Θ = 36.9°, P_R = 6.4 watts, P_X = 4.8 VARS, P_A = 8 VA, PF = .8
4. Z = 1118 ohms, I = .089 amps, V_R = 89.4 volts, V_L = 44.7 volts, Θ = 26.6°, P_R = 8 watts, P_X = 4 VARS, P_A = 8.94 VA, PF = .894
5. Z = 22.4 ohms, I = 4.47 amps, V_R = 44.7 volts, V_L = 89.4 volts, Θ = 63.4°, P_R = 200 watts, P_X = 400 VARS, P_A = 446 VA, PF = .448
6. I_R = .2 amps, I_L = .2 amps, I_T = .283 amps, Z = 353 ohms, Θ = −45°, P_R = 20 watts, P_X = 20 VARS, P_A = 28.3 VA, PF = .707
7. I_R = 1 amp, I_L = .4 amps, I_T = 1.08 amps, Z = 92.8 ohms, Θ = −21.8°, P_R = 100 watts, P_X = 40 VARS, P_A = 108 VA, PF = .928
8. I_R = .1 amps, I_L = .133 amps, I_T = .166 amps, Z = 600 ohms, Θ = −53.1°, P_R = 10 watts, P_X = 13.3 VARS, P_A = 16.6 VA, PF = .6
9. I_R = .1 amps, I_L = .2 amps, I_T = .224 amps, Z = 447 ohms, Θ = −63.4°, P_R = 10 watts, P_X = 20 VARS, P_A = 22.4 VA, PF = .447
10. I_R = 10 amps, I_L = 5 amps, I_T = 11.2 amps, Z = 8.94 ohms, Θ = −26.6°, P_R = 1000 watts, P_X = 500 VARS, P_T = 1120 VA, PF = .894

page 150

1. 45°/div. . . . . PA 45°
2. 45°/div. . . . . PA 54°
3. 60°/div. . . . . PA 60°
4. 60°/div. . . . . PA 30°
5. 72°/div. . . . . PA 30°
6. 72°/div. . . . . PA 60°
7. 45°/div. . . . . PA 36°
8. 90°/div. . . . . PA 45°
9. 60°/div. . . . . PA 30°
10. 72°/div. . . . . PA 45°

Note With these practice problems, it is very difficult to be accurate. Answers are rounded off to the best accuracy possible. With an actual oscilloscope, the phase angle is quite easy to read accurately. The purpose of these drawings is to provide practice in calculating the angles.

CHAPTER 9

page 161

1. 159 kΩ 2. .1 μF 3. 10 kHz 4. .01 μF 5. 1 kHz
6. .106 Ω 7. infinity 8. app. 0 ohms 9. 159 ohms 10. infinity

page 168

1. C_T = .009 μF, X_{C1} = 2650 Ω, X_{C2} = 265,000 Ω, X_{C3} = 26,500 Ω, X_{CT} = 294,000 Ω
2. C_T = 26.4 μF, X_{C1} = 56.4 Ω, X_{C2} = 17.7 Ω, X_{C3} = 26.5 Ω, X_{CT} = 100 Ω
3. Z = 141 Ω, I = .71A, V_R = 70.7 V, V_C = 70.7 V, Θ = −45°

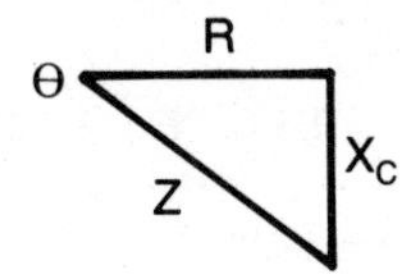

4. Z = 55.9 Ω, I = 1.79A, V_R = 44.7 V, V_C = 89.4 V, Θ = −63.4°

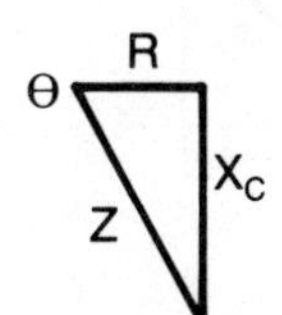

5. Z = 100.5 Ω, I = .99A, V_R = 9.9 V, V_C = 99.5 V, Θ = −84.3°

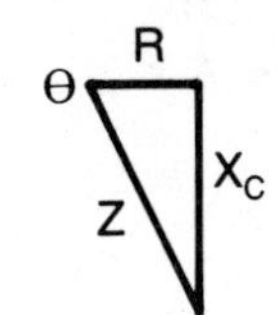

6. Z = 75.2 Ω, I = 1.33A, V_R = 99.8 V, V_C = 6.65 V, Θ = −3.8°

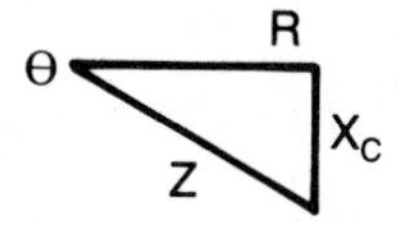

7. Z = 79 Ω, I = 1.26A, V_R = 94.9 V, V_C = 31.6 V, Θ = −18.4°

R
Θ
Xc
Z

8. Z = 50 Ω, I = 2A, V_R = 0 V, V_C = 100 V, Θ = −90°

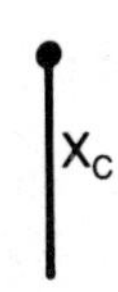

9. $R = 80\ \Omega$, $X_L = 160\ \Omega$, $Z = 179\ \Omega$, $V_a = 44.7$ V, $\Theta = -63.4°$

10. $R = 75\ \Omega$, $X_L = 45\ \Omega$, $Z = 87.5\ \Omega$, $V_a = 29.2$ V, $\Theta = -31°$

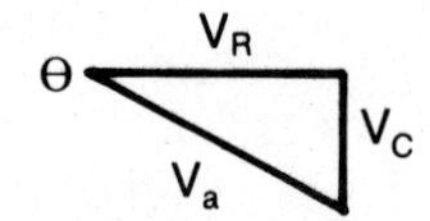

page 175

1. $C_T = 6\ \mu F$, $X_{C1} = 1590\ \Omega$, $X_{C2} = 796\ \Omega$, $X_{C3} = 531\ \Omega$, $X_{CT} = 265\ \Omega$
2. $C_T = 225\ \mu F$, $X_{C1} = 31.8\ \Omega$, $X_{C2} = 15.9\ \Omega$, $X_{C3} = 21.2\ \Omega$, $X_{CT} = 7.07\ \Omega$
3. $I_R = 1A$, $I_C = 1A$, $I_T = 1.4A$, $\Theta = 45°$, $Z = 70.7$ ohms

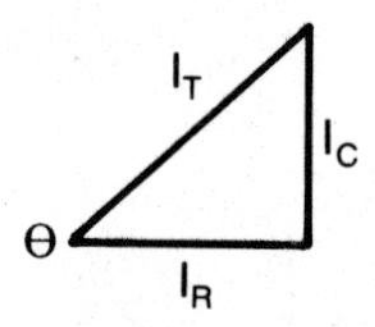

4. $I_R = 4A$, $I_C = 2A$, $I_T = 4.5A$, $\Theta = 26.6°$, $Z = 22$ ohms

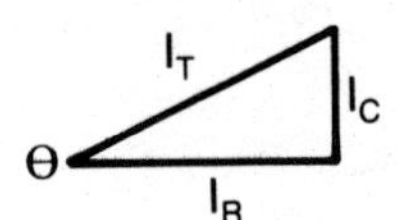

5. $I_R = 10A$, $I_C = 1A$, $I_T = 10.05A$, $\Theta = 5.7°$, $Z = 10$ ohms

6. $I_R = 1.3A$, $I_L = 20A$, $I_T = 20.04$, $\Theta = 86.3°$, $Z = 5$ ohms

7. $I_R = 1.3A$, $I_C = 4A$, $I_T = 4.2A$, $\Theta = 72°$, $Z = 23.8$ ohms

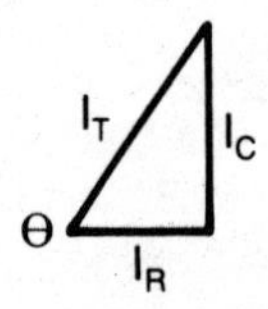

8. I_R = 12A, I_C = 20A, I_T = 28.3A, Θ = 45°, Z = 3.5 ohms

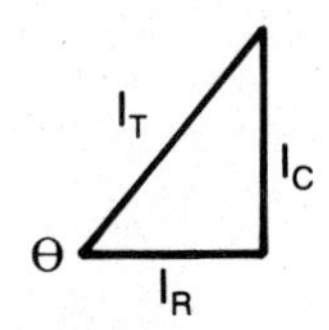

9. R = 100 Ω, X_C = 33.3 Ω, I_T = 3.16A, Θ = 71.6°, Z = 31.6Ω

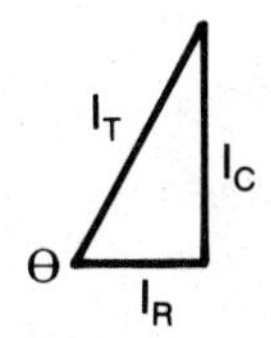

10. R = 4 kΩ, X_C = 10 kΩ, I_T = 27 mA, Θ = 21.8°, Z = 3.7 kΩ

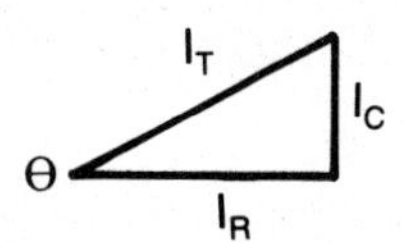

page 176

1. Z = 707 ohms, I = .141 amps, V_R = 70.7 volts, V_C = 70.7 volts, Θ = −45°, P_R = 10 watts, P_X = 10 VARS, P_A = 14.1 VA, PF = .707
2. Z = 269 ohms, I = .371 amps, V_R = 37.1 volts, V_C = 92.9 volts, Θ = −68.2°, P_R = 13.8 watts, P_X = 34.5 VARS, P_A = 37.1 VA, PF = .371
3. Z = 1250 ohms, I = .08 amps, V_R = 80 volts, V_C = 60 volts, Θ = −36.9°, P_R = 6.4 watts, P_X = 4.8 VARS, P_A = 8 VA, PF = .8
4. Z = 1118 ohms, I = .089 amps, V_R = 89.4 volts, V_C = 44.7 volts, Θ = −26.6°, P_R = 8 watts, P_X = 4 VARS, P_A = 8.94 VA, PF = .894
5. Z = 22.4 ohms, I = 4.47 amps, V_R = 44.7 volts, V_C = 89.4 volts, Θ = −63.4°, P_R = 200 watts, P_X = 400 VARS, P_A = 447 VA, PF = .447
6. I_R = .2 amps, I_C = .2 amps, I_T = .283 amps, Z = 354 ohms, Θ = 45°, P_R = 20 watts, P_X = 20 VARS, P_A = 28.3 VA, PF = .707
7. I_R = 1 amp, I_C = .4 amps, I_T = 1.08 amps, Z = 92.6 ohms, Θ = 21.8°, P_R = 100 watts, P_X = 40 VARS, P_A = 108 VA, PF = .928
8. I_R = .1 amps, I_C = .133 amps, I_T = .167 amps, Z = 600 ohms, Θ = 53.1°, P_R = 10 watts, P_X = 13.3 VARS, P_A = 16.7 VA, PF = .6
9. I_R = .1 amps, I_C = .2 amps, I_T = .224 amps, Z = 447 ohms, Θ = 63.4°, P_R = 10 watts, P_X = 20 VARS, P_A = 22.4 VA, PF = .448
10. I_R = 10 amps, I_C = 5 amps, I_T = 11.2 amps, Z = 8.94 ohms, Θ = 26.6°, P_R = 1000 watts, P_X = 500 VARS, V_T = 1120 VA, PF = .894

CHAPTER 10

page 190

1. Z = 90 ohms, I_T = .11A, Θ = +56.3°, net L
2. Z = 180 ohms, I_T = 55.6 mA, Θ = −56.3°, net C
3. Z = 7.8 kilohms, I_T = 1.3 mA, Θ = −39.8°, net C
4. Z = 22.4 ohms, I_T = .45A, Θ = −26.6°, net C

5. $Z = 283$ ohms, $I_T = 35$ mA, $\Theta = +45°$, net L
6. $I_T = 1.12A$, $Z = 8.9$ ohms, $\Theta = +26.6°$, net C
7. $I_T = .112A$, $Z = 89$ ohms, $\Theta = -26.6°$, net L
8. $I_T = 18$ mA, $Z = 556$ ohms, $\Theta = +56.3°$, net C
9. $I_T = 1.14A$, $Z = 8.76$ ohms, $\Theta = +28.8°$, net C
10. $I_T = 1.16A$, $Z = 8.62$ ohms, $\Theta = -64.5°$, net L

page 200

1. $Z = 178\angle 47.7$ $I = 2.47\angle -47.7$
2. $I = 10\angle 53.1$ $Z = 10\angle -53.1$ $V_R = 60\angle 53.1$ $V_C = 80\angle -36.9$
3. $Z = 318\angle -56.6$ $I = .472\angle 56.6$
4. $Z = 123\angle -35.8$ $I = 1.79\angle 35.8$ $V_R = 179\angle 35.8$ $V_L = 236\angle 125.8$ $V_C = 356\angle -54.2$
5. $Z = 660\angle -83.5$ $I = .33\angle 83.5$
6. $I_R = 6\angle 0$ $I_L = 8\angle -90$ $I_T = 10\angle -53.1$ $Z = 12\angle 53.1$
7. $I_R = 6\angle 0$ $I_C = 8\angle 90$ $I_T = 10\angle 53.1$ $Z = 12\angle -53.1$
8. $Z = 1560\angle 38.7$
9. $Z = 44\angle 22.9$
10. $Z = 77.2\angle -29.2$
11. $Z = 53.2\angle 20$
12. $Z = 144\angle 1.52$
13. $Z = 40.2\angle 41.5$
14. $Z = 39.2\angle 25$
15. $Z = 38.2\angle 12.5$

CHAPTER 11

page 213

1. $f_r = 12.6$ Hz
2. $f_r = 1.624$ MHz
3. a. $Q = 100$
   b. $X_C = X_L = 1500$ ohms
   c. $I = 1$ mA
   d. $V_C = 1.5$ V
4. $L = 254\ \mu H$ at 1 MHz
   $L = 15.9\ \mu H$ at 4 MHz
5. $C_{max} = 32.7$ pF
   $C_{min} = 21.7$ pF

CHAPTER 12

page 238

1. $V_{dc} = 28.8$ volts, $I_T = 275$ mA, $V_Z = 12$ volts, $I_Z = 275$ mA, $P_Z = 3.3$ watts, $V_{RS} = 16.8$ volts, $R_s = 61.1$ ohms, $P_{Rs} = 4.62$ watts
2. $V_{dc} = 21.6$ volts, $I_T = 55$ mA, $V_Z = 9$ volts, $I_Z = 55$ mA, $P_Z = 49.5$ mW, $V_{RS} = 12.6$ volts, $R_s = 229$ ohms, $P_{Rs} = 69.3$ mW
3. $V_{dc} = 14.4$ volts, $I_T = 825$ mA, $V_Z = 6$ volts, $I_Z = 825$ mA, $P_Z = 4.95$ watts, $V_{Rs} = 8.4$ volts, $R_s = 10.2$ ohms, $P_{Rs} = 6.93$ watts
4. $V_{dc} = 36$ volts, $I_T = 1.65$ amps, $V_Z = 15$ volts, $I_Z = 1.65$ amps $P_Z = 24.75$ watts, $V_{RS} = 21$ volts, $R_s = 12.7$ ohms, $P_{Rs} = 34.65$ watts
5. $V_{dc} = 72$ volts, $I_T = 165$ mA, $V_Z = 24$ volts, $I_Z = 165$ mA, $P_Z = 3.96$ watts, $V_{RS} = 48$ volts, $R_s = 291$ ohms, $P_{Rs} = 7.92$ watts

Index